Visible Light Communications

IEEE Press
445 Hoes Lane
Piscataway, NJ 08854

Visible Light Communications

Modulation and Signal Processing

Zhaocheng Wang

Qi Wang

Wei Huang

Zhengyuan Xu

Published by John Wiley & Sons, Inc., Hoboken, New Jersey.
Published simultaneously in Canada.

Library of Congress Cataloging-in-Publication Data is available.

ISBN: 978-1-119-33138-4

Printed in the United States of America.

10 9 8 7 6 5 4 3 2 1

Contents

Preface

This book presents the state-of-the-art of visible light communication (VLC) focusing on the modulation and signal processing aspects. VLC has many advantages, such as wide unregulated bandwidth, high security and low cost over its traditional radio frequency counterpart. It has attracted increasing attention from both academia and industry, and is considered as a promising complementary technology in the fifth generation (5G) wireless communications and beyond, especially in indoor applications. This book provides for the first time a systematical and advanced treatment of modulation and signal processing for VLC and optical camera communication (OCC) systems. Example designs are presented and the analysis of their performance is detailed. In addition, the book includes a bibliography of current research literature and patents in this area.

Visible Light Communications: Modulation and Signal Processing endeavours to provide topics from VLC models to extensive coverage of the latest modulation and signal processing techniques for VLC systems. Major features of this book include a practical guide to design of VLC systems under lighting constraints, and the combination of the theoretical rigor and practical examples in present OCC systems.

Although it contains some introductory materials, this book is intended to serve as a useful tool and a reference book for communication and signal processing professionals, such as engineers, designers and developers with VLC related projects. For university undergraduates majoring in communication and signal processing, this book can be used as a supplementary tool in their design projects. Graduate students and researchers working in the field of modern communications will also find this book of interest and valuable. The book is organized as follows.

Chapter 1 provides an overview of the history of VLC, its advantages, applications, related modulation and signal processing techniques, and standardization progresses.

Chapter 2 investigates optical channel models and channel capacity subject to lighting constraints from light emitting diode (LED), where chromaticity control, dimming control and flicker mitigation are also discussed. The link characteristics including shadowing, direct versus indirect lighting and natural light are introduced. Typical optical channel models are addressed in detail. In addition, channel capacity under different lighting constraints is derived to achieve tight upper and lower bounds.

Chapter 3 reviews carrierless, single carrier modulations and some coding schemes for VLC systems. Modulation and coding techniques for dimming control and flicker mitigation are also introduced to satisfy illumination requirements.

Chapter 4 briefly reviews conventional optical orthogonal frequency division multiplexing (OFDM) schemes and then focuses on recent developments on optical OFDM including performance enhancement, spectrum- and power- efficient optical OFDM, and optical OFDM under lighting constraints. Comprehensive comparisons of the existing and proposed modulation techniques are provided as well.

Chapter 5 discusses multicolor modulation schemes under illumination requirements. The LED colorimetry is introduced as a measure for illumination quality, and various modulation schemes are explored to support both communication and high quality illumination.

Chapter 6 explains optical multiple-input multiple-output (MIMO) techniques for imaging and non-imaging VLC systems, including modern optical MIMO, optical spatial modulation, optical space shift keying, and optical MIMO-OFDM. Furthermore, multiuser precoding techniques for VLC systems are also introduced under lighting constraints.

Chapter 7 addresses the signal processing and optimization issues for VLC systems including pre- and post-equalization, interference mitigation and capacity maximization. The hybrid visible light communication and wireless fidelity (VLC-WiFi) system is also introduced to provide better coverage, and the system optimization problem is formulated and solved.

Chapter 8 introduces OCC fundamentals. It describes a typical OCC link, from the optical signal source, propagation path, to optical lens, filters, pixelated image sensors and the receiver. Different noise models such as ambient noise, temporal noise and fixed pattern noise are also addressed. Inter-pixel interference in the active pixel sensor, optical crosstalk due to diffraction and light diffusion, and the distortion due to perspective are introduced.

Chapter 9 discusses OCC modulation schemes and system design aspects. It also introduces various system impairment factors and mitigation techniques, including tracking and coding techniques to achieve synchronization. The off-line and real-time prototypes as well as the potential applications of smartphone cameras are illustrated.

This work was supported by National Key Basic Research Program of China under Grant No. 2013CB329200.

The authors also wish to thank Mr. Rui Jiang at Tsinghua University, China for his contributions to Chapter 2, Mr. Jiandong Tan at Tsinghua University, China for his help with writing Chapter 4.

The authors are indebted to anonymous reviewers for their detailed and insightful constructive comments, as well as many researchers for their published works serving as rich reference sources in the book. The help provided by Mary Hatcher and other staff members from John Wiley & Sons is most appreciated.

1 Introduction to Visible Light Communications

1.1 History

Visible light communication (VLC) is an age-old technique which uses visible light to transmit messages from one place to another. In ancient China, communication by flames was an effective way to relay signals from border sentry stations to distant command offices on the Great Wall. Similarly, lighthouses were distributed along seashore or on islands to navigate the cargo ships on oceans. Nowadays, visible lights are also mounted on modern skyscrapers to not only indicate its presence at particular locations, but also provide reference signals to pilots flying a plane.

Along with the evolution of telecommunication science and technology, using visible lights instead of other electromagnetic waves to transmit information started to attract attentions from scientists, tracing back to the famous photophone experiment by Alexander Graham Bell in 1880 [1]. In his experiment, the voice signal was modulated onto the sunlight and the information was transmitted over a distance of about 200 m. Efforts to explore natural lights and artificial lights for communication continued for decades. In 1979, F. R. Gfeller and G. Bapst demonstrated the technical feasibility of indoor optical wireless communication using infrared light emitting diodes (LEDs) [2]. Built upon fluorescent lamps, VLC at low data rates was investigated in [3]. As LED illumination industry advanced, the fast switching characteristic of visible light LEDs prompted active researches on high-speed VLC. A concept was first proposed by Pang *et al.* in 1999 [4], using the traffic light LED as the optical signal transmitter. Later on, a series of fundamental studies were carried out by S. Haruyama and M. Nakagawa at Keio University in Japan. They investigated the possibility of providing concurrent illumination and communication using white LEDs for VLC systems [5, 6]. Meanwhile, they not only discussed and analyzed effects of light reflection and shadowing on the system performance, but also explored VLC applications at relatively low rates [7, 8]. Using LED traffic lights to transmit traffic information was experimented based on avalanche photodiode (APD) and two-dimensional image sensor receiver, respectively [9, 10]. VLC and power-line communication (PLC) were coherently integrated to provide a network capability [11], where the performance of an advanced orthogonal frequency division

Visible Light Communications: Modulation and Signal Processing. First edition. Zhaocheng Wang, Qi Wang, Wei Huang, and Zhengyuan Xu.
Published 2017 by John Wiley & Sons, Inc.

multiplexing (OFDM) modulation format was evaluated [12]. Applications were extended to brightness control [13] and high-accuracy positioning [14] in addition to communications.

As mobile broadband grows rapidly, the demand for high-speed data services also increases dramatically. VLC emerges as an alternative to alleviate radio spectrum crunch. Higher rate VLC has attracted global research attentions, in particular, from European researchers at the beginning, by maximally exploring the LED capabilities and increasing the spectral efficiency. Using a simple first-order analogue equalizer, a data rate of 100 Mbps was realized with on-off keying non-return-to-zero (OOK-NRZ) modulation in 2009 [15]. Meanwhile, 125 Mbps over 5 m using OOK and 200 Mbps over 0.7 m using OFDM were reported by Vucic *et al.* [16, 17], where photodiodes (PDs) were used in those VLC systems to detect optical signals. By adopting a 2×1 array of white LEDs and an imaging receiver consisting of a 3×3 photodetector array, a multiple-input multiple-output OFDM (MIMO-OFDM) system could deliver a total transmission rate of 220 Mbps over a range of 1 m [18]. The data rate can be further increased if APD is adopted. In 2010, the data rate of the OOK-based system reached 230 Mbps [19] and the data rate of the OFDM-based system approached 513 Mbps with bit- and power-loading [20]. In 2012, the highest data rate of a single LED-based VLC system achieved 1 Gbps with OFDM [21]. Additionally, carrierless amplitude and phase modulation (CAP) was introduced into VLC systems, and a data rate of 1.1 Gbps was achieved [22]. Using an MIMO structure, a 4×9 VLC system achieving 1.1 Gbps was presented, where the parallel streams were transmitted by 4 individual LEDs and detected by a 3×3 receiver array [23].

In the previous studies, a phosphor-converted LED (pc-LED) was adopted as optical signal transmitter. The bandwidth of a pc-LED is however limited by slow response of the phosphorescent component. In 2014, a post-equalization circuit consisting of two passive equalizers and one active equalizer was proposed to extend the bandwidth from tens of MHz to around 150 MHz [24]. If other types of LEDs having higher bandwidth are employed, it has potential to increase the throughput significantly. For example, using micro LEDs as transmitters in VLC systems could be firstly attributed to McKendry *et al.* and a data rate of 1 Gbps was reported at a price of low luminous efficiency [25]. Multicolor LEDs, radiating particularly red, green, and blue lights, can provide high-rate transmission by wavelength division multiplexing (WDM). Data were simultaneously conveyed in parallel by different colors such as red, green, and blue lights. In principle, the data rate could be tripled in the absence of color crosstalk. An OFDM-based VLC system using a multicolor LED was realized supporting a data rate of 803 Mbps over 0.12 m [26]. Using multicolor LED as the transmitter and APD as the receiver, the data rate of OFDM-based VLC systems was increased from 780 Mbps over 2.5 m to 3.4 Gbps over 0.3 m, where WDM and bit- and power-loading techniques were jointly applied [27–29]. In another study [30], the bandwidths of multicolor LED chips were extended to 125 MHz and modulated by 512 quadrature amplitude modulation (QAM) and 256WDM, respectively, and the frequency domain equalization based VLC system finally reached a data rate of 3.25 Gbps. The data rate of CAP-based VLC systems using multicolor LEDs was increased up to 3.22 Gbps, also benefiting from WDM technology [31].

It is well known that lighting LEDs typically serve as transmitters for downlink information transmission to mobile devices. In 2013, an asynchronous bidirectional VLC system was demonstrated in [32] where a 575 Mbps downlink transmission was realized by red and green LEDs, and a 225 Mbps uplink transmission by a single blue LED. From a network perspective, a spectrum reuse scheme based on different colors was proposed for different cells in an indoor optical femtocell, where multiple users can share the spectrum and access the network simultaneously [33]. User-centric cluster formation methods were proposed for interference-mitigation in [34]. A VLC system can also be combined with a wireless fidelity (WiFi) system to provide seamless coverage after a judicious handover scheme was designed and applied [35].

In multicolor LED-based VLC systems, signals from three color light sources were transmitted independently in most experiments, leaving room for capacity increase. In 2015, Manousiadis *et al.* used a polymer-based color converter to generate red, green, and blue lights emitted by blue micro LEDs [36]. Three color lights were modulated and mixed for white light illumination. The aggregate data rate from three colors was 2.3 Gbps. Techniques to explore spatial and temporal capabilities of devices were also investigated. A MIMO VLC system employing different field of view (FOV) detectors in order to improve signal-to-noise ratio (SNR) was analyzed in [37]. An optical diversity scheme was proposed, where the original data and its delayed versions were simultaneously transmitted over orthogonal frequencies [38]. Data rate can be significantly enhanced by employing different degrees of freedom. Combining with WDM, high-order CAP, and post-equalization techniques, Chi *et al.* showed that a multicolor LED based VLC system could provide a data rate of 8 Gbps [39]. A novel layered asymmetrically clipped optical OFDM scheme was proposed to make a tradeoff between complexity and performance of an intensity-modulated direct-detection (IM/DD) VLC system [40]. Under lighting constraints, DC-informative modulation and system optimization techniques were proposed [41–43]. Some receiver design issues were particularly addressed in weak illuminance environments and several bidirectional real-time VLC systems with low complexity were reported [44, 45].

Besides individual research groups, there are also many large scale organizations and research teams worldwide that have contributed to the development and standardization of VLC technology. In Europe, the HOME Gigabit Access (OMEGA) project was launched in 2008 to develop a novel indoor wireless access network, providing gigabit data rates for home users [46]. The project members included France Telecom, Siemens, University of Oxford, University of Cambridge, and many other companies and universities. This project finally demonstrated a real-time VLC system using 16 white LEDs on the ceiling to transfer HD video streams at 100 Mbps. Another organization called OPTICWISE was funded by the European Science Foundation under an action of the European Cooperation in Science and Technology (COST), which allowed coordination of nationally funded VLC researches across European countries. Significant research results and professional activities were reported from its various groups [47].

In Japan, Visible Light Communication Consortium (VLCC) consisting of many Hi-tech enterprises and manufacturers in the areas of illumination and communication, such as Casio, NEC, and Panasonic, was founded in 2003. It was devoted to marketing investigation, application promotion, and technology standardization. After years of development, it evolved to Visible Light Communications Association (VLCA) in 2014 to collaborate various industries closely for realizing the visible light communication infrastructure, from telecommunication to lighting, social infrastructure, Internet, computer, semiconductor, etc.

In the United States, the Ubiquitous Communication by Light Center (UC-Light), Center on Optical Wireless Applications (COWA), and Smart Lighting Engineering Research Center (ERC), are notable VLC research groups. UC-Light focuses on efficient lighting, communication, and navigation technologies by LEDs, and aims to create new technological innovations, economic activities, and energy-saving benefits. COWA is dedicated to the optical wireless applications of communications, networking, imaging, positioning, and remote sensing. ERC concentrates on LED communication systems and networks, supporting materials and lighting devices, and applications for detection of biological and biomedical hazards.

In China, two sizable teams were built in 2013 to focus on the research of optical wireless communications over broad spectra, including visible light communication. One was funded by National Key Basic Research Program of China (973 Program), including about 30 researchers from top universities and research institutes. The other was funded by National High Technology Research and Development Program of China (863 Program). Both project teams have made tremendous efforts on theory breakthrough, technology development, and real-time VLC system demonstrations. The real-time data rate has reached 1.145 Gbps at 2.5 m to deliver multimedia services, and the highest off-line data rate of 50 Gbps was achieved at a shorter distance. To jointly prompt commercialization of VLC technologies, Chinese Visible Light Communications Alliance (CVLCA) was founded in 2014, which attracted universities and industries in lighting, telecommunication, energy, consumer electronics, and financing agencies.

1.2 Advantages and applications

Visible light communication has many attractive advantages compared to its radio frequency (RF) counterpart, which include but are not limited to the following aspects.

(1) Wide spectrum: As the demand for high-speed wireless services is increasing dramatically, RF spectrum is getting congested. The radio wave spectrum is limited, from 3 kHz to 300 GHz, while the visible light spectrum is at least 1000 times greater, which is from 400 THz to 780 THz [48].

(2) No electromagnetic interference: Since light does not cause any electromagnetic interference, VLC is suitable for communications in the electromagnetic interference immunity (EMI) environments, such as hospitals, nuclear power plants, and airplanes.

(3) Easy implementation: VLC modules can be made small and compact, so that they can be easily implemented into the existing lighting infrastructure. The modulation unit, digital-to-analog converter, and driving circuit can be integrated into LEDs. The photodiode, analog-to-digital converter, and other signal processing units can be manufactured as a portable external receiver, or embedded into the lighting infrastructure.

(4) Low cost: The implementation of a VLC system is relatively simple. Instead of designing an entire wireless communication system, it reuses the ubiquitous lighting infrastructure, and only a few additional modules are added to the lighting system. As LED industry is rapidly developing, the cost of massively producing VLC transceivers is expected to decrease.

(5) High energy efficiency: As green lighting devices, LEDs have been recognized as the next generation lighting devices, which can reduce the energy consumption of traditional lighting sources by 80% [49]. If all the lighting sources are replaced by LEDs, the global electricity consumption is expected to reduce by as much as 50% [50]. According to a recent report from the U.S. Department of Energy, by the year of 2025, it is possible to save the amount of energy up to 217 terawatt-hours (TWh) with the adoption of LED lighting technology [51].

(6) Health safety: Unlike infrared LED and laser having concentrated optical power within a narrow beam, lighting LED is a diffusive light source. Therefore, it is intrinsically safe for many application scenarios with large emitted optical power. Since lighting LED does not generate radiation as radio frequency or microwave devices do, no obvious health hazard is incurred to the environment and end users.

(7) Information security: Security is an important issue to RF communication because radio waves can penetrate walls, causing information leakage. Since light cannot penetrate opaque objects, VLC can be confined in an indoor, enclosed space and more secure communication links are ensured.

The aforementioned features help to yield various indoor and outdoor VLC applications. The most desirable application, perhaps, is indoor high-speed Internet access for smart phones and computers. People usually spend much more time staying indoors than outdoors, in offices and homes for study, work, entertainment, etc. It would be convenient to access the Internet by simply using LED lighting devices on the ceiling. The inherent modulation bandwidth of LEDs (orders of MHz to hundreds of MHz) is able to provide much higher data rate than WiFi and existing mobile networks. Equipped with advanced techniques, such as multicarrier modulation, wavelength multiplexing, and equalization, the VLC data rate can be increased up to gigabit per second.

Besides offices and homes, electromagnetic sensitive environments also require safe and reliable wireless services. Visible light does not cause any electromagnetic interference to the existing electrical equipment, and is thus ideal for communication in those environments. In a hospital, for example, some sophisticated and expensive medical equipment, such as magnetic resonance imaging equipment, must be insulated from electromagnetic interference. The electronic devices radiating the electromagnetic waves are prohibited in an airplane cabin during takeoff and landing because those waves might cause equipment malfunction. In a nuclear plant, it is also very restrictive to use a mobile phone. It is evident that VLC becomes a safe technology for communications in such EMI environments.

In some cases, users would like to directly communicate to each other at high speed, without routing messages through a network, such as machine-to-machine (M2M) and device-to-device (D2D) communications. Two VLC transceivers such as smart phones or laptops can realize point-to-point communication directly. Light communication becomes a feasible solution as well.

It is well known that LED is a natural transmitter and can easily broadcast information, which can be embedded in LED displays and screens in different public areas, such as waiting hall at the airports and train stations, and sent to passengers. If an image sensor in a camera is used as signal detector, optical camera communication (OCC) could receive the broadcasting information [52]. Also, in shopping malls and outlets, merchandise and advertisement information can be broadcasted to customers through lighting LEDs or signage. Exhibitions, galleries, and museums are also ideal places to use LEDs for seamless information broadcasting.

Besides that, people could take the advantage of densely distributed LEDs for location references and use triangularization algorithms to forecast device positions. As a result, highly accurate indoor positioning and navigation come true by LEDs, like GPS in outdoors by satellites. LEDs could also send control signals to an intelligent robot and guide its precise movement along a route to reach its predefined destination [53].

Since there are a large number of LEDs deployed/used outdoors as well, street lights, traffic lights, and vehicle lights are also applicable for establishing VLC wireless links among vehicles, vehicle and roadside lighting infrastructure, vehicle and traffic lights [9, 10, 54]. Since the vehicle is usually equipped with an image sensor array, it can predict its relative motion together with data transmission [55–57]. Underwater VLC is also a competitive communication technology for ocean exploration.

The aforementioned indoor and outdoor applications span a variety of fields, which could gradually penetrate different markets for various services, from low rate communication and positioning, to high-rate communication, and intelligent transportation. As words "visible light" indicate, VLC will have a bright future in our modern life.

1.3 Overview of modulation and signal processing

For VLC systems, LEDs and photodiodes are used as alternative transceivers to convey information via visible light. Accordingly, modulation and signal processing for

VLC systems possess new features and new challenges, compared to their RF counterparts. Normally, LED works under a forward bias while photodiode is driven by a reverse voltage. Since LED is used for lighting and communication simultaneously, its chromaticity and nonlinearity have to be investigated in VLC systems. As for the photodiode, key parameters such as absorption coefficient, quantum efficiency, and responsivity are considered in the system model. Based on whether there exists a line-of-sight (LOS) link between the transmitter and the receiver, optical wireless propagation links can be classified into two categories: LOS link and non-line-of-sight (NLOS) link. Besides, noise from other devices and surrounding environment should be considered. Based on the dominant noise in practical scenarios, three common optical wireless channel models are discussed, i.e., free-space optical intensity channel, discrete-time Poisson channel, and improved free-space intensity channel. Since there are no analytic expressions of channel capacity, several upper and lower bounds have been illustrated. Considering these specific channel models of VLC systems, several modulation and signal processing schemes have been demonstrated.

Single carrier modulation and carrierless modulation schemes are addressed firstly. Pulse amplitude modulation (PAM) is a simple modulation format widely used in VLC systems. When multipath channel is considered, PAM together with frequency-domain equalization is utilized to combat inter-symbol interference (ISI). Besides, several implementation schemes are introduced in order to overcome the effect of LED nonlinearity, i.e., PAM can be implemented with multiple LEDs, where each LED is modulated by OOK. Pulse position modulation (PPM) is another simple modulation format for VLC systems and PPM together with decision feedback equalization could eliminate the ISI. Since PPM has low data rate with only one pulse in a single symbol duration, several modified schemes have been proposed including differential PPM, multipulse PPM, overlapping PPM, and variable PPM. Besides, CAP is also adopted in VLC systems due to its high spectral efficiency and simple implementation, which can also be extended to multi-dimensional CAP. Meanwhile, various modified modulation and coding schemes have been proposed for dimming control in single carrier VLC systems, which could support communication and illumination simultaneously.

Optical OFDM techniques have been investigated in order to realize broadband and high-rate transmission. Since IM/DD methodology is used in VLC systems, the amplitude of optical OFDM signals is constrained to be real-valued and non-negative. Therefore, the conventional OFDM method is not feasible for intensity modulation and several optical OFDM schemes have been proposed to satisfy the specific signal constraints in VLC systems, such as DC-biased optical OFDM (DCO-OFDM), asymmetrically clipped optical OFDM (ACO-OFDM), pulse-amplitude-modulated discrete multitone (PAM-DMT), and unipolar OFDM (U-OFDM). Similar to conventional RF systems, optical OFDM suffers from high peak-to-average power ratio (PAPR), which might introduce severe nonlinear distortion and impair the performance of VLC systems. There are several techniques to enhance the performance of optical OFDM by optimizing DC bias and scaling factor, mitigating the nonlinear effect of LED, and PAPR reduction. Besides, some recently proposed power- and spectral-efficient optical OFDM methodologies, such as hybrid optical OFDM, enhanced U-

OFDM, and layered ACO-OFDM have shown great potential for future VLC systems. In addition, seamless integrations of OFDM modulation and dimming control are discussed, including pulse width modulation, reverse polarity optical OFDM and asymmetrical hybrid optical OFDM, which have shown that dimmable OFDM can support a wide dimming range with a relatively small throughput fluctuation.

Multicolor modulation is an interesting candidate for VLC systems, compared to the traditional RF modulation methods. White LEDs are usually classified into single-chip LEDs and RGB-type LEDs. The single-chip LEDs use a single blue LED that excites a yellow phosphor to create an overall white emission, while the RGB-type LEDs combine light from LEDs of three primary colors of red, green, and blue. They are preferable to single-chip LEDs since the transmission rate can be improved owing to their faster response time. Moreover, three wavelengths corresponding to the three primary colors can be used to carry multiple data streams independently and thus offer the possibility of WDM. Accordingly, multicolor modulation schemes under illumination requirements for VLC systems with RGB-type LEDs have been illustrated, whereby color shift keying (CSK) is developed and adopted in the IEEE 802.15.7 standard. Furthermore, the optimal design rules of CSK constellation as well as Qual-LED CSK are provided to achieve superior capacity, while CSK with coded modulation is introduced for practical scenarios. Moreover, WDM system combined with channel coding is detailed, and a receiver-side predistortion is proposed before channel decoding, which has shown significant performance gain.

Despite the fact that the spectrum of visible light is as wide as several THz, the bandwidth of off-the-shelf LED is limited, which makes it very challenging to achieve high-rate transmission. Meanwhile, in order to provide sufficient illumination, multiple LED units are usually installed in a single room. In such scenarios, MIMO techniques can be naturally employed in indoor VLC schemes to boost the data rate. Typically, there are two optical MIMO approaches for VLC systems, namely non-imaging MIMO and imaging MIMO. For non-imaging MIMO systems, each receiver collects the surrounding light with its own optical concentrator, and optical MIMO, optical spatial modulation, and optical space shift keying can be used. For imaging MIMO systems, an imaging diversity receiver is utilized to distinguish the light from different transmitters. Meanwhile, in order to support data transmission for multiple users simultaneously, precoding techniques are employed to eliminate the inter-user interference under the lighting constraints in VLC systems. Moreover, MIMO-OFDM is introduced for single-user and multiuser VLC systems, which provides high spectral efficiency and robust reception.

Due to the special characteristics of transceivers and channels for VLC systems, several signal processing and optimization issues for VLCs have been discussed. For multi-chip-based multiple-input single-output VLC system, an electrical and optical power allocation scheme is introduced to maximize the multi-user sum-rate in consideration of the luminance, chromaticity, amplitude, and bit error rate constraints. Considering the vulnerability of VLC LOS links, heterogeneous VLC-WiFi systems offer a solution for future indoor communications that combines VLC to support high-data-rate transmission and RF to support reliable connectivity. In such heterogeneous systems, vertical handover is critical to improve the system performance and a dynamic approach is adopted to obtain a tradeoff between the switching cost and the delay requirement, where the vertical handover is formulated as a Markov decision process problem.

For VLC systems with narrow FOV, the PD shot noise modeled by Poisson statistics is signal-dependent since it originates from the quantum nature of the received optical energy rather than external noises, which is in contradiction to the conventional signal-independent additive white Gaussian noise model. Therefore, novel signal processing and estimation techniques are illustrated to guarantee the transmission performance. OCC is a new form of visible light communication, which employs pervasive image sensors assembled in consumer electronic devices as the receiver. The advantages of OCC include the wide spectrum compared to the conventional VLC systems, the pervasive optical light sources including illumination LED, display and traffic light, and the pervasive consumer cameras having natural multicolor sensitivity, the feasibility of massive MIMO and anti-interference image-sensor-based receivers. With these advantages, OCC combined with mobile computing could realize novel forms of sensing and communication applications, such as indoor location, intelligent transportation system, screen-camera communication, and privacy protection. However, there exist also challenging issues to be addressed, including the limited frame rate, synchronization issue, non-negligible shot noise, perspective distortion, pixel misalignment, and blur effect.

To investigate the channel characteristics and system performances of OCC systems, the pixel-sensor structure and its operation procedure for CMOS image sensors have been addressed and the noise composition, including photo shot noise, dark current shot noise, fixed-pattern noise, source follower noise, sense node reset noise, and quantization noise at high illumination, is illustrated and analyzed. A plurality of experimental results demonstrate that the noise in a CMOS image-sensor-based receiver can be modeled as Gaussian noise, such as signal-independent electrical thermal noise as well as the signal-dependent and signal-independent shot noise. Based on these noise models, the SNR in OCC systems should be redefined, and accordingly, a unified communication model is proposed for OCC systems. Moreover, channel capacity of OCC systems has been investigated and the asymptotic upper bound and the tight lower bound with peak and average power constraints have been addressed. The capacity bounds indicate that a spectral efficiency of 8–11 bit/s/Hz is achievable under an ideal channel with diversity structure, and there is room for improvement using the today's OCC prototypes.

According to specific OCC channel characteristics, the modulation schemes, synchronization issues and several technical challenges in a real-time OCC system have been addressed. Based on the signal-dependent noise model, a capacity-achieving discrete nonuniform signaling scheme has been designed for OCC systems. However, it requires the feedback link, which possesses high complexity. Alternative modulation schemes which convey signal on different domains are adopted in OCC systems, including the under-sample-based modulation schemes in time/frequency domains, the rolling-shutter-effect-based modulation schemes in time/frequency domains, color-intensity modulation (CIM) in color space, and the spatial OFDM/WDM in spatial/frequency domains. Moreover, the effect of nonideal factors, such as linear misalignment, geometry distortion, blur effect and vignetting, and the corresponding mitigating schemes, are discussed, including equalization, perspective correction, adaptive coding, and modulation. For a practical OCC wireless link, synchronization is important and several methodologies have been discussed. The per-line tracking, inter-frame coding, and rateless coding could tackle the synchronization issues by decoding imperfect frames and recovering any lost frames.

Furthermore, a real-time CIM-MIMO OCC prototype has been realized, which utilizes spatial, color, and intensity dimensions to generate a high-dimensional signal constellation and parallel wireless links, leading to an increased data rate and improved bit error rate performance. Several technical challenges including unstable frame rate, joint nonlinearity and crosstalk, flicker noise, and rolling shutter, have been tackled.

For a real-time OCC system, commercial CMOS cameras are used as receivers. The corresponding products can be used in near-field screen-camera communications and indoor visible light positioning. If the sensor is equipped with an external optical lens, the transmission distance between the light source and the sensor can be significantly extended, which makes the system suitable for other applications, for example, capturing signals from a distant traffic light, or information broadcasting displays in a public area such as shopping mall and transportation hub.

1.4 Standards

With rapid evolution of VLC technologies, it is imperative to develop the corresponding standards to harmonize the physical layer (PHY) protocols and media access control (MAC) layer protocols, and help to transfer technologies into applications promptly, which has attracted much attention from various international and national standardization bodies.

The first international VLC standard, that is IEEE 802.15.7, was published by IEEE 802.15.7 working group for wireless personal area networks in 2011 [58]. The standard clearly specifies the PHY and MAC layers for short-range optical wireless communications using visible light for indoor and outdoor applications. IEEE 802.15.7 accommodates three different PHY layer types, i.e., PHY I, PHY II, and PHY III, respectively. PHY I supports lower rate (11.6–266.6 kb/s) and long-distance outdoor applications, PHY II supports higher rate (1.25–96 Mb/s) systems working in indoor infrastructures and point-to-point applications, and PHY III is designed to support the same rate (1.25–96 Mb/s) with multicolor light sources/detections. PHY I and PHY II adopt OOK and VPPM, which is a combination of two-pulse position modulation and pulse width modulation (PWM). A color shift keying modulation format, generated by using three-color light sources out of the seven-color bands, is also defined. Different forward error correction (FEC) schemes and run length limited (RLL) codes are added to meet various channel conditions and to guarantee the lighting brightness. In the MAC layer, IEEE 802.15.7 supports three different topologies, namely star, peer-to-peer, and broadcast. The MAC layer is also responsible for the following major tasks: initiating/maintaining procedures, association/disassociation procedures, color-function support mechanism, illumination and dimming support mechanism, mobility support mechanism, color stabilization, etc.

In 2014, a new working group 802.15.7r1 was formed to make revisions on the previous standard. The new standard, called as IEEE 802.15.7r1, is expected to be published in 2017 [59]. IEEE 802.15.7r1 will specify the following three different application scenarios depending on various data rates and devices. First, LED-ID is low-rate photodiode-based communication sending identification information through various LEDs. Second, OCC is an image-sensor-based communication which offers positioning/localization, message broadcasting, etc. Accordingly, three different source types have been defined, i.e., discrete source (15 bps–4 kbps), surface source (90 bps–8 kbps) and two-dimensional screen source (40 bps–64 kbps). At current stage, the modulation formats are still under on-going discussions. As a related application, a new interest group, called as IEEE 802.15 Vehicular Assistant Technology (VAT), was formed in January 2017 for OCC-based long range vehicular applications. Smart automotive lighting in vehicle safety systems has been also investigated in [60]. Third, light fidelity (LiFi) is high-rate photodiode-based communication that can support Gbps data stream, bidirectional and multiple access, mobility, and handover. The technical specifications focusing on modulation, coding, bandwidth, and optical clock rate have been intensively discussed. Although IEEE 802.15.7r1 has not been finalized, the endorsed reference channel models were presented in [61], where four different reference scenarios, including work place, office room with secondary light, living room, and manufacturing cell, are emulated by a powerful software Zemax to describe the channel impulse responses.

Besides IEEE 802.15.7 and IEEE 802.15.7r1, International Telecommunication Union (ITU), established a study group (named as SG15) to standardize the VLC technology within the G.vlc framework in September 2015. Research community together with key industrial members, such as Huawei and Marvell, are constructively and jointly developing a high-speed VLC standard. So far, G.vlc has been specifying VLC modulation format, dimming control, channel and source models, band plans, and network topology. Recently, SG15 decided to start a new G.occ framework (Gbps OCC) in order to cover various aspects of optical wireless applications.

In addition to international efforts, there are also national organizations focusing on VLC standardization. In Japan, VLCC was established in November 2003, whose members were major electronic companies and research centers. VLCC tried to merge VLC technology into LED lightings in offices and homes, commercial displays, traffic signals, and small lamps on home appliances. The Visible Light ID System was standardized by Japan Electronics and Information Technology Industries Association (JEITA), for commercial applications including indoor navigation and POS/client data exchange. In 2014, VLCA was established as the successor to VLCC, to facilitate various industrial collaborations and further develop the application and business of VLC technology.

Globally, China becomes the largest LED manufacturer and consumer market, and owns the most complete LED industry chain. Its VLC technology has bloomed in the recent decade, where lighting, wireless communication and automobile industries are all actively participating in the technology development and standardization of VLC systems. In March 2017, Smart Visible Light Industrial Technology Innovation Association was established in Guangdong Province, China, with over 20 industry members, including ZTE, Philips Lighting, and Audi. Its main goal is to publicize, popularize, and standardize the VLC technology in various industrial and commercial sectors. The Chinese VLC standard is being drafted by China Electronics Standardization Institute (CESI), and its first version will be released soon.

The above on-going standardization activities will prompt successful and rapid applications of various VLC technologies, which span from positioning, accurate control, low rate communication, to information broadcasting, and high-speed indoor and outdoor communications, for mobile devices, robotics, vehicles, and even new forms of terminals and applications such as drones, unmanned underwater vehicles, and virtual/augmented reality [62].

References

1 A. G. Bell, W. G. Adams, Tyndall, and W. H. Preece, "Discussion on the photophone and the conversion of radiant energy into sound," *J. Soc. Telegraph Eng.*, vol. 9, no. 34, pp. 375–383, 1880.

2 F. R. Gfeller and U. Bapst, "Wireless in-house data communication via diffuse infrared radiation," *Proc. IEEE*, vol. 67, no. 11, pp. 1474–1486, Nov. 1979.

3 D. Jackson, T. Buffaloe, and S. Leeb, "Fiat lux: A fluorescent lamp digital transceiver," *IEEE Trans. Ind. Appl.*, vol. 34, no. 3, pp. 625–630, May/Jun. 1998.

4 G. Pang, T. Kwan, C. H. Chan, and H. Liu, "LED traffic light as a communications device," in *Proc. IEEE/IEEJ/JSAI International Conference on Intelligent Transportation Systems 1999* (Tokyo, Japan), Oct. 5–8, 1999, pp. 788–793.

5 Y. Tanaka, S. Haruyama, and M. Nakagawa, "Wireless optical transmissions with white colored led for wireless home links," in *Proc. IEEE International Symposium on Personal Indoor and Mobile Radio Communications (PIMRC) 2000* (London, United Kindom), Sep. 18–21, 2000, vol. 2, pp. 1325–1329.

6 Y. Tanaka, T. Komine, S. Haruyama, and M. Nakagawa, "Indoor visible light data transmission system utilizing white LED lights," *IEICE Trans. Commun.*, vol. 86, no. 8, pp. 2440–2454, Aug. 2003.

7 T. Komine and M. Nakagawa, "Fundamental analysis for visible-light communication system using LED lights," *IEEE Trans. Consum. Electron.*, vol. 50, no. 1, pp. 100–107, Feb. 2004.

8 T. Komine and M. Nakagawa, "A study of shadowing on indoor visible-light wireless communication utilizing plural white LED lightings," in *Proc. International Symposium on Wireless Communication Systems (ISWCS) 2004* (Mauritius), Sep. 20–22, 2004, pp. 36–40.

9 M. Akanegawa, Y. Tanaka, and M. Nakagawa, "Basic study on traffic information system using LED traffic lights," *IEEE Trans. Intell. Transp. Syst.*, vol. 2, no. 4, pp. 197–203, Dec. 2001.

10 H. B. C. Wook, T. Komine, S. Haruyama, and M. Nakagawa, "Visible light communication with LED-based traffic lights using 2-dimensional image sensor," in *Proc. IEEE Consumer Communications and Networking Conference (CCNC) 2006* (Las Vegas, USA), Jan. 8–10, 2006, pp. 243–247.

11 T. Komine and M. Nakagawa, "Integrated system of white LED visible light communication and powerline communication," *IEEE Trans. Consum. Electron.*, vol. 49, no. 1, pp. 71–79, Feb. 2003.

12 T. Komine, S. Haruyama, and M. Nakagawa, "Performance evaluation of narrowband OFDM on integrated system of power line communication and visible light wireless communication," in *Proc. International Symposium on Wireless Pervasive Computing (ISWPC) 2006* (Phuket, Thailand), Jan. 16–18, 2006, pp. 1–6.

13 H. Sugiyama, S. Haruyama, and M. Nakagawa, "Brightness control methods for illumination and visible-light communication systems," in *Proc. International Conference on Wireless and Mobile Communications (ICWMC) 2007*

(Guadeloupe, France), Mar. 4–9, 2007, pp. 78–83.

14 M. Yoshino, S. Haruyama, and M. Nakagawa, "High-accuracy positioning system using visible LED lights and image sensor," in *Proc. IEEE Radio and Wireless Symposium 2008* (Orlando, FL), Jan. 22–24, 2008, pp. 439–442.

15 H. L. Minh, D. O'Brien, and G. Faulkner, "100-Mb/s NRZ visible light communications using a postequalized white LED," *IEEE Photon. Technol. Lett.*, vol. 21, no. 15, pp. 1063–1065, Aug. 2009.

16 J. Vucic, C. Kottke, and S. Nerreter, "125 Mbit/s over 5m wireless distance by use of OOK-modulated phosphorescent white LEDs," in *Proc. European Conference on Optical Communication (ECOC) 2009* (Vienna, Austria), Sep. 20–24, 2009, pp. 1–2.

17 J. Vucic, C. Kottke, S. Nerreter, and A. Buttner, "White light wireless transmission at 200Mb/s net data rate by use of discrete-multitone modulation," *IEEE Photon. Technol. Lett.*, vol. 21, no. 20, pp. 1511–1513, Oct. 2009.

18 A. H. Azhar, T. Tuan-Anh, and D. O'Brien, "Demonstration of high-speed data transmission using mimo-ofdm visible light communications," in *Proc. IEEE Global Communications Conference (GLOBECOM) Workshops 2010* (Miami, FL), Dec. 5–10, 2010, pp. 1052–1056.

19 J. Vucic, C. Kottke, and S. Nerreter, "230 Mbit/s via a wireless visible light link based on OOK modulation of phosphorescent white LEDs," in *Proc. Optical Fiber Communication Conference and Exposition and the National Fiber Optic Engineers Conference (OFC/NFOEC) 2010* (San Diego, CA), Mar. 21–25, 2010, pp. 1–3.

20 J. Vucic, C. Kottke, S. Nerreter, K. Langer, and J. W. Walewski, "513 Mbit/s visible light communications link based on DMT-modulation of a white LED," *J. Lightw. Technol.*, vol. 28, no. 24, pp. 3512–3518, Dec. 2010.

21 A. M. Khalid, G. Cossu, R. Corsini, P. Choudhury, and E. Ciaramella, "1-Gb/s transmission over a phosphorescent white LED by using rate-adaptive discrete multitone modulation," *IEEE Photon. J.*, vol. 4, no. 5, pp. 1465–1473, Oct. 2012.

22 F. M. Wu, C. T. Lin, and C. C. Wei, "1.1-Gb/s white-LED-based visible light communication employing carrier-less amplitude and phase modulation," *IEEE Photon. Technol. Lett.*, vol. 24, no. 19, pp. 1730–1732, Oct. 2012.

23 A. Azhar, T. Tran, and D. O'Brien, "A gigabit/s indoor wireless transmission using MIMO-OFDM visible-light communications," *IEEE Photon. Technol. Lett.*, vol. 25, no. 2, pp. 171–174, Jan. 2013.

24 H. Li, X. Chen, B. Huang, D. Tang, and H. Chen, "High bandwidth visible light communications based on a post-equalization circuit," *IEEE Photon. Technol. Lett.*, vol. 26, no. 2, pp. 119–122, Jan. 2014.

25 J. McKendry, R. Green, and A. Kelly, "High speed visible light communications using individual pixels in a micro light emitting diode array," *IEEE Photon. Technol. Lett.*, vol. 22, no. 18, pp. 1346–1348, Sep. 2010.

26 J. Vucic, C. Kottke, K. Habel, and K. D. Langer, "803 Mbit/s visible light WDM link based on DMT modulation of a single RGB LED luminary," in *Proc. Optical Fiber Communication Conference and Exposition and the National Fiber Optic Engineers Conference (OFC/NFOEC) 2011* (Los Angeles, CA), Mar. 6–10, 2011, pp. 1–3.

27 G. Cossu, A. M. Khalid, P. Choudhury, R. Corsini, and E. Ciaramella, "Long distance indoor high speed visible light communication system based on RGB LEDs," in *Proc. Asia Communications and Photonics Conference (ACP) 2012* (Guangzhou, China), Nov. 7–10, 2012, pp. 1–3.

28 G. Cossu, A. M. Khalid, P. Choudhury, R. Corsini, and E. Ciaramella, "2.1 Gbit/s visible optical wireless transmission," in *Proc. European Conference and Exhibition on Optical Communication (ECOC) 2012* (Amsterdam, Netherlands), Sep. 16–20, 2012, pp. 1–4.

29 G. Cossu, A. M. Khalid, P. Choudhury, R. Corsini, and E. Ciaramella, "3.4 Gbit/s visible optical wireless transmission based on RGB LED," *Opt. Exp.*, vol. 20, no. 26, pp. B501–B506, Dec. 2012.

30 Y. Wang, R. Li, Y. Wang, and Z. Zhang,

"3.25-Gbps visible light communication system based on single carrier frequency domain equalization utilizing an RGB LED," in *Proc. Optical Fiber Communications Conference and Exhibition (OFC) 2014* (San Francisco, CA), Mar. 9–13, 2014, pp. 1–3.

31 F. M. Wu, C. T. Lin, and C. C. Wei, "3.22-Gb/s WDM visible light communication of a single RGB LED employing carrier-less amplitude and phase modulation," in *Proc. Optical Fiber Communication Conference and Exposition and the National Fiber Optic Engineers Conference (OFC/NFOEC) 2013* (Anaheim, CA), Mar. 17–21, 2013, pp. 1–3.

32 Y. Wang, Y. Wang, N. Chi, J. Yu, and H. Shang, "Demonstration of 575-Mb/s downlink and 225-Mb/s uplink bi-directional SCM-WDM visible light communication using RGB LED and phosphor-based LED," *Opt. Exp.*, vol. 21, no. 1, pp. 1203–1208, Jan. 2013.

33 K. Cui, J. Quan, and Z. Xu, "Performance of indoor optical femtocell by visible light communication," *Opt. Commun.*, vol. 298–299, pp. 59–66, Jul. 2013.

34 X. Li, F. Jin, R. Zhang, J. Wang, Z. Xu, and L. Hanzo, "Users first: User-centric cluster formation for interference-mitigation in visible-light networks," *IEEE Trans. Wirel. Commun.*, vol. 15, no. 1, pp. 39–53, Jan. 2016.

35 F. Wang, Z. Wang, C. Qian, L. Dai, and Z. Yang, "MDP-based vertical handover scheme for indoor VLC-WiFi systems," in *Proc. OptoElectronics and Communications Conference (OECC) 2015* (Shanghai, China), Jun. 28–Jul. 2, 2015, pp. 1–3.

36 P. Manousiadis, H. Chun, and S. Rajbhandari, "Demonstration of 2.3 Gb/s RGB white-light VLC using polymer based colour-converters and GaN micro-LEDs," in *Proc. IEEE Summer Topicals Meeting Series (SUM) 2015* (Nassau, Bahamas), Jul. 13–15, 2015, pp. 222–223.

37 A. Sewaiwar, P. P. Han, and Y. H. Chung, "3-Gbit/s Indoor visible light communications using optical diversity schemes," *IEEE Photon. J.*, vol. 7, no. 6, p. 7904609, Dec. 2015.

38 C. He, T. Q. Wang, and J. Armstrong, "Performance of optical receivers using photodetectors with different fields of view in a MIMO ACO-OFDM system," *J. Lightw. Technol.*, vol. 33, no. 23, pp. 4957–4967, Dec. 2015.

39 Y. Wang, L. Tao, X. Huang, J. Shi, and N. Chi, "8-Gb/s RGBY LED-based WDM VLC system employing high-order CAP modulation and hybrid post equalizer," *IEEE Photon. J.*, vol. 7, no. 6, p. 7904507, Dec. 2015.

40 Q. Wang, C. Qian, X. Guo, Z. Wang, D. Cunningham, and I. White, "Layered ACO-OFDM for intensity-modulated direct-detection optical wireless transmission," *Opt. Exp.*, vol. 23, no. 9, pp. 12382–12393, May 2015.

41 C. Gong, S. Li, Q. Gao, and Z. Xu, "Power and rate optimization for visible light communication system with lighting constraints," *IEEE Trans. Signal Process.*, vol. 63, no. 16, pp. 4245–4256, Aug. 2015.

42 Q. Gao, R. Wang, Z. Xu, and Y. Hua, "DC-informative joint color-frequency modulation for visible light communications," *J. Lightw. Technol.*, vol. 33, no. 11, pp. 2181–2188, Jun. 2015.

43 Q. Gao, C. Gong, S. Li, and Z. Xu, "DC-informative visible light communications under lighting constraints," *IEEE Wirel. Commun.*, vol. 22, no. 2, pp. 54–60, Apr. 2015.

44 X. Liu, C. Gong, S. Li, and Z. Xu, "Signal characterization and receiver design for visible light communication under weak illuminance," *IEEE Commun. Lett.*, vol. 20, no. 7, pp. 1349–1352, Jul. 2016.

45 H. Chen, C. Wu, H. Li, X. Chen, Z. Gao, S. Cui, and Q. Wang, "Advances and prospects in visible light communications," *J. Semicond.*, vol. 37, no. 1, p. 011001, Jan. 2016.

46 D. C. O'Brien, G. Faulkner, H. L. Minh, O. Bouchet, M. E. Tabach, M. Wolf, J. W. Walewski, S. Randel, S. Nerreter, M. Franke, K. D. Langer, J. Grubor, and T. Kamalakis, "Home access networks using optical wireless transmission," in *Proc. IEEE International Symposium on Personal, Indoor and Mobile Radio Communications (PIMRC) 2008* (Cannes, France), Sep. 15–18, 2008, pp. 1–5.

47 V. Jungnickel, M. Uysal, N. Serafimovski,

T. Baykas, D. O'Brien, E. Ciaramella, Z. Ghassemlooy, R. Green, H. Haas, P. A. Haigh, V. P. G. Jimenez, F. Miramirkhani, M. Wolf, and S. Zvanovec, "A European view on the next generation optical wireless communication standard," in *Proc. IEEE Conference on Standards for Communications and Networking (CSCN) 2015* (Tokyo, Japan), Oct. 28–30, 2015, pp. 106–111.

48 H. Parikh, J. Chokshi, N. Gala, and T. Biradar, "Wirelessly transmitting a grayscale image using visible light," in *Proc. International Conference on Advances in Technology and Engineering (ICATE) 2013* (Mumbai, India), Jan. 23–25, 2013, pp. 1–6.

49 C. W. Chow, C. H. Yeh, Y. F. Liu, and Y. Liu, "Improved modulation speed of LED visible light communication system integrated to main electricity network," *Electron. Lett.*, vol. 47, no. 15, pp. 867–868, Jul. 2011.

50 M. Kavehrad, "Sustainable energy-efficient wireless applications using light," *IEEE Commun. Mag.*, vol. 48, no. 12, pp. 66–73, Dec. 2010.

51 N. Bardsley *et al.*, "Solid-state lighting research and development: Multi-year program plan," U.S. Dept. Energy, Washington, DC, USA, Tech. Rep., 2014, [online], *http://www1.eere.energy.gov/buildings/ssl/techroadmaps.html.*

52 W. Huang, P. Tian, and Z. Xu, "Design and implementation of a real-time CIM-MIMO optical camera communication system," *Opt. Exp.*, vol. 24, no. 21, pp. 24567–24579, Oct. 2016.

53 J. Hu, C. Gong, and Z. Xu, "Demonstration of a robot controlling and positioning system based on visible light," in *Proc. International Conference on Wireless Communications & Signal Processing (WCSP) 2016* (Yangzhou,China), Oct. 13–15, 2016, pp. 1–6.

54 K. Cui, G. Chen, Z. Xu, and R. D. Roberts, "Traffic light to vehicle VLC channel characterization," *Appl. Opt.*, vol. 51, no. 27, pp. 6594–6605, Sep. 2012.

55 T. Yamazato, I. Takai, H. Okada, T. Fujii, T. Yendo, S. Arai, M. Andoh, T. Harada, K. Yasutomi, K. Kagawa, and S. Kawahito, "Image-sensor-based visible light communication for automotive applications," *IEEE Commun. Mag.*, vol. 52, no. 7, pp. 88–97, Apr. 2014.

56 T. Yamazato, M. Kinoshita, S. Arai, E. Souke, T. Yendo, T. Fujii, K. Kamakura, and H. Okada, "Vehicle motion and pixel illumination modeling for image sensor based visible light communication," *IEEE J. Sel. Area. Commun.*, vol. 33, no. 9, pp. 1793–1805, Sep. 2015.

57 Y. Goto, I. Takai, T. Yamazato, H. Okada, T. Fujii, S. Kawahito, S. Arai, T. Yendo, and K. Kamakura, "A new automotive VLC system using optical communication image sensor," *IEEE Photon. J.*, vol. 8, no. 3, p. 6802716, Jun. 2016.

58 IEEE Std. 802.15.7-2011, *Part 15.7: Short-Range Wireless Optical Communication Using Visible Light*, Sep. 2011.

59 "The IEEE 802.15.7r1 Study Group," [online], *http://www.ieee802.org/15/.*

60 S. H. Yu, O. Shih, H. M. Tsai, N. Wisitpongphan, and R. D. Roberts, "Smart automotive lighting for vehicle safety," *IEEE Commun. Mag.*, vol. 51, no. 12, pp. 50–59, Dec. 2013.

61 M. Uysal, F. Miramirkhani, O. Narmanlioglu, T. Baykas, and E. Panayirci, "IEEE 802.15.7r1 reference channel models for visible light communications," *IEEE Commun. Mag.*, vol. 55, no. 1, pp. 212–217, Jan. 2017.

62 S. Arnon, Visible Light Communication, *Cambridge University Press*, 2015.

2
Visible Light Communications: Channel and Capacity

In this chapter, the channel and capacity of visible light communication (VLC) are introduced. Specifically, the characteristics of light emitting diode (LED) as the transmitter and photodiode as the receiver are described in Section 2.1. When LED is employed for lighting and communication simultaneously, its nonlinearity and lighting constraints are investigated in Section 2.2. Besides, absorption coefficient, quantum efficiency, and responsivity of the photodiode are demonstrated in Section 2.3. Furthermore, different propagation links between the transmitter and the receiver are analyzed in Section 2.4. Since the dominant noise might be different in various application scenarios, three optical wireless channels are addressed including free-space optical intensity channel, discrete-time Poisson channel, and improved free-space intensity channel in Sections 2.5 and 2.6. Considering there exist no analytic expressions of the channel capacity, the state-of-the-art upper and lower bounds are presented in Section 2.6 as well.

2.1
LED characteristics

One of the first red LEDs was developed in 1962 based on GaAsP [1]. Compared to conventional lighting sources such as fluorescent and incandescent lights, LEDs have many advantages including energy efficiency, light density, lifetime, and reliability. Benefiting from the refinement of III-V alloy and the development of the epitaxy methods, LEDs have gained significant performance improvement over the last fifty years. The efficiency of commercial LEDs has been dramatically increased from 0.1 lm/W to a level that is above 100 lm/W. Currently, LEDs can emit the light covering all visible spectrum from short wavelength (i.e., violet) to long wavelength (i.e., red). As a result, LEDs have been widely applied in our daily lives, such as general lighting, traffic lights, and flat panel display. The market share of LEDs in global commercial lighting is continuously growing and the revenue from commercial LEDs sales would exceed 20 billion US dollars in the coming years. Although the price of LEDs is relatively higher than conventional light sources at present, it is foreseeable that the commercialization of LED associated with the advancement of

Visible Light Communications: Modulation and Signal Processing. First edition. Zhaocheng Wang, Qi Wang, Wei Huang, and Zhengyuan Xu.
Published 2017 by John Wiley & Sons, Inc.

the fabrication technique would further reduce their costs.

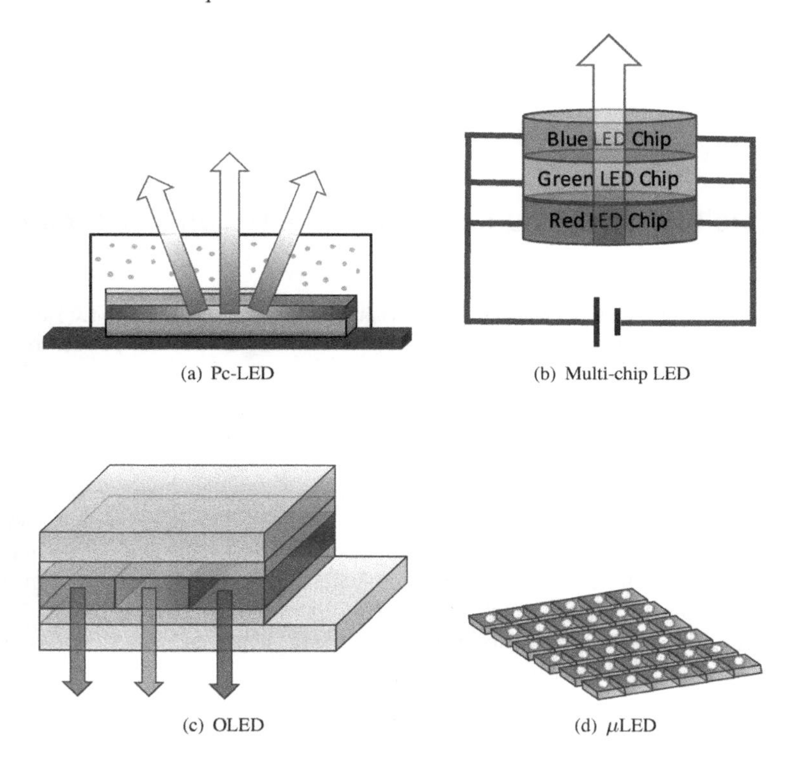

(a) Pc-LED

(b) Multi-chip LED

(c) OLED

(d) μLED

Figure 2.1 Different LED types.

Until now, there are various types of LEDs such as phosphor-converted LED (pc-LED), multi-chip LED, organic LED (OLED), and micro LED (μLED), which are shown in Fig. 2.1. Pc-LED and multi-chip LED are two common types of white LEDs for lighting, which use two or more different wavelength lights to generate the white light. In pc-LED packages, one or more visible light-emitting phosphors are coated on an LED chip emitting short-wavelength light. The pc-LEDs employ some of the short-wavelength light to pump the phosphors and produce long-wavelength light while the rest of the short-wavelength light is leaked out. By mixing these different wavelength lights together, the white light could be generated. Typical commercial pc-LEDs utilize the cerium doped yttrium aluminum garnet (Ce:YAG) phosphor to produce the yellow light and mix it with the blue light emitted by the gallium nitride based LED chip [2]. Due to the development of modern manufacturing technology, the luminous efficacy of pc-LED has been improved to above 150 lm/W [3]. However, the intrinsic modulation bandwidth of pc-LED is limited to several MHz due to the slow relaxation time of the phosphor [4]. On the other hand, multi-chip LEDs exploit three or more LED chips to emit different monochromatic lights and mix them together according to the predefined ratio to produce the white light, i.e., red-green-blue (RGB) LEDs. Multi-chip LEDs can provide variable color points and control the white light dynamically. The color rendering of mutli-chip LEDs is excellent (color rendering index > 95). With the help of external detectors including thermal, electrical or optical sensors, the undesirable high variability in the color point can be reduced considerably. Although multi-chip LEDs are more complex and expensive, their intrinsic modulation bandwidth is several times larger than that of pc-LEDs [5].

The basic structure of OLEDs is thin-film organic semiconductors sandwiched between the anode and the cathode. The luminescence mechanism for OLEDs is different from inorganic LEDs. In the recombination of electron-hole pair, a high-energy molecular state called singlet or triplet exciton is formed. The exciton would emit the light and its wavelength is related to the emitting layer material rather than the band gap. Organic LEDs based on small-molecular or polymer (SMOLEDs or PLEDs) are usually used in flat panel display. Compared to liquid crystals, OLEDs possess several advantages in energy efficiency, contrast ratio, refresh rate, and the capacity of vibrant color rendering.

μLED is an emerging type of LEDs that can be used in self-emissive micro-displays, multi-site photostimulation and hybrid inorganic/organic devices. Unlike liquid crystal displays, μLED displays are self-luminescent and power-efficient. Besides, they can support wide-angle viewing without color shift and degradation. Different from the organic materials of OLEDs which are chemically unstable, μLEDs inherit the advantages of inorganic LEDs and have a longer lifetime. Usually, μLED display integrates massive yet small LED elements. The size of each element is only $\mu m \times \mu m$ or smaller. The common method of fabricating μLED arrays is to arrange several microchip elements onto a substrate, and the elements of μLED are addressed individually, which increases the layout complexity. An alternative solution is to address the elements either row by row or column by column.

2.1.1 Operation principles

Common LEDs are generally based on the theory of p-n junction, a boundary between two types of semiconductor materials (i.e., p-type and n-type). For the p-type region, the holes are major carriers and the electrons are minor carriers. While for the n-type region, the electrons are major carriers and the holes are minor carriers. Without a bias voltage, the holes as the major carriers in the p-type region would diffuse into the n-type region and recombine with the electrons, leaving positively charged ions behind near the boundary of the n-type region. On the other side, the electrons in the n-type region would diffuse into the p-type region and recombine with the holes, leaving the negatively charged ions behind near the boundary of the p-type region. As a result, a built-in electric field known as diffusion voltage is formed in the boundary between the p-type region and the n-type region, which is

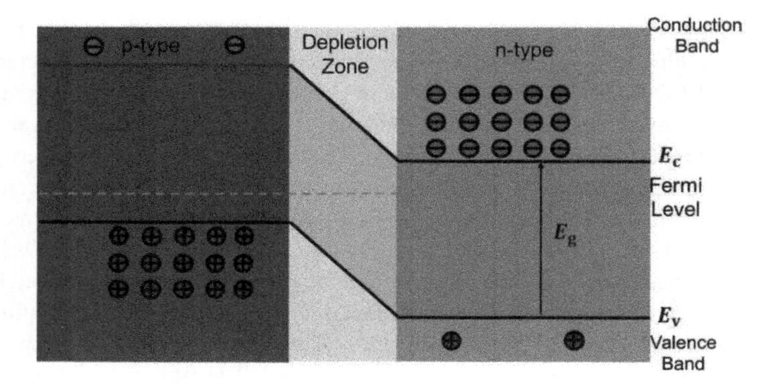

Figure 2.2 p-n junction under an equilibrium condition.

expressed as

$$V_D = \frac{kT}{q} \ln \frac{N_A N_D}{n_i^2}, \tag{2.1}$$

where q is the elementary charge, k is the Boltzmann constant, T is the absolute temperature, N_A and N_D are the acceptor concentration at p-type region and the donor concentration at n-type region, respectively, and n_i is the intrinsic carrier concentration of the semiconductor. The built-in voltage would obstruct the diffusion of the major carriers from the p-type and n-type regions and an equilibrium state is reached, which is illustrated in Fig. 2.2. The built-in space charge region between p-type region and n-type region is also called the depletion zone.

As shown in Fig. 2.3, when a forward bias voltage V is loaded on the p-n junction, where positive electrode is connected to p-type region and negative electrode is connected to n-type region, the holes in p-type region and the electrons in n-type region are injected into the opposite side, resulting in the width decrease of the depletion zone, which can be approximately given by

$$W_D \approx \sqrt{\frac{2\epsilon}{q} \frac{N_A + N_D}{N_A N_D} (V_D - V)}, \tag{2.2}$$

where ϵ is the dielectric permittivity of the semiconductor. If the depletion zone is thin enough, the electrons would cross the p-n junction into the neutral p-type region and recombine with the holes. Due to the energy/band gap between the electrons in the conduction band and the holes in the valence band, the recombination of the electron-hole pair could cause a photon emission. The energy/band gap (E_g) between the energy at the top of the valence band (E_v) and that at the bottom of the conduction band (E_c) can be expressed as

$$E_g = E_c - E_v. \tag{2.3}$$

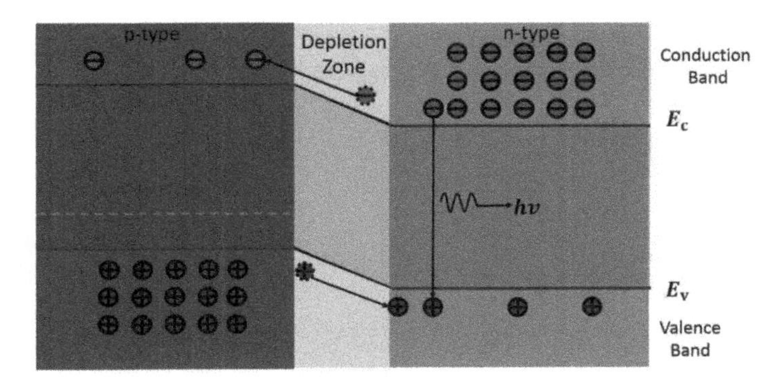

Figure 2.3 p-n junction under forward bias.

Here, the energy of the electrons in the conduction band is given by

$$E_e = E_c + \frac{h^2\hat{k}^2}{2m_e^*}, \tag{2.4}$$

and the energy of the holes in the valence band is given by

$$E_h = E_v - \frac{h^2\hat{k}^2}{2m_h^*}, \tag{2.5}$$

where h is Planck's constant, $\hat{k}$ is the wavenumber, m_e^* and m_h^* are the electron and hole effective masses, respectively. Since the energy of the electrons in the conduction band is higher than the energy of the holes in the valence band, during the spontaneous recombination of the electron-hole pair, the emitted photon energy can be expressed as

$$hv = E_e - E_h = E_g + \frac{h^2\hat{k}^2}{2m_r^*}, \tag{2.6}$$

where m_r^* is the reduced effective mass, which satisfies $\frac{1}{m_r^*} = (\frac{1}{m_e^*} + \frac{1}{m_h^*})$.

Since different materials have distinct band gaps, the emitted lights could have different wavelengths and present different colors. For example, the band gap of InGaP is ~1.9 eV (~650 nm, deep red). For $Al_xGa_{1-x}N$, its band gap varies from ~0.7 eV (~1800 nm, infrared) to ~3.4 eV (~365 nm, UV-A).

2.1.2
LED nonlinearity

For an ideal LED system model, the input (forward bias) is linear to the output (emitted optical power). In practice, LEDs always suffer from nonlinear distortion, which would degrade the system performance considerably. The classical Shockley ideal diode equation was proposed in the early 1950s to describe this nonlinearity effect between the current and the voltage, which is given by [6]

$$I = I_{\mathrm{S}}(e^{qV/kT} - 1), \tag{2.7}$$

where I_{S} is the saturation current, which is expressed as

$$I_{\mathrm{S}} = qA(\sqrt{\frac{D_{\mathrm{p}}}{\tau_{\mathrm{p}}}\frac{n_i^2}{N_{\mathrm{A}}}} + \sqrt{\frac{D_{\mathrm{n}}}{\tau_{\mathrm{n}}}\frac{n_i^2}{N_{\mathrm{D}}}}), \tag{2.8}$$

where A is the junction area, D_{p} and D_{n} denote the electron and hole diffusion constants, and τ_{p} and τ_{n} are the minority carrier lifetimes of electrons and holes. Since the diffusion constants, the minority carrier lifetimes, and the intrinsic carrier concentration are all temperature-dependent, the saturation current is not constant for a specific LED.

As forward voltage is typically much larger than thermal voltage (i.e., $\frac{kT}{e}$), the Shockley ideal diode equation can be simply approximated as

$$I = I_{\mathrm{S}} e^{qV/kT}. \tag{2.9}$$

In addition, the current-optical power (I–P) conversion is also nonlinear, which can be modeled as either memory-less model or memory model. A typical memory-less LED model is a polynomial model. Based on Taylor series, a polynomial approximation for the nonlinear transfer function can be obtained as

$$P = \sum_{n=0}^{N} \alpha_n (I - I_{\mathrm{DC}})^n, \tag{2.10}$$

where α_n is the coefficient of the nth order power of the nonlinear transfer function and I_{DC} denotes the direct current (DC). As the nonlinear transfer function is modeled to be static, the polynomial approximation is only valid when the modulation frequency is below 3-dB bandwidth of the LEDs [7].

Since an LED's capacitance and conductance are frequency-dependent, the polynomial model is not capable of describing the dynamics and memory effects of the LEDs accurately. Instead, a Volterra model, combining the nonlinearity and the memory effects together, is employed. The current-optical power conversion based on Volterra series for the continuous-time system can be expressed as

$$P(t) = P_0 + \sum_{n=1}^{\infty} \frac{1}{n!} P_n(t), \tag{2.11}$$

where P_0 is DC component of the optical power, and $P_n(t)$ is the nth order component of $P(t)$, which can be further detailed as

$$P_n(t) = \int_{-\infty}^{+\infty} \cdots \int_{-\infty}^{+\infty} h_n(\tau_1, ..., \tau_n) \prod_{k=1}^{n} I(t - \tau_k) d_{\tau_k}, \tag{2.12}$$

where $h_n(\tau_1, ..., \tau_n)$ is the nth order Volterra Kernel of the nonlinear system, which can be obtained from the measurement data [8]. The nth order Volterra Kernel $h_n(\tau_1, ..., \tau_n)$ can be regarded as the higher-order impulse response of the nonlinear system.

For the discrete-time case, the nth order component $P_n(m)$ can be expressed as

$$P_n(m) = \sum_{k_1=0}^{+\infty} \cdots \sum_{k_n=0}^{+\infty} h_n(k_1, ..., k_n) \prod_{j=1}^{n} I(m - k_j). \tag{2.13}$$

In practice, the series is truncated, i.e., the order of (2.12) and (2.13) is set to be a specific value instead of the infinity. The drawback of the Volterra model comes from its high complexity in estimating the Volterra coefficients.

2.2
LED lighting constraints

LED lighting constraints are crucial to modulation and signal processing for VLC systems, which include dimming control, chromaticity control, and flicker-free communication.

2.2.1
Dimming control

To describe the light brightness perceived by human eyes, spectral luminous efficiency function $V(\lambda)$ is defined by International Commission on Illumination (CIE), which indicates that the human visual system is more sensitive to the light with middle wavelengths compared to either short or long wavelengths. The perceived light power is measured as luminous flux, which is given by [9]

$$\Phi = K_\mathrm{m} \int_\lambda P(\lambda)V(\lambda)d\lambda, \tag{2.14}$$

where K_m is a constant of 683 lm/W to convert irradiance to illuminance and $P(\lambda)$ is the power spectral distribution. Accordingly, the luminous intensity is defined as

$$I_\mathrm{t} = \frac{d\Phi}{d\Omega}, \tag{2.15}$$

where Ω denotes the spatial angle.

Since LEDs are specific semiconductor devices that emit incoherent light when driven by current, the information to be conveyed is usually modulated into the instantaneous optical power of the LEDs. In indoor VLC systems, the brightness of LED light should be dimmed for the convenience of illumination. Usually, the driver circuit has a set of transistors that combine the dimming signal with the biased modulating signal and switch the LEDs.

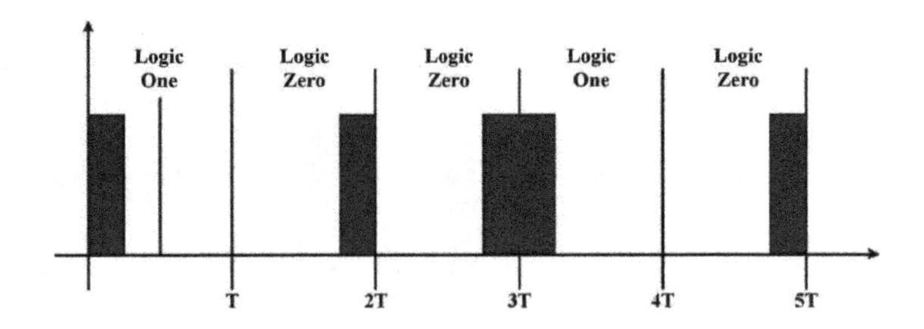

Figure 2.4 VPPM signal with 25% pulse width.

In recent times, VLC is mainly aimed at supporting high-rate transmission. However, lighting quality and power consumption, which are also crucial aspects of VLC systems, had mostly been overlooked. The lighting requirements for indoor scenarios are generally application specific. Bedroom/living room might require lighting levels as low as 1% of the maximum illumination for aesthetic and comfort purposes. An illuminance level of 300 lux (lumen per square meter) is preferred for reading and writing purpose, whereas 30 lux is sufficient for computer task [10]. Other locations such as corridors and stairwells have a flexible dimming requirement where life time and energy-saving are the primary considerations of the LEDs. The brightness of an LED is adjusted by controlling the forward current, which can be classified into analog dimming, digital dimming, and hybrid dimming. Analog dimming adjusts the current amplitude linearly to the radiated optical flux. In digital dimming, pulse width modulation (PWM) scheme is usually adopted, where the time period (T) of the PWM signal is fixed and the duty cycle varies proportionally to the required dimming level. Hybrid dimming combines both analog dimming and digital dimming for further reduction of perceived chromaticity shifts [11].

IEEE 802.15.7 standard uses both on-off keying (OOK) and variable pulse position modulation (VPPM) for VLC links. OOK dimming can be realized by adjusting the light intensity of both "on" and "off" status, or the light intensity could remain unchanged whereby the average duty cycle of the waveform can be adjusted by the insertion of "compensation" time into the modulation waveform. During the "compensation" time, the light source is fully turned on or off which allows a DC component to be added to the waveform. VPPM changes the duty cycle of each optical symbol based on the required dimming level. It is similar to 2-PPM when the duty cycle is 50%. The logic zero and logic one symbols are pulse width modulated depending on the dimming duty cycle requirements. The pulse width ratio of pulse position modulation (PPM) can be adjusted to produce the required duty cycle for supporting dimming. Figure 2.4 shows an example waveform indicating how VPPM can attain a 25% dimming duty cycle, where both logic zero and logic one have a 25% pulse width [12].

2.2.2 Chromaticity control

When multi-chip LEDs are used, chromaticity is a critical issue which presents the quality of a color regardless of its luminance. Since human color perception is determined by three types of cones in the retina of human eyes, which are sensitive to the light of long, middle, and short wavelengths, respectively, three tristimulus values (R, G, B) associated with their color matching functions $(\bar{r}(\lambda), \bar{g}(\lambda)$, and $\bar{b}(\lambda))$ are utilized to describe any color perception based on red/green/blue primaries. However, some portions of these color matching functions might be negative. As a result, a linear transformation is performed to obtain alternative positive tristimulus values (X, Y, Z) and color matching functions $(\bar{x}(\lambda), \bar{y}(\lambda)$, and $\bar{z}(\lambda))$, which are expressed as

$$X = K_{\mathrm{m}} \int_{\lambda} P(\lambda)\bar{x}(\lambda)d\lambda, \tag{2.16a}$$

$$Y = K_{\mathrm{m}} \int_{\lambda} P(\lambda)\bar{y}(\lambda)d\lambda, \tag{2.16b}$$

$$Z = K_{\mathrm{m}} \int_{\lambda} P(\lambda)\bar{z}(\lambda)d\lambda. \tag{2.16c}$$

Then, the chromaticity of the color can be represented by two coordinate points x and y in the CIE 1931 color space chromaticity diagram, which are defined as

$$x = \frac{X}{X+Y+Z}, \tag{2.17a}$$

$$y = \frac{Y}{X+Y+Z}. \tag{2.17b}$$

Good design of VLC systems shall guarantee that there is no color mismatching from human eyes' point of view [13]. Color shift keying (CSK), already adopted in the IEEE 802.15.7 standard, is an instance that considers the chromaticity control in signal modulation. CSK is similar to frequency shift keying whereby the bit patterns are encoded according to different color combinations [12]. The modulation scheme relies on the x-y color coordinates in the CIE 1931 color space chromaticity diagram to realize color matching. Specifically, in order to provide various colors for conveying the data information, the IEEE 802.15.7 standard breaks the spectrum into seven color bands to support multiple LED color choices for visible light communications. Figure 2.5 indicates the center of the seven color bands on the x-y color coordinates defined by CIE 1931 where the 3-bit values indicate each of the seven color bands. For example, in 4-CSK (two bits per symbol), the light source is wavelength keyed such that one of four possible wavelengths (colors) is transmitted per bit pair combination. Different wavelengths (colors) are generated by the three color light sources out of the seven color bands. The three vertices of the CSK constellation triangle are decided by the center wavelength of the three color bands on the x-y color coordinates. The final output color (e.g., white) is guaranteed by the color coordinates shown in Fig. 2.5.

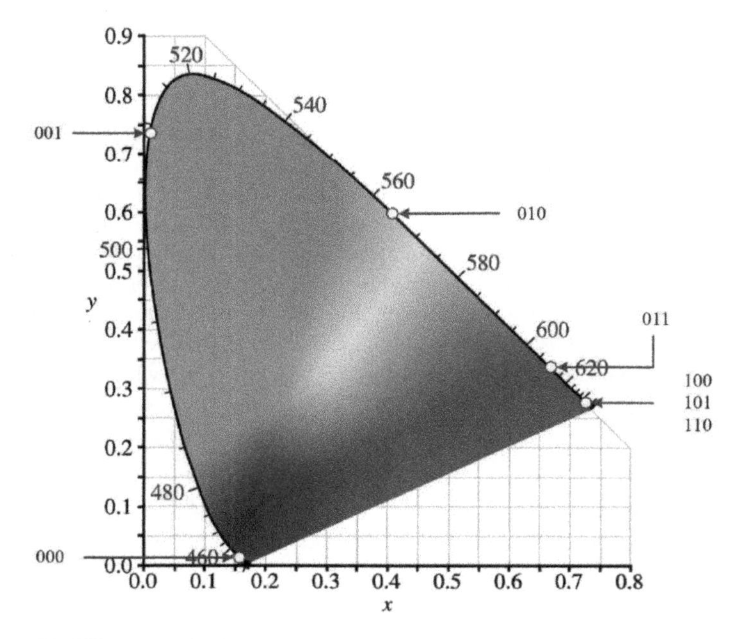

Figure 2.5 CIE 1931 x-y color coordinates, where x and y are the chromaticity values. The outer curve is the spectral locus with wavelengths shown in nm. The three-digit values refer to the center wavelength of the seven bands defined in the IEEE 802.15.7 standard.

2.2.3 Flicker-free communication

Flicker is defined as the periodic or non-periodic output power (brightness) fluctuation which human eyes can perceive. It can fatigue the eyes quickly and deteriorate the eye sight if they are exposed to a noticeable flicker for a long period of time. According to statistical analysis, about 1 in 4,000 people are highly susceptible to flashing lights cycling in the range from 3 to 70 Hz. Less well known is the fact that long time exposure to higher frequency and unintentional flicker in the range from 70 to 160 Hz could also cause malaise, headaches, and visual impairments. Unless human beings stay in natural daylight, they are likely to be exposed to such kind of flicker, since fluorescent lamps or LEDs are subject to flicker. Accordingly, many efforts have been carried out to design good-quality and fast-switching LED drivers in order to reduce the negative flicker effect.

Apparently, flicker mitigation technology is crucial in VLC systems. To facilitate flicker-free VLC, it is important to have a DC-free signal so that the average light intensity does not change. To achieve DC-free properties, DC-free modulation codes can be employed. A simple and commonly used modulation code is the binary Manchester code, where bit 1 is converted into symbols [+1,-1] and bit -1 is converted into symbols [-1,+1]. Flicker mitigation technology is classified into intra-frame flicker mitigation and inter-frame flicker mitigation in the IEEE 802.15.7 standard [12]. Intra-frame flicker mitigation aims to eliminate the flicker within the transmission of a data frame. For OOK and VPPM modulations, it is implemented by using the dimmed OOK mode and run length limited (RLL) line coding. RLL line codes are adopted to avoid long runs of 1's and 0's which could potentially cause flicker and clock recovery problems. Various RLL line codes such as Manchester, 4B6B, and 8B10B codes are defined in the IEEE 802.15.7 standard, and provide tradeoffs between coding overhead and ease of implementation. For CSK modulation, it is implemented by ensuring constant average power across multiple light sources along with scrambling and high optical clock rates. Inter-frame flicker mitigation applies to both data transmission and idle periods. While idling, visibility patterns or idle patterns may be used to ensure that light emission by the VLC transmitters have the same average brightness over adjacent maximum flicker time period.

2.3 Photodiode characteristics

A photodiode is used as the optical receiver to convert the optical signal to the electrical signal in visible light communications. When a photon with enough energy is absorbed in the photodiode, an electron moves from the valence band to the conduction band, resulting in the generation of an electron-hole pair. In this process, the energy of the photon hv should not be less than the energy gap between the valence band and the conduction band E_g, i.e., $hv \geq E_\mathrm{g}$.

Usually, the photodiode is driven by the reverse voltage, where the anode connects to the negative terminal while the cathode connects to the positive terminal. If the absorption occurs in the depletion zone, the built-in electric field would impel the separation of the electron-hole pairs. The holes would drift toward the anode and the electrons drift toward the cathode. Consequently, the photocurrent is generated. The reverse voltage can strengthen the built-in field in the depletion zone to accelerate the drift of the photon-induced carriers and enlarge the length of the depletion zone as well as decrease the capacitance so that the response time is shortened.

Both direct-bandgap (InGaAs, GaAs, etc) and indirect-bandgap (Si, Ge, etc) semiconductors can be used for photodiodes. Compared to the indirect-bandgap semiconductors, direct-bandgap counterparts often have higher absorption coefficients. When absorbing the same amount of the light, the absorption region of the direct-bandgap semiconductors is thinner than their indirect-bandgap counterparts. In practice, the materials such as Ge and InGaAs are chosen to fabricate the photodiodes for receiving long-wavelength light (infrared spectral range). As for short-wavelength light (200–1600 nm), the receiver material is preferred to be Si. Thus, Si-based photodiodes are commonly used in VLC systems.

There are various types of photodiodes such as PN photodiode, PIN photodiode and avalanche photodiode (APD). PN photodiode consists of a thin p-type, highly doped layer and an n-type substrate. Its frequency response exhibits a double cutoff: the lifetime cutoff (MHz) due to the lifetime of the carrier in the diffusion regions at p-type and n-type sides, and the RC cutoff (GHz) due to the transit time and capacitance effects. The performance of PN photodiode is usually limited by the carrier lifetime with a maximum cutoff frequency of 100–200 MHz.

Compared to PN photodiode, an additional intrinsic region is sandwiched between the p-type region and the n-type region in PIN photodiode in order to improve the frequency response and the high-frequency efficiency. In PIN photodiode, the depletion region is much larger than the carriers' diffusion region. As a result, the photocurrent due to carrier diffusion at the p-type and n-type region can be ignored and the cutoff frequency is increased to the order of GHz. The frequency response of PIN photodiode is limited by the photodiode capacitance and the transit time of the carriers drifting through the depletion region.

As a highly sensitive semiconductor, APD utilizes the impact ionization process to detect and amplify the current. In APDs, the photon-generated carriers produce more electron-hole pairs by collision with bounded electrons, i.e., the impact ionization. The photon-generated carriers occur in the generation region and the avalanche multiplication happens in the multiplication region. For conventional APDs, these two regions are the same. While for a sperate absorption and multiplication APD (SAM-APD), these two regions are physically separated. Compared to PIN photodiodes, APD has higher sensitivity.

A. Absorption coefficient

According to the Beer-Lambert law, the received radiant flux Φ_r in the photodiodes is given by

$$\Phi_{\mathrm{r}} = \Phi_{\mathrm{t}} \cdot e^{-\tau}, \tag{2.18}$$

where Φ_t is the radiant flux of the light penetrating into the surface of the photodiode and τ is the optical depth, which is defined as

$$\tau = \int_0^d \alpha(z)dz, \tag{2.19}$$

where d is the thickness of the photodiode that the light goes through and $\alpha(\cdot)$ is the attenuation/absorption coefficient, which describes how far the light with specific wavelength penetrates into the photodiode before it is totally absorbed. A higher absorption coefficient is usually in favor of the improvement of the photodiode's quantum efficiency.

If the attenuation is uniformly distributed, (2.18) can be rewritten as

$$\Phi_{\mathrm{r}} = \Phi_{\mathrm{t}} \cdot e^{-\alpha d}. \tag{2.20}$$

At this time, the coefficient α is also referred to as the linear attenuation coefficient, which is related to the photodiode material as well as the wavelength of the light.

B. Quantum efficiency

Quantum efficiency describes the photodiode's sensitivity to the light. Two different types of quantum efficiency are often discussed, which are internal quantum efficiency and external quantum efficiency, respectively. The internal quantum efficiency η_i is defined as the ratio of the number of the charge carriers (corresponding to the photocurrent) to the number of the photons (corresponding to the light absorbed by the photodiode), i.e.,

$$\eta_i = \frac{I_{pc}/q}{(\Phi_t - \Phi_r)/hv}, \tag{2.21}$$

where I_{pc} is the photocurrent.

One typical internal quantum efficiency model for the silicon photodiode is expressed as [14]

$$\eta_i = c + \frac{(1-c)}{\alpha D}(1 - e^{-\alpha D}) - \frac{W_n e^{-\alpha W_{p,d}}}{\alpha L^2}, \tag{2.22}$$

where c denotes the collection efficiency at the silicon-silicon dioxide interface, D is the depth where the collection efficiency is unity, $W_{p,d}$ is the width of the p-type region and the depletion region, W_n is the width of the n-type region and L defines the depth at which the minority carriers diffuse in the n-type region. In (2.22), the first two terms describe the collection efficiency from the oxide interface to the point $W_{p,d}$ while the third term is an approximation of the collection efficiency in the n-type region with a uniform dopant concentration.

In addition, the external quantum efficiency is defined as the ratio of the number of the charge carriers (corresponding to the photocurrent) to the number of the photons (corresponding to the incident light Φ_i at the surface of the photodiode), i.e.,

$$\eta_e = \frac{I_{pc}/q}{\Phi_i/hv}. \tag{2.23}$$

C. Responsivity

Similar to the quantum efficiency, responsivity is defined in units of amperes per watt of radiant power (A/W), which depends on the wavelength of the incident light. The conversion from the external quantum efficiency to the responsivity is given by

$$R = \eta_e \frac{q}{hv}. \tag{2.24}$$

2.4 Propagation links

Based on whether there exists a line-of-sight (LOS) path between the transmitter and the receiver, the propagation links can be classified into two categories: LOS link and non-line-of-sight (NLOS) link. Since the power of the LOS path dominates and the power of the reflected paths is much lower, an LOS wireless system usually has a higher power efficiency. However, the optical channel vulnerability is a tricky problem. When the LOS path is blocked by moving objects, the system performance would be rapidly deteriorated, even leading to communication interruption. While for a NLOS wireless system, the lights radiated by the transmitter are reflected by the surfaces of the ceiling or walls within a room. Compared to the LOS link scenarios, multipath propagation improves the robustness of the NLOS link based VLC systems. Even when barriers exist between the transmitter and the receiver, the signals via reflected paths can still be detected.

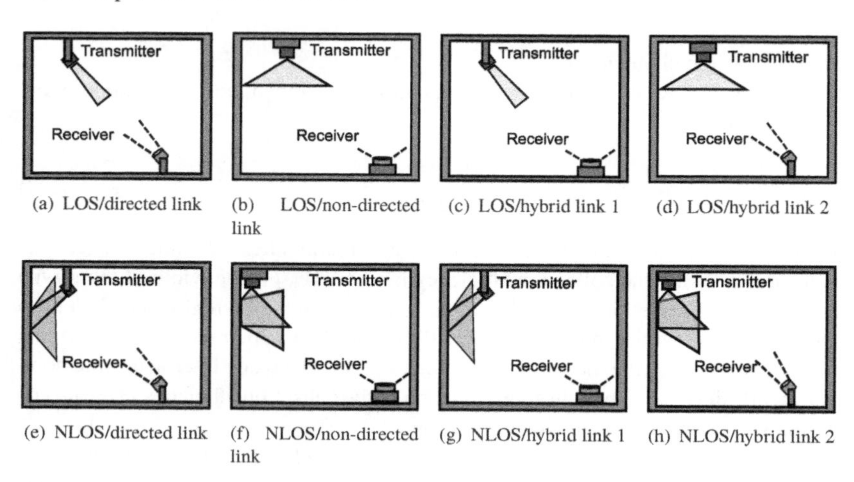

(a) LOS/directed link (b) LOS/non-directed link (c) LOS/hybrid link 1 (d) LOS/hybrid link 2

(e) NLOS/directed link (f) NLOS/non-directed link (g) NLOS/hybrid link 1 (h) NLOS/hybrid link 2

Figure 2.6 Classifications of the propagation links [15].

Considering the directionality of the transmitter and the receiver, the VLC propagation links could be also classified into three categories: directed link, non-directed link, and hybrid link. For the directed link, the transmitter and the receiver directly point to each other with narrow semiangle and field of view (FOV). Thus, the directed link based system has a high power efficiency. While in the non-directed link, both the transmitter and the receiver have wide semiangles for ease of use. As for the hybrid link, the transmitter and the receiver have different directionality (narrow semiangle transmitter in combination with wide FOV receiver or wide semiangle transmitter in combination with narrow FOV receiver). The classifications of these propagation links are shown in Fig. 2.6.

In conventional radio frequency communications, multipath propagation could cause the variation of the magnitude of the received electromagnetic signals and inter-symbol interference. By contrast, VLC is usually free from multipath fading because the physical detection area of the photodiode is much larger than the square wavelength of the light. The inherent size of the photodiode could be treated as a two-dimensional antenna array, which provides spatial diversity to eliminate the multipath fading effect. However, although multipath fading is neglected in VLC systems, time spread due to multipath propagation is still an issue for signal detection, especially for the NLOS link case.

2.4.1 LOS link

For most LEDs, the generalized Lambert law is applied, which indicates that the radiant intensity is relevant to the cosine of the angle θ between the emitted light and the normal to the LED surface. Thus, the radiant intensity can be expressed as [16]

$$R(\theta) = \frac{m+1}{2\pi} \cos^m(\theta) P_{\mathrm{LED}}, \tag{2.25}$$

where P_{LED} is the total radiated power, which is given by

$$P_{\mathrm{LED}} = \int_{\lambda} P(\lambda) d\lambda, \tag{2.26}$$

where $P(\lambda)$ is the spectral power distribution and m is the order of Lambertian emission, which depends on the semiangle at half illuminance of the LED $\Phi_{1/2}$, i.e.,

$$m = -\frac{\ln 2}{\ln \cos \Phi_{1/2}}. \tag{2.27}$$

The coefficient $\frac{m+1}{2\pi}$ ensures that the radiant intensity integration over the surface of a hemisphere equals the total optical power.

When the distance is defined as d, the irradiance can be given by

$$E_{\mathrm{e}}(d) = \frac{R(\theta)}{d^2}. \tag{2.28}$$

For the optical receiver, the detected optical power is proportional to the effective signal-collection area, which is given by

$$A_{\mathrm{e}}(\psi) = \begin{cases} A f(\psi) g(\psi) \cos(\psi), & \text{if } 0 \leq \psi \leq \Psi_{\mathrm{c}}, \\ 0, & \text{if } \psi > \Psi_{\mathrm{c}}, \end{cases} \tag{2.29}$$

where ψ is the incident angle, A is the photodetector physical area, Ψ_{c} is the concentrator FOV, $f(\psi)$ and $g(\psi)$ denote the optical filter gain and the concentrator gain, respectively. For a non-imaging concentrator, $g(\psi)$ can be expressed as

$$g(\psi) = \begin{cases} \frac{n^2}{\sin^2(\Psi_{\mathrm{c}})}, & \text{if } 0 \leq \psi \leq \Psi_{\mathrm{c}}, \\ 0, & \text{if } \psi > \Psi_{\mathrm{c}}, \end{cases} \tag{2.30}$$

where n is the internal refractive index. Thus the received optical power can be obtained by

$$P_{\mathrm{r}} = E_{\mathrm{e}}(d) A_{\mathrm{e}}(\psi). \tag{2.31}$$

For VLC channel, the frequency response is relatively flat near DC and only the LOS path is considered to calculate the DC gain. From (2.28) and (2.29), it can be expressed as

$$H_{\mathrm{LOS}} = \begin{cases} \frac{(m+1)A}{2\pi d^2}\cos^m(\theta)f(\psi)g(\psi)\cos(\psi), & \text{if } 0 \leqslant \psi \leqslant \Psi_c, \\ 0, & \text{if } \psi > \Psi_c. \end{cases} \tag{2.32}$$

It can be observed from (2.32) that if photodetector physical area A, the optical filter gain $f(\psi)$ and the concentrator gain $g(\psi)$ are fixed, the DC gain depends on the distance d between the transmitter and the receiver, the irradiation angle θ and the incident irradiation angle ψ. If the VLC receiver is moving in the room, it is important to adjust some of those factors to guarantee the system performance.

2.4.2 NLOS link

In the NLOS link case, the DC gain is calculated via the multiple reflections from the surfaces of a room. The impulse response of multiple bounces is expressed as

$$h(t) = \sum_{k=0}^{\infty} h^{(k)}(t; P(\lambda)), \tag{2.33}$$

where k is the number of bounces. For the ith reflected path, the $(i-1)$th bounce point is regarded as a virtual lighting source obeying the Lambert law and the ith bounce point is treated as a virtual receiver. After k bounces, the impulse response can be recursively calculated, which is given by [17]

$$\begin{aligned} h^{(k)}(t; P(\lambda)) = & \frac{1}{P_{\mathrm{LED}}}\int_s [L_1 L_2 \cdots L_{k+1} P_n^{(k)} \mathrm{rect}(\frac{\psi_{k+1}}{\Psi_c}) \times \\ & \delta(t - \frac{d_1 + d_2 + \cdots + d_{k+1}}{c})]dA_s, \end{aligned} \tag{2.34}$$

where dA_s is a small reflection area, $\mathrm{rect}(\cdot)$ is the rectangular function, i.e.,

$$\mathrm{rect}(x) = \begin{cases} 1, & \text{if } |x| \leq 1, \\ 0, & \text{if } |x| > 1, \end{cases} \tag{2.35}$$

$\delta(\cdot)$ is the delta function, c is the light velocity, $d_1, d_2, \cdots$, and d_{k+1} are the distance of each path during k bounces, $L_1, L_2, \cdots, L_{k+1}$ are the path loss/DC gain of each path during k bounces, which are expressed as

$$L_1 = \frac{(m+1)A_{\mathrm{ref}}}{2\pi d_1^2}\cos^m(\theta_1)\cos(\psi_1), \tag{2.36}$$

$$L_2 = \frac{A_{\mathrm{ref}}}{\pi d_2^2}\cos(\theta_2)\cos(\psi_2), \tag{2.37}$$

$$\vdots$$

$$L_{k+1} = \frac{A_{\mathrm{PD}}}{\pi d_{k+1}^2} \cos(\theta_{k+1}) f(\psi_{k+1}) g(\psi_{k+1}) \cos(\psi_{k+1}), \tag{2.38}$$

and $P_n^{(k)}$ denotes the optcial power of the reflected light after k-bounces, which is given by

$$P_n^{(k)} = \int_\lambda P(\lambda)\gamma_1(\lambda)\gamma_2(\lambda)\cdots\gamma_k(\lambda)d\lambda, \tag{2.39}$$

where $\gamma_i(\lambda)$ is the spectral reflectance of the surface at the ith bounce.

When the spectral reflectance is assumed to be wavelength-independent, i.e., $\gamma(\lambda) = \gamma$, the DC gain based on the first reflection is given by

$$H_{\mathrm{ref}}^{(1)} = \int_s \frac{(m+1)A}{2\pi^2 d_1^2 d_2^2} \gamma \cos^m(\theta_1) \cos(\theta_2) f(\psi) g(\psi) \cos(\psi_1) \cos(\psi_2) dA_s. \tag{2.40}$$

2.5 Noise in VLC systems

The noise existing in VLC systems can be classified into two categories: noise from the light including the quantum noise (or photon fluctuation noise) from the optical signal itself, and the background radiation noise from ambient light, and noise from the receiver devices such as dark current noise, thermal noise, and $1/f$ noise. Many types of noise can be regarded as shot noise in the wireless optical link, such as dark current noise, quantum noise, and background radiation noise. Specifically, dark current noise is due to random generation of electrons and holes within the depletion region without photon-induced excitation, which is signal-independent. Quantum noise is produced by the random arrival rate of photons from the optical source, which is signal-dependent. On the other hand, background radiation noise is caused by the reception of the photons from ambient light, which is signal-independent and can be modeled as being additive, white, and Gaussian due to its high intensity. For many application scenarios, the received signal-to-noise ratio (SNR) is limited by the background radiation noise, which is much stronger than the quantum noise from the optical source as well as other noise sources.

Next, various noise types in VLC systems are detailed as below.

A. Quantum noise

Quantum noise or photon fluctuation noise is caused by the discrete nature of the photons from the optical source. When the optical power from the light source is kept unchanged, the number of arrival photons is statistically constant during a long period. However, in a short time interval, the number of photons follows a Poisson distribution, i.e.,

$$P(n = k) = \frac{\lambda^k}{k!} e^{-\lambda}, \quad k = 1, 2, \cdots, \tag{2.41}$$

where λ is the average number of arrival photons per interval and n is the number of arrival photons in a given time interval.

Since intensity modulation is usually employed in VLC systems, the quantum noise always appears to be shot noise, which has a one-sided power spectral density in unit of A^2/Hz as

$$\sigma^2 = 2qi_{\text{pc}}, \tag{2.42}$$

where q is the electronic charge, i_{pc} denotes the photocurrent and we have $i_{\text{pc}} = RP_{\text{LED}}$, where R is the photodiode responsivity and P_{LED} is the light source power.

B. Background radiation noise

Background radiation noise or ambient light noise is caused by the reception of the photons from the environment. Ambient light sources include the sun, the sky, incandescent lamps, and fluorescent lamps. Background radiation noise is signal-independent and can be modeled as being additive, white, and Gaussian due to its high intensity. In the NLOS link case where a wide FOV receiver is employed, the received SNR is limited by the background radiation noise that is much stronger than the quantum noise from the optical source as well as other noise sources even with the adoption of the optical filters.

When the spectral radiance L_{e} ($\mathrm{W{\cdot}m^{-2}{\cdot}sr^{-1}{\cdot}Hz^{-1}}$) is assumed to be independent of the wavelength, the received background noise power can be expressed as [18]

$$P_{\text{bg}} = L_{\text{e}} \Omega_{\text{s}} A T_0 g(\psi) B_{\text{opt}} \frac{\cos(\psi)}{\cos(\theta)}, \tag{2.43}$$

where Ω_{s} is part of the FOV subtended by the background source at the receiver, T_0 is the transmittance of the atmosphere, and B_{opt} is the bandwidth of the optical filter. From (2.43), the background noise power strongly depends on the FOV and the optical bandwidth of the receiver, and its variance is given by

$$\sigma_{\text{bg}}^2 = 2qB_{\text{pd}} R P_{\text{bg}}, \tag{2.44}$$

where B_{pd} is the electrical bandwidth of the photodiode.

C. Thermal noise

Thermal noise or Jonson-Nyquist noise is caused by the random fluctuation of the charge carriers (usually electrons) in any conducting medium at a temperature higher than the absolute zero temperature. The power spectral density of thermal noise remains constant ("white") in a wide frequency range up to the near-infrared frequency. Considering the independent agitation of massive charge carriers, the thermal noise obeys Gaussian distribution according to the central limit theorem. The variance of the thermal noise in the noisy resistor in $A^2 \cdot Hz^{-1}$ is given by

$$\sigma_{\text{thermal}}^2 = \frac{4KT}{R_F}, \tag{2.45}$$

where R_F is the resistance.

Thermal noise in the noisy resistor can be modeled as a voltage source ($V_{\text{eq}} = \sqrt{4KTR_F}$) in series or a current source ($I_{\text{eq}} = \sqrt{\frac{4KT}{R_F}}$) in parallel with a noiseless resistor with a resistance R_F. In both cases, the sources generate the Gaussian white noise.

D. $1/f$ noise

$1/f$ noise is an intermediate between white noise and Brownian noise caused by Brownian motion, whose power spectral density is given by

$$S_{1/f}(f) = \frac{c}{f^\alpha}, \tag{2.46}$$

where c is a constant and α denotes the exponent satisfying $0 < \alpha < 2$ (usually close to 1). $1/f$ noise is not white and becomes strong at low frequencies.

E. Dark current noise

Dark current is an electric current, which exists in the photodiode even when there is no incident light. The dark current in the p-n junction based devices consists of the surface and bulk currents which are caused by the random generation of the electron-hole pairs thermally or tunneling between the conduction band and the valence band. Thus, it is related to the loaded bias voltage and photodiode temperature.

The dark current can be divided into two categories: the surface dark current and the bulk current. The surface current contains the surface generation-combination current and the surface leakage shunt current, while the bulk current is made of the bulk diffusion current, the bulk generation-combination current and the bulk tunneling current. Since the dark current causes random fluctuations of the average photocurrent, it usually exhibits as shot noise with a variance of

$$\sigma_d^2 = 2qB_{\text{pd}}i_d, \tag{2.47}$$

where i_d is the dark current.

2.6 Channel capacity

For intensity-modulated direct-detection (IM/DD)-based VLC systems, the classic Shannon channel capacity formula is considered, i.e.,

$$C = \log(1 + \text{SNR}). \tag{2.48}$$

However, it might not be suitable for actual optical wireless channels due to the following reasons:

- In IM/DD-based VLC systems, the intensity/amplitude of the signal would determine the voltage loaded on LEDs and the instantaneous emitted optical power. Thus, the signal is constrained to be nonnegative and real-valued. Moreover, LEDs have a maximum permissible forward current, resulting in a limited maximum amplitude value of the signal. As a result, the transmitted signal is confined to be unipolar and amplitude-limited.
- Traditionally, if the electrical power of the signal is constrained, the signal following a Gaussian distribution would approach Shannon channel capacity. However, since LEDs are primarily used for illumination, the signal is subject to the average optical power instead of the electrical power. When the electrical power constraint is not applied, the input signal distribution to approach Shannon channel capacity does not necessarily follow Gaussian distribution in optical wireless channels.

In recent years, the capacities for different optical wireless channel models (e.g., free-space optical intensity channel, discrete-time Poisson channel, and improved free-space optical intensity channel) have been investigated. Several capacity bounds are derived based on three constraints:

1) The nonnegative constraint, which is illustrated as

$$P(x < 0) = 0; \tag{2.49}$$

2) The peak-power constraint, which is given by

$$P(x > A) = 0; \tag{2.50}$$

3) The average-power constraint, which is expressed as

$$E(x) \leq P. \tag{2.51}$$

Moreover, for analysis, the average-to-peak power ratio is defined as

$$\alpha = \frac{P}{A}. \tag{2.52}$$

Next, different channel models in actual VLC systems are described. Afterward, the corresponding capacity bounds are introduced.

2.6.1 Channel models

The optical wireless channels can be divided into three categories: free-space optical intensity channel, discrete-time Poisson channel, and improved free-space intensity channel.

A. Free-space optical intensity channel

Free-space optical intensity channel is commonly employed in visible light communications where the noise exhibits white and Gaussian features and is independent of the signal [19–24]. This channel model can be described as

$$Y = X + Z, \tag{2.53}$$

where X and Y are the input and output signals, respectively, and Z is Gaussian noise with zero mean and variance of σ^2. Without loss of generality, the channel gain is assumed to be unit in the following analysis. The conditional probability distribution of the output signal is given by

$$W(y|x) = \frac{1}{\sqrt{2\pi\sigma^2}} e^{-\frac{(y-x)^2}{2\sigma^2}}. \tag{2.54}$$

(2.54) is valid when the background radiation noise and the thermal noise are dominant in VLC systems. However, if the power of data-carrying optical source is larger than ambient lighting and the noise power in electronic devices, this model cannot describe the channel characteristics precisely since the dominant noise source might be the quantum noise caused by the discrete nature of the photons from the optical source.

B. Discrete-time Poisson channel

Discrete-time Poisson channel is an alternative channel model widely investigated in VLC systems [25, 26], where the discrete nature of the photons is taken into consideration. The transmission signal is modeled as a Poisson counting process, where the number of arrival photons remains statistically constant during a long period but varies in a short time interval. The arrival rate of the photons from the data-carrying optical source is proportional to its average optical power. At the receiver, a photon counter is employed to measure the number of the received photons. Usually, the received photons consist of the photons from the data-carrying optical source and the photons from the background radiation noise. The conditional probability distribution of the output signal is given by

$$W(y|x) = \frac{(x+\lambda)^y}{y!} e^{-(x+\lambda)}, \tag{2.55}$$

where λ is the arrival rate of the photons from the background radiation noise.

C. Improved free-space optical intensity channel

Improved free-space optical intensity channel is discussed in [27], where the output signal follows Gaussian distribution but its variance is signal-dependent. In this model, quantum noise, background radiation noise, and thermal noise are considered and assumed to be Gaussian. The quantum noise is signal-dependent and its variance is proportional to the signal strength while the background radiation noise and the thermal noise are signal-independent. The output signal is given by

$$Y = X + \sqrt{X}Z_1 + Z_0, \tag{2.56}$$

where we have $Z_1 \sim N(0, \varsigma^2\sigma^2)$ and $Z_0 \sim N(0, \sigma^2)$. ς denotes the ratio of the power of the signal-dependent noise to the signal-independent noise. Accordingly, the conditional probability distribution of the output signal can be expressed as

$$W(y|x) = \frac{1}{\sqrt{2\pi\sigma^2(1+\varsigma^2 x)}} e^{-\frac{(y-x)^2}{2\sigma^2(1+\varsigma^2 x)}}. \tag{2.57}$$

2.6.2
Capacity bounds for free-space optical intensity channel

A. Upper bounds

The capacity upper bounds of free-space optical intensity channel have been widely investigated. From the geometric point of view, based on the well-known sphere packing method [28], tight closed-form upper bounds are derived in [21, 22] using the Steiner-Minkowski formula when only an average-power constraint is imposed. Nevertheless, obvious gaps exist between these upper bounds and the lower bounds in the high optical signal-to-noise ratio (OSNR) region due to the mathematical approximation of the intrinsic volumes of the simplex. To narrow the gap between the upper and the lower bounds in the high OSNR region, alternative mathematical approximation for the intrinsic volumes of the simplex has been proposed in [23]. In [24], an even tighter sphere-packing upper bound is derived using a recursive approach. However, the notable gap in the low OSNR region is still unsolved.

In [19], tight upper bounds are derived using a dual expression for channel capacity with different ratios of the peak power to the average power. When both peak-power and average-power constraints are applied and the ratio of the peak power to the average power is fixed, they would coincide with the lower bound as the average power tends to infinity. The drawback of these upper bounds comes from their higher complexity than the sphere-packing bounds.

1) ***Sphere packing***

When m independent transmitted symbols are denoted as $\mathbf{x} = [x_1\, x_2\, \cdots\, x_m]$ and the average optical power in a VLC channel is defined as P, the admissible set of the transmitted symbols is given by

$$T(P) = \left\{ \mathbf{x} \,\middle|\, \forall i, x_i \in \mathbb{R}^+, \frac{1}{m}\sum_{i=1}^{m} x_i \le P \right\}. \tag{2.58}$$

These elements in the admissible set form a regular m-simplex [29], and the received signal is calculated by

$$\mathbf{y} = \mathbf{x} + \mathbf{z}. \tag{2.59}$$

From the geometrical point of view, each element in the signal vector $\mathbf{x}$ is surrounded by a small uncertainty region caused by the noise $\mathbf{z}$ [28]. When the noise is white and Gaussian, the perturbations of the received samples are independent. If m is large enough, the perturbation would lie within an m-dimensional sphere $B_m(\rho)$ with the radius ρ and centered at the original signal point. The volume of the m-dimensional sphere $B_m(\rho)$ is expressed as [28]

$$V(B_m(\rho)) = \frac{\pi^{m/2}}{\Gamma(\frac{m}{2}+1)}\rho^m, \tag{2.60}$$

where $\Gamma(\cdot)$ is gamma function and $\rho = \sqrt{m}\sigma$.

In other words, the received signal would fall into the set $O(P, \rho)$ defined as the outer parallel body to the admissible set $T(P)$ at distance ρ with the probability near to one, which is illustrated as

$$O(P, \rho) = \left\{ \mathbf{y} \,\middle|\, \mathbf{y} = \mathbf{x} + \mathbf{z}, \mathbf{x} \in T(P), \mathbf{z} \in B_m(\rho) \right\}. \tag{2.61}$$

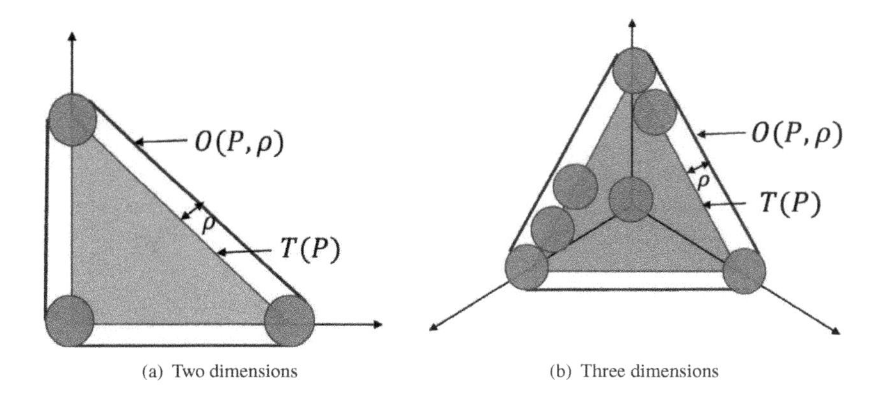

Figure 2.7 Illustrations of m-simplex and its outer parallel body.

The two-dimensional and three-dimensional illustrations of m-simplex and its outer parallel body are shown in Fig. 2.7(a) and Fig. 2.7(b), respectively. Based on the outer parallel body $O(P, \rho)$, the number of nonoverlapping spheres packed in $O(P, \rho)$ is the maximum number of the different and distinguishable transmitted symbols, and can be regarded as the upper bound on channel capacity, which is expressed as [28]

$$C \leq \lim_{m\to\infty} \frac{1}{m} \log \frac{V(O(P, \rho))}{V(B_m(\rho))}. \tag{2.62}$$

Likewise, when only peak-power constraint is applied, the admissible set is defined as

$$T(A) = \left\{ \mathbf{x} \,\middle|\, \forall i, x_i \in \mathbb{R}^+, x_i \leq A \right\}. \tag{2.63}$$

It forms an m-dimensional cube and its outer parallel is given by

$$O(A, \rho) = \left\{ \mathbf{y} \,\middle|\, \mathbf{y} = \mathbf{x} + \mathbf{z}, \mathbf{x} \in T(A), \mathbf{z} \in B_m(\rho) \right\}. \tag{2.64}$$

The upper bound on channel capacity is expressed as

$$C \leq \lim_{m \to \infty} \frac{1}{m} \log \frac{V(O(A, \rho))}{V(B_m(\rho))}. \tag{2.65}$$

The two-dimensional representations of the m-dimensional cube and its outer parallel body are illustrated in Fig. 2.8(a). In Fig. 2.8(b), the method of the sphere packing is presented.

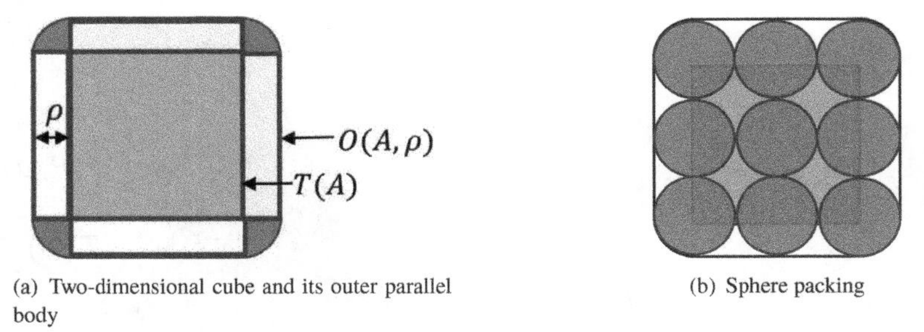

(a) Two-dimensional cube and its outer parallel body

(b) Sphere packing

Figure 2.8 Two-dimensional cube with its outer parallel body and sphere packing.

To obtain the closed-form expressions for the upper bound, two methods are used, which are illustrated as Steiner-Minkowski formula and recursive approach.

- ***Steiner-Minkowski formula***

The volume of the outer parallel body $V(O(P, \rho))$ is expressed as [30]

$$V(O(P, \rho)) = \sum_{k=0}^{m} V_k(P) V(B_{m-k}(\rho)), \tag{2.66}$$

where the intrinsic volume $V_k(P)$ is given by

$$V_k(P) = \eta_k (mP)^k, \qquad k = 0, 1, ..., m, \tag{2.67}$$

and

$$\eta_k = \begin{cases} \frac{1}{m!}, & \text{if } k = m, \\ \binom{m}{k} \frac{1}{k! 2^{m-k}} + \binom{m}{k+1} \frac{\sqrt{k+1}}{k! \sqrt{\pi}^{m-k}} \\ \times \int_0^{+\infty} e^{-x^2} (\int_{-\infty}^{x/\sqrt{k+1}} e^{-y^2} \, dy)^{m-k-1} \, dx, \\ & \text{if } 0 \leq k < m. \end{cases} \tag{2.68}$$

Substituting (2.66), (2.67), and (2.68) into (2.62), the upper bound can be rewritten as

$$\begin{aligned} C &\leq \lim_{m\to\infty} \frac{1}{m} \log \sum_{k=0}^{m} \frac{\Gamma(\frac{m}{2}+1)(mP)^k \eta_k}{\Gamma(\frac{m-k}{2}+1)\pi^{\frac{k}{2}} \rho^k} \\ &\leq \lim_{m\to\infty} \frac{1}{m} \log \left[(m+1) \sup_{k\in\chi} \frac{\Gamma(\frac{m}{2}+1)(mP)^k \eta_k}{\Gamma(\frac{m-k}{2}+1)\pi^{\frac{k}{2}} \rho^k} \right] \\ &= \lim_{m\to\infty} \frac{1}{m} \log \sup_{k\in\chi} \frac{\Gamma(\frac{m}{2}+1)(mP)^k \eta_k}{\Gamma(\frac{m-k}{2}+1)\pi^{\frac{k}{2}} \rho^k}, \end{aligned} \tag{2.69}$$

where $\chi = \{0, 1, ..., m\}$.

To derive a simple closed-form upper bound, the coefficient η_k is considered. For every $0 \leq k < m$, there is an inequality for η_k, that is

$$\eta_k < \binom{m+1}{k+1} \frac{k+1}{k!(m-k)(\frac{k}{\beta}+1)} \gamma^{-\frac{k}{\beta}} \frac{2^{m-k}-1}{2^{m-k}}, \tag{2.70}$$

where γ and β are positive real-valued coefficients of the tight lower bound (i.e., $\gamma \exp(-\beta x^2)$) on the complementary error function $\mathrm{erfc}(\mathrm{x})$.

According to (2.69) and (2.70), the upper bound on channel capacity is calculated by [23]

$$C < \sup_{\alpha\in[0,1]} \left\{ \log \left[\frac{(\frac{e}{2\pi})^{\frac{\alpha}{2}} \gamma^{-\frac{\alpha}{\beta}}}{(\alpha)^{2\alpha}(1-\alpha)^{\frac{3}{2}(1-\alpha)}} (\frac{P}{\sigma})^{\alpha} \right] \right\}. \tag{2.71}$$

Another upper bound based on the Steiner-Minkowski formula (with different approximation value of η_k) is given by [22]

$$C \leq \sup_{\alpha\in[0,1]} \left\{ \alpha \log \left(\sqrt{\frac{e^2}{4\pi}} \frac{P}{\sigma} \right) - \log \left((1-\alpha)^{\frac{(1-\alpha)}{2}} (1-\frac{\alpha}{2})^{1-\frac{\alpha}{2}} \alpha^{\frac{3}{2}\alpha} \right) \right\}. \tag{2.72}$$

When only peak-power constraint is applied, by way of packing sphere in the cube with Steiner-Minkowski formula, the upper bound on channel capacity is further given by [24]

$$C \leq \sup_{\alpha\in[0,1]} \left\{ \alpha \log \left(\frac{1}{\sqrt{2\pi e}} \frac{A}{\sigma} \right) - \log \left((1-\alpha)^{\frac{3(1-\alpha)}{2}} \alpha^{\alpha} \right) \right\}. \tag{2.73}$$

- ***Recursive approach***

The recursive approach proposed in [24] utilizes the geometry of m-dimensional spheres $B_m(\rho)$ to obtain a tighter upper bound at the high OSNR region. When only average-power constraint is imposed, the total volume of the spheres packed in m-simplex is given by

$$\hat{V}(P,\rho) = V_{\text{in}}(T(P)) + V_{\text{out}}(T(P)), \tag{2.74}$$

where $V_{\text{in}}(T(P))$ denotes the volume of the spheres and the portions of spheres inside the m-simplex and $V_{\text{out}}(T(P))$ denotes the volume of the portions of spheres outside the m-simplex. $V_{\text{in}}(T(P))$ can be upper-bounded by the volume of the m-simplex, while the portions outside the m-simplex are distributed among the $(m-1)$-dimensional faces of the m-simplex, which would lead to packing $(m-1)$-dimensional spheres on the $(m-1)$-dimensional faces of the m-simplex. m-simplex has $\binom{m+1}{m}$ $(m-1)$-dimensional faces, and the $(m-1)$-dimensional face is an $(m-1)$-simplex $T_{m-1}(P)$. By repeating the above procedure for the $(m-i)$-dimensional faces ($i = 0, 1, ..., m$), the upper bound of $\hat{V}(P,\rho)$ can be expressed as

$$\hat{V}(P,\rho) \leq \sum_{i=0}^{m} \binom{m+1}{m-i+1} V(T_{m-i}(P)) \frac{V(B_m(\rho))}{V(B_{m-i}(\rho))}. \tag{2.75}$$

Then, the upper bound on channel capacity can be obtained by

$$C \leq \log\Big(\sum_{i=0}^{m} \binom{m+1}{m-i+1} \frac{V(T_{m-i}(P))}{V(B_{m-i}(\rho))}\Big), \tag{2.76}$$

where the $(m-i)$-simplex $V(T_{m-i}(P))$ is given by

$$V(T_{m-i}(P)) = \frac{(mP)^{m-i}}{(m-i)!}\sqrt{m-i+1}. \tag{2.77}$$

Similar to (2.69), the upper bound can be obtained by [24]

$$C \leq \sup_{\alpha\in[0,1]} \left\{ \alpha \log\Big(\sqrt{\frac{e}{2\pi}}\frac{P}{\sigma}\Big) - \log\Big((1-\alpha)^{1-\alpha}\alpha^{\frac{3}{2}\alpha}\Big) \right\}. \tag{2.78}$$

Likewise, when only peak-power constraint is considered, by packing sphere in the cube with recursive approach, the upper bound on channel capacity is further given by [24]

$$C \leq \sup_{\alpha\in[0,1]} \left\{ \alpha \log\Big(\frac{1}{\sqrt{2\pi e}}\frac{A}{\sigma}\Big) - \log\Big((1-\alpha)^{(1-\alpha)}\alpha^{\frac{\alpha}{2}}2^{\alpha-1}\Big) \right\}. \tag{2.79}$$

2) ***Dual expression of channel capacity***

When a dual expression of channel capacity is applied, the channel capacity under both peak-power and average-power constraints is given by [19]

$$C \leq \sup_{Q} E_Q[D(W(\cdot|X)\|R(\cdot))], \tag{2.80}$$

where $W(\cdot|\cdot)$ is the conditional output distribution, Q denotes any channel input distribution, $R(\cdot)$ is the channel output distribution, and $D(\cdot\|\cdot)$ is the relative entropy.

When the channel output has a probability distribution $p(y)$, then (2.80) can be rewritten as

$$C \leq \sup_{Q} E_Q[-\int_{-\infty}^{\infty} p(y)dW(y|X)] - \frac{1}{2}\log 2\pi e\sigma^2]. \tag{2.81}$$

The upper bounds in different cases are given as follows [19].

- Case I: Peak-power and average-power constraints are considered with $\alpha \in (0, 1/2)$,

$$C \leq \frac{1}{2}\log\left(1 + \alpha(1-\alpha)\frac{A^2}{\sigma^2}\right) \tag{2.82}$$

and

$$\begin{aligned} C \leq \left(1 - Q\Big(\frac{\delta + \alpha A}{\sigma}\Big) - Q\Big(\frac{\delta + (1-\alpha)A}{\sigma}\Big)\right)\log\Big(\frac{A}{\sigma}\frac{e^{\frac{\nu\delta}{A}} - e^{-\nu(1+\frac{\delta}{A})}}{\sqrt{2\pi}\nu(1-2Q(\frac{\delta}{\sigma}))}\Big) \\ -\frac{1}{2} + Q\Big(\frac{\delta}{\sigma}\Big) + \frac{\delta}{\sqrt{2\pi}\sigma}e^{-\frac{\delta^2}{2\sigma^2}} + \frac{\sigma}{A}\frac{\nu}{\sqrt{2\pi}}\Big(e^{-\frac{\delta^2}{2\sigma^2}} - e^{-\frac{(A+\delta)^2}{2\sigma^2}}\Big) \\ +\nu\alpha\Big(1 - 2Q\Big(\frac{\delta + \frac{A}{2}}{\sigma}\Big)\Big) \end{aligned} \tag{2.83}$$

where ν and δ are positive parameters given in [19, (14) & (15)].

- Case II: Peak-power and average-power constraints are considered with $\alpha \in [1/2, 1]$,

$$C \leq \frac{1}{2}\log\Big(1 + \frac{A^2}{4\sigma^2}\Big) \tag{2.84}$$

and

$$\begin{aligned} C \leq \left(1 - 2Q\Big(\frac{\delta + \frac{A}{2}}{\sigma}\Big)\right)\log\left(\frac{A + 2\delta}{\sqrt{2\pi}\sigma\Big(1 - 2Q(\frac{\delta}{\sigma})\Big)}\right) \\ -\frac{1}{2} + Q\Big(\frac{\delta}{\sigma}\Big) + \frac{\delta}{\sqrt{2\pi}\sigma}e^{-\frac{\delta^2}{2\sigma^2}} \end{aligned} \tag{2.85}$$

where δ is a positive parameter given in [19, (21)].

- Case III: Only the average-power constraint is considered,

$$C \leq \log\left(\beta e^{-\frac{\delta^2}{2\sigma^2}} + \sqrt{2\pi}\sigma Q\left(\frac{\delta}{\sigma}\right)\right) - \log\left(\sqrt{2\pi}\sigma\right) - \frac{\delta P}{2\sigma^2}$$
$$+\frac{\delta^2}{2\sigma^2}\left(1 - Q\left(\frac{\delta}{\sigma}\right) - \frac{P}{\delta}Q\left(\frac{\delta}{\sigma}\right)\right)$$
$$+\frac{1}{\beta}\left(P + \frac{\sigma}{\sqrt{2\pi}}\right), \quad \text{if } \delta \leq -\frac{\sigma}{\sqrt{e}} \tag{2.86}$$

and

$$C \leq \log\left(\beta e^{-\frac{\delta^2}{2\sigma^2}} + \sqrt{2\pi}\sigma Q\left(\frac{\delta}{\sigma}\right)\right) + \frac{1}{2}Q\left(\frac{\delta}{\sigma}\right) + \frac{\delta}{2\sqrt{2\pi}\sigma}e^{-\frac{\delta^2}{2\sigma^2}}$$
$$+\frac{\delta^2}{2\sigma^2}\left(1 - Q\left(\frac{\delta + P}{\sigma}\right)\right) + \frac{1}{\beta}\left(P + \delta + \frac{\sigma}{\sqrt{2\pi}}e^{-\frac{\delta^2}{2\sigma^2}}\right)$$
$$-\frac{1}{2}\log 2\pi e\sigma^2, \quad \text{if } \delta \geq 0 \tag{2.87}$$

where β and δ are free parameters given in [19, (29), (30), (31) & (32)].

B. Lower bounds

With a proper choice of the input distribution, a tight lower bound can be obtained to approach the channel capacity or upper bound. Two specific lower bounds on channel capacity for free-space optical intensity channel are detailed as below.

- ***Entropy power inequality***

For any specific input distribution, based on the entropy power inequality (EPI), we have

$$\begin{aligned} C \geq& I(X;Y) \\ =& H(X+Z) - H(Z) \\ \geq& \frac{1}{2}\log\left(e^{2H(X)} + e^{2H(Z)}\right) - H(Z) \\ =& \frac{1}{2}\log\left(1 + \frac{e^{2H(X)}}{2\pi e\sigma^2}\right). \end{aligned} \tag{2.88}$$

The maximum differential entropy of the input signals, i.e., $\max H(X)$, would result in a tight lower bound [19].

- Case I: When peak-power and average-power constraints are considered with $\alpha \in (0, 1/2)$, the input distribution approaching maximum differential entropy is given by

$$p(x) = \frac{\nu}{A(1 - e^{-\nu})}e^{-\frac{\nu x}{A}}, \quad 0 \leq x \leq A, \tag{2.89}$$

where ν satisfies

$$\frac{1}{\nu} - \frac{e^{-\nu}}{1 - e^{-\nu}} = \alpha. \tag{2.90}$$

In this case, the lower bound is given by

$$C \geq \frac{1}{2} \log \left(1 + A^2 \frac{e^{2\alpha\nu}}{2\pi e \sigma^2} \left(\frac{1 - e^{-\nu}}{\nu} \right)^2 \right). \tag{2.91}$$

- Case II: When peak-power and average-power constraints are considered with $\alpha \in [1/2, 1]$, the input distribution approaching maximum differential entropy is given by

$$p(x) = \frac{1}{A}, \quad 0 \leq x \leq A, \tag{2.92}$$

and the lower bound is given by

$$C \geq \frac{1}{2} \log \left(1 + \frac{A^2}{2\pi e \sigma^2} \right). \tag{2.93}$$

- Case III: When only the average-power constraint is considered, the input distribution approaching maximum differential entropy is given by

$$p(x) = \frac{1}{P} e^{-\frac{x}{P}}, \quad x \geq 0, \tag{2.94}$$

and the lower bound is given by

$$C \geq \frac{1}{2} \log \left(1 + \frac{P^2 e}{2\pi \sigma^2} \right). \tag{2.95}$$

The upper bounds and lower bounds for Case I and Case II are shown in Fig. 2.9 and Fig. 2.10, respectively.

- ***Discrete input distribution***

Under the average-power constraint, all probability distributions on the input X form the set given by

$$\mathscr{P} = \left\{ p(x) \mid p(x < 0) = 0, E(x) \leq P, \int p(x) dx = 1 \right\}. \tag{2.96}$$

To find a capacity-achieving input distribution, the distributions of equidistant mass points are considered. The set consisting of these discrete distributions is illustrated as

$$\mathscr{P}_l = \left\{ p_l(x) \in \mathscr{P} \mid a_k \geq 0, l > 0, p_l(x) = \sum_{k=0}^{\infty} a_k \delta(x - kl) \right\}, \tag{2.97}$$

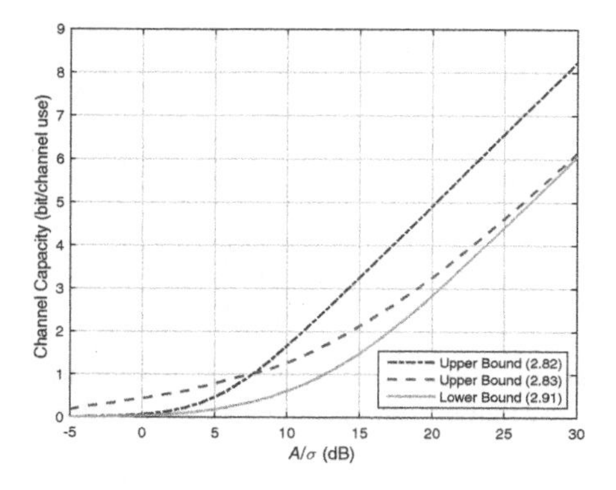

Figure 2.9 Comparison of free-space optical intensity channel capacity bounds when both peak-power and average-power constraints are applied with $\alpha = 0.1$.

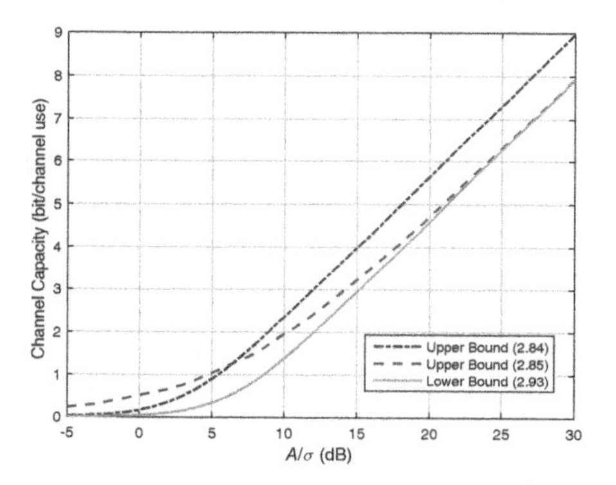

Figure 2.10 Comparison of free-space optical intensity channel capacity bounds when both peak-power and average-power constraints are applied with $\alpha \in [1/2, 1]$.

where a_k is the amplitude of the kth mass point and l is the distance between two mass points. Intuitively, the input distribution with maximum entropy under the average-power constraint may achieve the channel capacity, which is given by [22]

$$\begin{aligned}
p_l^* &= \arg \max_{p_l(x)\in\mathscr{P}_l} H(p_l(x)) \\
\text{s.t.} \quad & \sum_{k=0}^{\infty} a_k k l \leq P, \\
& \sum_{k=0}^{\infty} a_k = 1, \\
& a_k \geq 0.
\end{aligned} \tag{2.98}$$

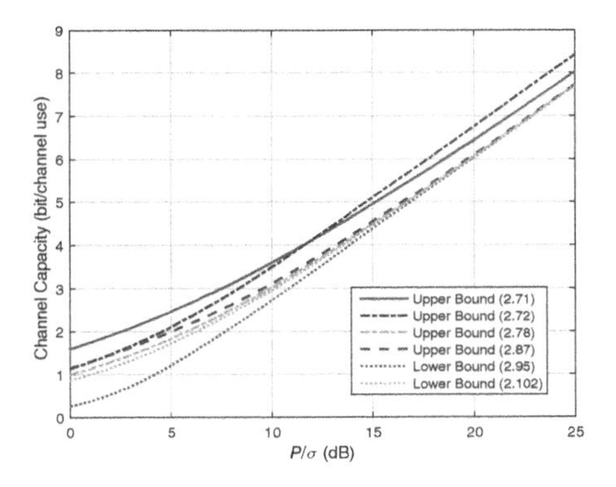

Figure 2.11 Comparison of free-space optical intensity channel capacity bounds when only the average-power constraint is applied.

Based on (2.98), the input signal with maximum entropy follows a geometric distribution given by

$$p_l^* = \sum_{k=0}^{\infty} \frac{l}{l+P}(\frac{P}{l+P})^k \delta(x - kl). \tag{2.99}$$

Afterwards, the mutual information can be calculated by

$$I_{p_l^*}(X;Y) = -\int f_W(w) \log f_W(w) dw - \frac{1}{2}\log(2\pi e \beta^2), \tag{2.100}$$

where

$$f_W(w) = \sum_{k=1}^{\infty} \frac{1}{1+\frac{P}{l}} \left(\frac{\frac{P}{l}}{1+\frac{P}{l}}\right)^k \frac{l}{\sqrt{2\pi\sigma^2}} e^{-(w-k)^2 l^2/2\sigma^2}. \tag{2.101}$$

Since the mutual information $I_{p_l^*}(X;Y)$ is a function of the argument l, a tight lower bound can be obtained by [22]

$$C \geq C_L = \max_l I_{p_l^*}(X;Y). \tag{2.102}$$

Although this method does not provide a closed-form expression, the lower bound can be numerically calculated from (2.102). When only the average-power constraint is applied, the upper and lower bounds are shown in Fig. 2.11.

2.6.3 Capacity bounds for discrete-time Poisson channel

In the discrete-time Poisson channel, the conditional output signal follows a Poisson distribution as shown in (2.55). Like the free-space optical intensity channel, there is no analytical expression for channel capacity subject to the average-power and peak-power constraints. Thus, only capacity bounds on the discrete-time Poisson channel are presented.

A. Upper bounds

Similar to free-space optical intensity channel, upper bounds on the discrete-time Poisson channel are derived using the dual expression of the channel capacity for the following three different cases [25].

- Case I: Peak-power and average-power constraints are imposed and $\alpha \in (0, 1/3)$,

$$C \leq \frac{1}{2}\log A - (1-\alpha)\nu - \log\left(\frac{1}{2} - \alpha\nu\right) - \frac{1}{2}\log 2\pi e + \mathrm{o}_A(1), \quad (2.103)$$

where $\mathrm{o}_A(1)$ is defined as a function that tends to zero as the peak power and average power tend to infinity, and ν satisfies

$$\frac{1}{2\nu} - \frac{e^{-\nu}}{\sqrt{\pi\nu}\mathrm{erf}(\sqrt{\nu})} = \alpha. \quad (2.104)$$

- Case II: Peak-power and average-power constraints are imposed and $\alpha \in [1/3, 1]$ (when $\alpha = 1$, only peak-power constraint is considered),

$$C \leq \frac{1}{2}\log A - \frac{1}{2}\log\frac{\pi e}{2} + \mathrm{o}_A(1). \quad (2.105)$$

- Case III: Only the average-power constraint is applied,

$$C \leq \frac{1}{2}\log P + \mathrm{o}_P(1), \quad (2.106)$$

where $\mathrm{o}_P(1)$ is defined as a function that tends to zero as the average power goes to infinity.

B. Lower bounds

The entropy of the output is lower-bounded by [25]

$$H(Y) \geq H(X) + (1 + E(X))\log(1 + \frac{1}{E(X)}) - 1, \quad (2.107)$$

and the conditional entropy $H(Y|X)$ is upper-bounded by

$$H(Y|X) \leq \frac{1}{2}E\Big(\log 2\pi e(X + \lambda + \frac{1}{12})\Big). \quad (2.108)$$

The lower bound on Poisson channel capacity can be given by

$$\begin{aligned} C &\geq I(X;Y) \\ &= H(Y) - H(Y|X) \\ &\geq H(X) + (1 + E(X))\log(1 + \frac{1}{E(X)}) - 1 - \frac{1}{2}E\Big(\log 2\pi e(X + \lambda\frac{1}{12})\Big). \end{aligned} \quad (2.109)$$

According to (2.109), the lower bound strongly depends on the entropy of input signal [25].

- Case I: When peak-power and average-power constraints are imposed with $\alpha \in (0, 1/3)$, the input distribution approaching the maximum entropy is given by

$$p(x) = \frac{\sqrt{\nu}}{\sqrt{A\pi x}\mathrm{erf}(\sqrt{\nu})} e^{-\frac{\nu}{A}x}, \quad 0 < x \leq A, \tag{2.110}$$

where ν satisfies (2.104). Thus, the lower bound is given by

$$\begin{aligned} C \geq & \frac{1}{2}\log A - (1-\alpha)\nu - \log(\frac{1}{2} - \alpha\nu) - e^{\nu}(\frac{1}{2} - \alpha\nu)\cdot \\ & \left(\log\left(1 + \frac{\lambda + \frac{1}{12}}{A}\right) + 2\sqrt{\frac{\lambda + \frac{1}{12}}{A}}\arctan\left(\frac{A}{\lambda + \frac{1}{12}}\right)\right) \\ & + (P+1)\log(1 + \frac{1}{P}) - 1 - \frac{1}{2}\log 2\pi e. \end{aligned} \tag{2.111}$$

- Case II: When peak-power and average-power constraints are imposed with $\alpha \in [1/3, 1]$, the input distribution approaching the maximum entropy is given by

$$p(x) = \frac{1}{2\sqrt{Ax}}, \quad 0 < x \leq A, \tag{2.112}$$

and the lower bound is given by

$$\begin{aligned} C \geq & \frac{1}{2}\log A - \left(1 + \frac{A}{3}\right)\log\left(1 + \frac{3}{A}\right) - 1 \\ & - \sqrt{\frac{\lambda + \frac{1}{12}}{A}}\left(\frac{\pi}{4} + \frac{1}{2}\log 2\right) - \frac{1}{2}\log\frac{\pi e}{2}. \end{aligned} \tag{2.113}$$

- Case III: When only the average-power constraint is applied, the input distribution approaching the maximum entropy is given by

$$p(x) = \frac{1}{\sqrt{2\pi P x}} e^{-\frac{x}{2P}}, \quad x > 0, \tag{2.114}$$

and the lower bound is given by

$$C \geq \frac{1}{2}\log P - \sqrt{\frac{\pi(\lambda + \frac{1}{12})}{2P}} + (P+1)\log(1 + \frac{1}{P}) - 1. \tag{2.115}$$

Figure 2.12 depicts the upper bounds and lower bounds on Poisson channel capacity for Case I and Case II, while Fig. 2.13 shows the upper bound and lower bound on Poisson channel capacity for Case III.

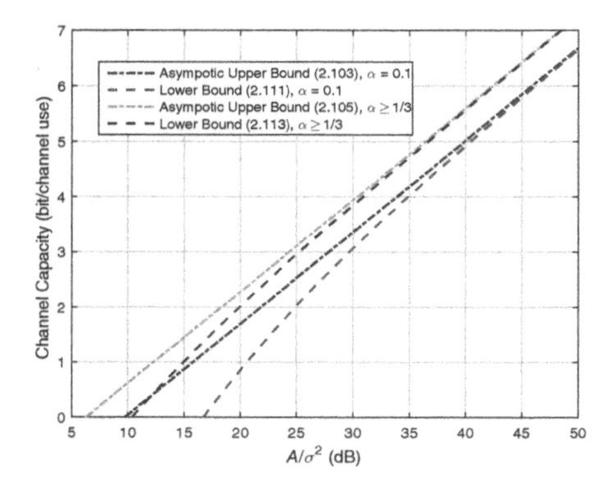

Figure 2.12 Comparison of Poisson channel capacity bounds when both peak-power and average-power constraints are applied.

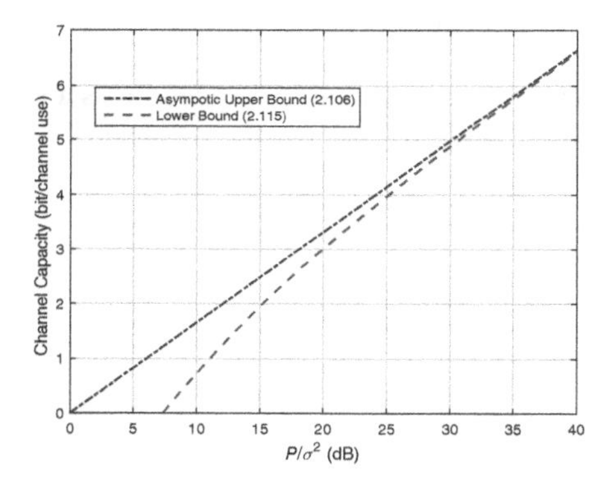

Figure 2.13 Comparison of Poisson channel capacity bounds when only the average-power constraint is applied.

2.6.4 Capacity bounds for improved free-space intensity channel

For improved free-space intensity channel, both the signal-dependent Gaussian noise and the signal-independent Gaussian noise are considered. Compared to the free-space optical intensity channel, the analysis of channel capacity becomes difficult when the signal-dependent Gaussian noise is included.

A. Upper bounds

By using the dual expression of the channel capacity, the upper bounds for improved free-space intensity channel are given as follows [27].

- Case I: Peak-power and average-power constraints are considered and $\alpha \in$

$(0, 1/3)$,

$$C \leq \frac{1}{2} \log \frac{A}{\sigma^2} - \frac{1}{2} \log 2\pi e \varsigma^2 - (1-\alpha)\mu - \log(\frac{1}{2} - \alpha\mu) + o_A(1), \tag{2.116}$$

where $\mu \in (0, \frac{1}{2\alpha})$ satisfies

$$\frac{1}{2\mu} - \frac{e^{-\mu}}{\sqrt{\pi\mu}\mathrm{erf}(\sqrt{\mu})} = \alpha. \tag{2.117}$$

- Case II: Peak-power and average-power constraints are considered and $\alpha \in [1/3, 1]$ (when $\alpha = 1$, only peak-power constraint is considered),

$$C \leq \frac{1}{2} \log \frac{A}{\sigma^2} - \frac{1}{2} \log \frac{\pi e \varsigma^2}{2} + o_A(1). \tag{2.118}$$

- Case III: Only the average-power constraint is considered,

$$C \leq \frac{1}{2} \log \frac{P}{\sigma^2} - \frac{1}{2} \log \varsigma^2 + o_P(1). \tag{2.119}$$

B. Lower bounds

The derivation of the lower bounds for improved free-space optical intensity channel is similar to Poisson channel. Based on the lower bound on the entropy of the output signal, we have

$$H(Y) \geq H(x) + \frac{1}{2} \log(1 + \frac{2\varsigma^2\sigma^2}{P}) + \frac{\sqrt{P(P + 2\varsigma^2\sigma^2)}}{\varsigma^2\sigma^2} - \frac{P + \varsigma^2\sigma^2}{\varsigma^2\sigma^2}, \tag{2.120}$$

and the lower bound on channel capacity can be given by

$$\begin{aligned} C &\geq I(X;Y) \\ &= H(Y) - H(Y|X) \\ &\geq H(X) + \frac{1}{2} \log(1 + \frac{2\varsigma^2\sigma^2}{P}) + \frac{\sqrt{P(P + 2\varsigma^2\sigma^2)}}{\varsigma^2\sigma^2} \\ &\quad - \frac{P + \varsigma^2\sigma^2}{\varsigma^2\sigma^2} - \frac{1}{2} E\Big(\log 2\pi e \sigma^2 (1 + \varsigma^2 X)\Big). \end{aligned} \tag{2.121}$$

The specific lower bounds for different cases are given as follows [27].

- Case I: When peak-power and average-power constraints are considered with $\alpha \in (0, 1/3)$, with the same distribution as (2.110), the lower bound is given by

$$C \geq \frac{1}{2} \log \frac{A}{\sigma^2} - \frac{1}{2} \log 2\pi e \varsigma^2 - (1-\alpha)\mu - \log(\frac{1}{2} - \alpha\mu) + \beta_{(A,\alpha,\mu,\sigma,\varsigma)}, \tag{2.122}$$

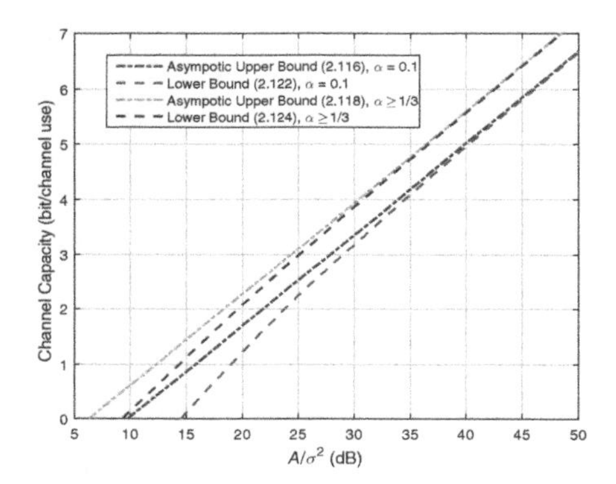

Figure 2.14 Comparison of improved free-space optical intensity channel capacity bounds when both peak-power and average-power constraints are applied with $\varsigma = 1$.

where

$$\beta_{(A,\alpha,\mu,\sigma,\varsigma)} = -e^{\mu}(\frac{1}{2} - \alpha\mu)(\frac{2}{\sqrt{\varsigma^2 A}} \arctan(\sqrt{\varsigma^2 A}) + \log(1 + \frac{1}{\varsigma^2 A})) + \frac{1}{2}\log(1 + \frac{2\varsigma^2\sigma^2}{\alpha A}) - \frac{\alpha A}{\varsigma^2\sigma^2} - 1 + \frac{\sqrt{\alpha A(\alpha A + 2\varsigma^2\sigma^2)}}{\varsigma^2\sigma^2}. \tag{2.123}$$

- Case II: When peak-power and average-power constraints are considered with $\alpha \in [1/3, 1]$, using the same distribution as (2.112), the lower bound is given by

$$C \geq \frac{1}{2}\log\frac{A}{\sigma^2} - \frac{1}{2}\log\frac{\pi e \varsigma^2}{2} + \beta_{(A,1/3,0,\sigma,\varsigma)}, \tag{2.124}$$

where $\beta_{(A,1/3,0,\sigma,\varsigma)}$ is defined as

$$\beta_{(A,\alpha,\mu,\sigma,\varsigma)} = -\frac{1}{\sqrt{\varsigma^2 A}} \arctan(\sqrt{\varsigma^2 A}) - \frac{1}{2}\log(1 + \frac{1}{\varsigma^2 A}) + \frac{1}{2}\log(1 + \frac{6\varsigma^2\sigma^2}{A}) - \frac{A}{3\varsigma^2\sigma^2} - 1 + \frac{\sqrt{A(A + 6\varsigma^2\sigma^2)}}{3\varsigma^2\sigma^2}. \tag{2.125}$$

- Case III: When only the average-power constraint is considered, using the same distribution as (2.114), the lower bound is given by

$$C \geq \frac{1}{2}\log\frac{P}{\sigma^2} - \log\varsigma^2 + \frac{1}{2}\log(1 + \frac{2\varsigma^2\sigma^2}{P}) - \frac{P}{\varsigma^2\sigma^2} - 1 + \frac{\sqrt{P(P + 2\varsigma^2\sigma^2)}}{\varsigma^2\sigma^2} - \sqrt{\frac{\pi}{2\varsigma^2 P}}. \tag{2.126}$$

The upper and lower bounds on improved free-space optical intensity channel capacity for Case I and Case II are shown in Fig. 2.14. In Fig. 2.15, the upper and lower bounds for Case III are illustrated.

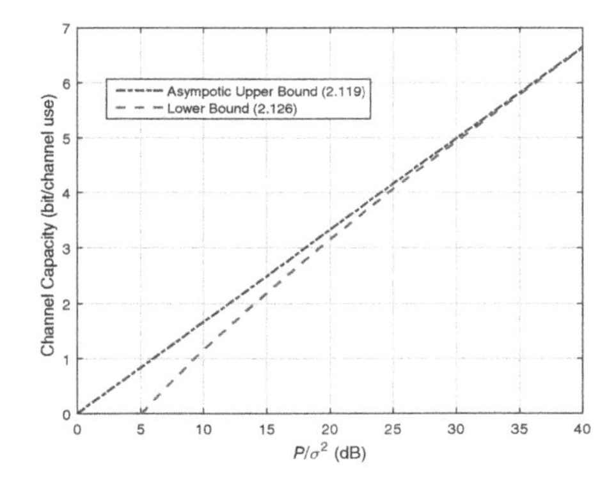

Figure 2.15 Comparison of improved free-space optical intensity channel capacity bounds when only the average-power constraint is applied with $\varsigma = 1$.

2.7 Conclusion

In this chapter, the characteristics of LED as the transmitter and photodiode as the receiver are described. Normally, LED works under a forward bias while photodiode is driven by a reverse voltage. When LED is employed for lighting and communication simultaneously, its nonlinearity and lighting constraints are investigated. Besides, absorption coefficient, quantum efficiency, and responsivity of the photodiode are demonstrated. Furthermore, different LOS and NLOS propagation links between the transmitter and the receiver are analyzed. Since the dominant noise might be different in various application scenarios, three optical wireless channels are addressed including free-space optical intensity channel, discrete-time Poisson channel, and improved free-space intensity channel. Considering there exist no analytic expressions of the channel capacity, the state-of-the-art upper and lower bounds are presented.

References

1 D. A. Steigerwald, J. C. Bhat, D. Collins, R. M. Fletcher, M. O. Holcomb, M. J. Ludowise, P. S. Martin, and S. L. Rudaz, "Illumination with solid state lighting technology," *IEEE J. Sel. Topics Quantum Electron.*, vol. 8, no. 2, pp. 310–320, Mar. 2002.
2 J. Wang, C. C. Tsai, W. C. Cheng, M. H. Chen, C. H. Chung, and W. H. Cheng, "High thermal stability of phosphor-converted white-lightemitting diodes employing Ce:YAG doped glass," *IEEE J. Sel. Topics Quantum Electron.*, vol. 17, no. 3, pp. 741–746, May/Jun. 2011.
3 S. Pimputkar, J. S. Speck, S. P. DenBaars, and S. Nakamura, "Prospects for LED lighting," *Nature Photon.*, vol. 3, pp. 179–181, Apr. 2009.
4 J. Grubor, S. Randel, K. D. Langer, and J. W. Walewski, "Broadband information broadcasting using LED-based interior lighting," *J. Lightw. Technol.*, vol. 26, no. 24, pp. 3883–3892, Dec. 2008.
5 D. Karunatilaka, F. Zafar, V. Kalavally, and R. Parthiban, "LED based indoor visible light communications: State of the art," *IEEE Commun. Surv. Tuts.*, vol. 17, no. 3, pp. 1649–1678, Mar. 2015.
6 A. Martí, J. L. Balenzategui, and R. F. Reyna, "Photon recycling and Shockley's diode equation," *J. Appl. Phys.*, vol. 82, no. 8, pp. 4067–4075, Oct. 1997.
7 I. Neokosmidis, T. Kamalakis, J. W. Walewski, B. Inan, and T. Sphicopoulos, "Impact of nonlinear LED transfer function on discrete multitone modulation: Analytical approach," *J. Lightw. Technol.*, vol. 27, no. 22, pp. 4970–4978, Nov. 2009.
8 T. Kamalakis, J. Walewski, G. Ntogari, and G. Mileounis, "Empirical Volterra-series modeling of commercial light-emitting diodes," *J. Lightw. Technol.*, vol. 29, no. 14, pp. 2146–2155, Jul. 2011.
9 G. Wyszecki and W. S. Stiles, *Color Science: Concepts and Methods, Quantitative Data and Formulae*, 2nd ed. New York: Wiley, 1982.
10 F. Zafar, D. Karunatilaka, and R. Parthiban, "Dimming schemes for visible light communications: The state of research," *IEEE Wirel. Commun.*, vol. 22, no. 2, pp. 29–35, Apr. 2015.
11 J. Gancarz, H. Elgala, and T. D. C. Little, "Impact of lighting requirements on VLC systems," *IEEE Commun. Mag.*, vol. 51, no. 12, pp. 34–41, Dec. 2013.
12 S. Rajagopal, R. D. Roberts, and S. K. Lim, "IEEE 802.15.7 visible light communication: Modulation schemes and dimming support," *IEEE Commun. Mag.*, vol. 50, no. 3, pp. 72–82, Mar. 2012.
13 R. Singh, T. O'Farrell, and J. P. R. David, "An enhanced color shift keying modulation scheme for high-speed wireless visible light communications," *J. Lightw. Technol.*, vol. 32, no. 14, pp. 2582–2592, Jul. 2014.
14 J. Geist, E. F. Zalewski, and A. R. Schaefer, "Spectral response self calibration and interpolation of silicon photodiodes," *Appl. Opt.*, vol. 19, no. 22, pp. 3795–3799, Nov. 1980.
15 J. M. Kahn and J. R. Barry, "Wireless infrared communications," *Proc. IEEE*, vol. 85, no. 2, pp. 265–298, Feb. 1997.
16 F. R. Gfeller and U. H. Bapst, "Wireless in-house data communication via diffuse

infrared radiation," *Proc. IEEE*, vol. 67, no. 11, pp. 1474–1486, Nov. 1979.

17 P. H. Pathak, X. Feng, P. Hu, and P. Mohapatra, "Visible light communication, networking, and sensing: A survey, potential and challenges ," *IEEE Commun. Surv. Tuts.*, vol. 17, no. 4, pp. 2047–2077, Nov. 2015.

18 N. S. Kopeika and J. Bordogna, "Background noise in optical communication systems," *Proc. IEEE*, vol. 58, no. 10, pp. 1571–1577, Oct. 1970.

19 A. Lapidoth, S. M. Moser, and M. A. Wigger, "On the capacity of free-space optical intensity channels," *IEEE Trans. Inf. Theory*, vol. 55, no. 10, pp. 4449–4461, Oct. 2009.

20 S. Hranilovic and F. R. Kschischang, "Capacity bounds for power- and band-limited optical intensity channels corrupted by Gaussian noise," *IEEE Trans. Inf. Theory*, vol. 50, no. 5, pp. 784–795, May 2004.

21 J. Wang, Q. Hu, J. Wang, M. Chen, and J. Wang, "Tight bounds on channel capacity for dimmable visible light communications," *J. Lightwave Technol.*, vol. 31, no. 23, pp. 3771–3779, Dec. 2013.

22 A. Farid and S. Hranilovic, "Capacity bounds for wireless optical intensity channels with Gaussian noise," *IEEE Trans. Inf. Theory*, vol. 56, no. 12, pp. 6066–6077, Dec. 2010.

23 R. Jiang, Z. Wang, Q. Wang, and L. Dai, "A tight upper bound on channel capacity for visible light communications," *IEEE Commun. Lett.*, vol. 20, no. 1, pp. 97–100, Jan. 2016.

24 A. Chaaban, J. Morvan, and M. Alouini, "Free-space optical communications: Capacity bounds, approximations, and a new sphere-packing perspective," *IEEE Trans. Commun.*, vol. 64, no. 3, pp. 1176–1191, Mar. 2016.

25 A. Lapidoth and S. M. Moser, "On the capacity of the discrete-time Poisson Channel," *IEEE Trans. Inf. Theory*, vol. 55, no. 1, pp. 303–322, Jan. 2009.

26 S. Shamai, "Capacity of a pulse amplitude modulated direct detection photon channel," *Proc. IEE*, vol. 137, no. 6, pp. 424–430, Dec. 1990.

27 S. M. Moser, "Capacity results of an optical intensity channel with input-dependent Gaussian noise," *IEEE Trans. Inf. Theory*, vol. 58, no. 1, pp. 207–223, Jan. 2012.

28 C. E. Shannon, "Communication in the presence of noise," *Proc. IRE*, vol. 37, no. 1, pp. 10–21, Jan. 1949.

29 W. Rudin, *Principle of Mathematical Analysis*. New York: McGraw-Hill, 1976.

30 U. Betke and M. Henk, "Intrinsic volumes and lattice points of crosspolytopes," *Monatsh. Math.*, vol. 115, no. 1, pp. 27–33, Mar. 1993.

3
Single Carrier/Carrierless Modulation and Coding

In this chapter, we present a review of carrierless and single carrier modulation schemes for visible light communication (VLC). In order to meet the illumination requirements, we also provide a brief introduction of modulation and coding techniques recently developed for dimming control and flicker mitigation.

In Section 3.1, pulse amplitude modulation (PAM), which is a simple modulation format widely used in VLC systems, is addressed. When a multipath channel is considered, frequency-domain equalization is preferred and its bit error rate (BER) is analyzed. Besides, several implementation considerations are brought to overcome the nonlinearity effect of light emitting diode (LED).

In Section 3.2, pulse position modulation (PPM) is illustrated as well as its corresponding decision feedback equalizers (DFE). PPM has low data rate with only one pulse in a single symbol duration. Several modified schemes are introduced including differential PPM (DPPM), multipulse PPM (MPPM), overlapping PPM (OPPM), and variable PPM (VPPM).

Carrierless amplitude phase modulation (CAP) is a variant of quadrature amplitude modulation (QAM), which has high spectral efficiency and simple implementation. In Section 3.3, the concept of CAP is detailed. After that, the design of multi-dimensional CAP is discussed as an extension of two-dimensional CAP.

In indoor VLC systems, since communication and illumination should be maintained at the same time, where the light may be dimmed to satisfy different illumination and energy requirements, modulation and coding schemes for the dimming control are discussed in Section 3.4.

3.1
Pulse amplitude modulation

PAM is a simple modulation methodology, widely used in VLC and other optical communication systems [1, 2]. On-off keying (OOK), which can be regarded as a special candidate of PAM with only two levels, has been adopted in the IEEE 802.15.7 standard [3]. Recently, gigabit/s VLC transmissions have been reported adopting either OOK or PAM [4–6].

Visible Light Communications: Modulation and Signal Processing. First edition. Zhaocheng Wang, Qi Wang, Wei Huang, and Zhengyuan Xu. Published 2017 by John Wiley & Sons, Inc.

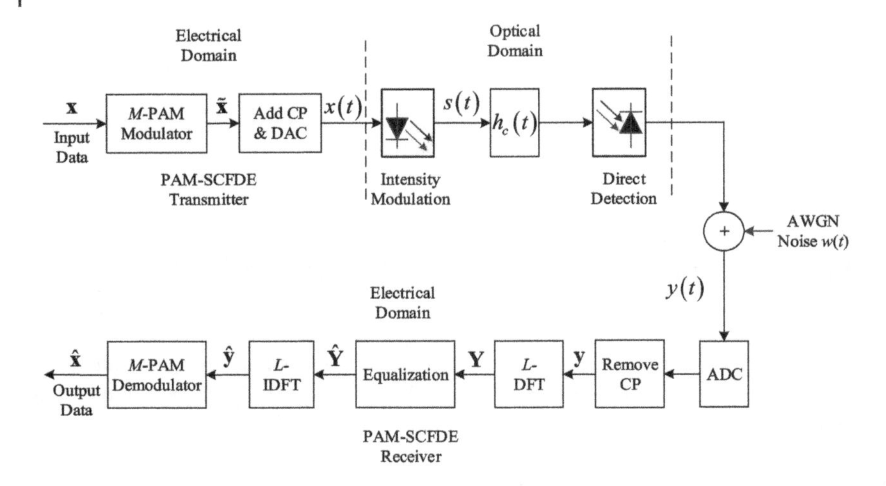

Figure 3.1 A PAM-SCFDE VLC system.

In order to remove the inter-symbol interference (ISI) for PAM systems, single-carrier modulation with frequency-domain equalization (SCFDE) is usually adopted, which is referred to as PAM-SCFDE [7]. A PAM-SCFDE VLC system diagram is illustrated in Fig. 3.1. The binary signal vector $\mathbf{x}$ is fed into the M-PAM modulator, where M denotes the modulation order. Since intensity modulation is preferred in VLC systems, the transmitted waveform is non-negative and direct current (DC) bias should be added to the bipolar PAM signal to generate a unipolar PAM waveform. Therefore, the output signal vector $\tilde{\mathbf{x}} = [\tilde{x}_0\ \tilde{x}_1\ \cdots\ \tilde{x}_{L-1}]^{\mathrm{T}}$ is real and non-negative, and we have $\tilde{x}_k \in \{0, 1, ..., M-1\}$ for $0 \leq k < L$, where L is the block length. A cyclic prefix (CP) with K elements is commonly added to mitigate the multipath dispersion and K is equal to or larger than the maximum number of taps in the multipath channel. After inserting the CP to the transmitted signal vector $\tilde{\mathbf{x}}$, the generated waveform is then sent to the digital-to-analog converter (DAC) to obtain $x(t)$, which is used to drive the LEDs and generate the optical signal radiated to the air.

At the receiver, the optical signal is directly detected by photodiode (PD), which converts the optical waveform into the electrical signal. The multipath channel dispersion can be modeled by $h_c(t)$. In addition, the shot and thermal noises at the receiver, denoted as $w(t)$, can be modeled as independent, identically distributed (i.i.d.) additive white Gaussian noise (AWGN) [7]. After analog-to-digital converter (ADC) and CP removal, we have

$$\mathbf{y} = \mathbf{H}\mathbf{x} + \mathbf{w}, \tag{3.1}$$

where $\mathbf{w} = [w_0\ w_1\ \cdots\ w_{L-1}]^{\mathrm{T}}$ is the sample vector of AWGN noise with $w_i \sim \mathcal{N}(0, N_0/2)$, which includes both shot noise and thermal noise. N_0 is the single-sided noise power spectral density, and $\mathbf{H}$ is a channel matrix with the size of $L \times L$.

If the line-of-sight (LOS) channel is considered, the channel matrix $\mathbf{H}$ is diagonal with the LOS channel coefficient h_0 as its diagonal elements. The equalized signal vector $\hat{\mathbf{y}}$ can be calculated by

$$\hat{\mathbf{y}} = h_0^{-1}\mathbf{y}, \tag{3.2}$$

which is used for the estimation of the transmitted signal vector $\hat{\mathbf{x}}$.

In a multipath VLC channel, $\mathbf{H}$ can be defined as an $L \times L$ circulant convolution matrix with the first column $[h_0\ h_1\ \cdots\ h_{K-1}\ 0\ \cdots\ 0]^{\mathrm{T}}$, and $\mathbf{h} = [h_0\ h_1\ \cdots\ h_{K-1}]$ is the K-tap discrete multipath channel coefficients with $h_i = h_c(i \cdot T_s)$ for $i \in [0, K-1]$, where $h_c(\cdot)$ is the channel response and T_s is the symbol duration. Thus, the channel matrix $\mathbf{H}$ can be diagonalized as

$$\mathbf{H} = \mathbf{F}^{\mathrm{H}}\mathbf{\Lambda}\mathbf{F}, \tag{3.3}$$

where $\mathbf{F}$ is an $L \times L$ discrete Fourier transform (DFT) matrix and $\mathbf{F}^{\mathrm{H}}$ denotes its Hermitian matrix. $\mathbf{\Lambda}$ denotes the $L \times L$ diagonal matrix, whose diagonal elements $\{\Lambda_1, \Lambda_2, \ldots, \Lambda_L\}$ are the eigenvalues of $\mathbf{H}$. When DFT is applied to (3.1), we have $\mathbf{Y} = \mathbf{\Lambda}\mathbf{X} + \mathbf{W}$, where $\mathbf{Y} = \mathbf{F}\mathbf{y}$, $\mathbf{X} = \mathbf{F}\mathbf{x}$ and $\mathbf{W} = \mathbf{F}\mathbf{w}$. As a result, frequency-domain equalization with the diagonal equalizer $\mathbf{V}_{\mathrm{EQ}}$ can be utilized, and the equalized signal vector can be calculated after inverse discrete Fourier transform (IDFT) as

$$\hat{\mathbf{y}} = \mathbf{F}^{\mathrm{H}}\mathbf{V}_{\mathrm{EQ}}\mathbf{Y} = \mathbf{F}^{\mathrm{H}}\mathbf{V}_{\mathrm{EQ}}\mathbf{\Lambda}\mathbf{F}\mathbf{x} + \mathbf{F}^{\mathrm{H}}\mathbf{V}_{\mathrm{EQ}}\mathbf{F}\mathbf{w}, \tag{3.4}$$

which is then used for M-PAM demodulation and the estimation of the transmitted signal vector $\hat{\mathbf{x}}$.

In a unipolar M-PAM system, the transmitted signal $x(t)$ can be rewritten as

$$x(t) = s_m p(t), \; m = 0, \ldots, M-1, \tag{3.5}$$

where $p(t)$ is a shaping pulse with a duration of T and the constellation point s_m is denoted as $s_m = m$ for $m = 0, \ldots, M-1$. Therefore, the transmitted energy is given by

$$\tau_m = \int_{-\infty}^{\infty} s_m^2 p^2(t)\, dt = m^2 \tau_p, \tag{3.6}$$

where τ_p is the energy of $p(t)$. The average energy of PAM signal is calculated as

$$\tau_{\mathrm{av}} = \frac{\tau_p}{M} \sum_{m=0}^{M-1} m^2 = \frac{\tau_p}{6}(M-1)(2M-1), \tag{3.7}$$

and the average electrical energy per bit, $E_{\mathrm{b(elec,av)}}$ can be defined as

$$E_{\mathrm{b(elec,av)}} = \frac{\tau_p}{6\log_2 M}(M-1)(2M-1). \tag{3.8}$$

When Gray mapping is used in the design of PAM constellation, the minimum Euclidean distance between different constellation points is given by

$$d_{\min} = \sqrt{\frac{6E_{\mathrm{b(elec,av)}}\log_2 M}{(M-1)(2M-1)}}. \tag{3.9}$$

Therefore, the BER of unipolar PAM systems in the LOS channel can be calculated as [7]

$$\begin{aligned} P_{\mathrm{LOS}}^{M-\mathrm{PAM}} &\approx \frac{2(M-1)}{M} Q\left(\frac{d_{\min}}{\sqrt{2N_0}}\right) \\ &= \frac{2(M-1)}{M} Q\left(\sqrt{\frac{3\gamma_{\mathrm{av}}\log_2 M}{(M-1)(2M-1)}}\right), \end{aligned} \tag{3.10}$$

where $Q(x) = \frac{1}{\sqrt{2\pi}} \int_x^{\infty} \exp\left(-\frac{u^2}{2}\right) du$ represents the standard tail probability function of the Gaussian distribution with zero mean and unit variance, and $\gamma_{\mathrm{av}} = E_{\mathrm{b(elec,av)}}/N_0$ is the ratio of average electrical energy per bit to noise power spectral density.

When a multipath VLC channel is considered, linear minimum mean squared error (LMMSE) equalization can be implemented to achieve better performance than its zero-forcing (ZF) counterpart in SCFDE systems [7, 8]. Its equalization matrix is given by

$$\mathbf{V}_{\mathrm{LMMSE}} = \left(\mathbf{\Lambda}^{\mathrm{H}}\mathbf{\Lambda} + \gamma^{-1}\mathbf{I}\right)^{-1}\mathbf{\Lambda}^{\mathrm{H}}, \tag{3.11}$$

which is used to replace $\mathbf{V}_{\mathrm{EQ}}$ in (3.4), where $\gamma = E_{\mathrm{b(elec)}}/N_0$ denotes the ratio of energy per bit to noise power spectral density, and $\mathbf{I}$ is the identity matrix.

When $z(\gamma, \mathbf{\Lambda})$ is defined as

$$z(\gamma, \mathbf{\Lambda}) \triangleq \frac{1}{L} \sum_{k=0}^{L-1} \frac{1}{1 + \gamma |\Lambda_k|^2}, \tag{3.12}$$

for the $L \times L$ diagonal matrix $\mathbf{\Lambda}$ in (3.3), γ_{av} can be expressed as [9]

$$\gamma_{\mathrm{av}} = z_{\mathrm{PAM}}^{-1} - 1, \tag{3.13}$$

where

$$z_{\mathrm{PAM}} = z(\gamma \log_2 M, \mathbf{\Lambda}). \tag{3.14}$$

According to (3.10), the BER of M-PAM-SCFDE with LMMSE equalization can be roughly given by

$$P_{\mathrm{LMMSE}}^{M-\mathrm{PAM}} \approx \frac{2(M-1)}{M} \mathcal{F}\left(\frac{3}{(M-1)(2M-1)}, z_{\mathrm{PAM}}\right), \tag{3.15}$$

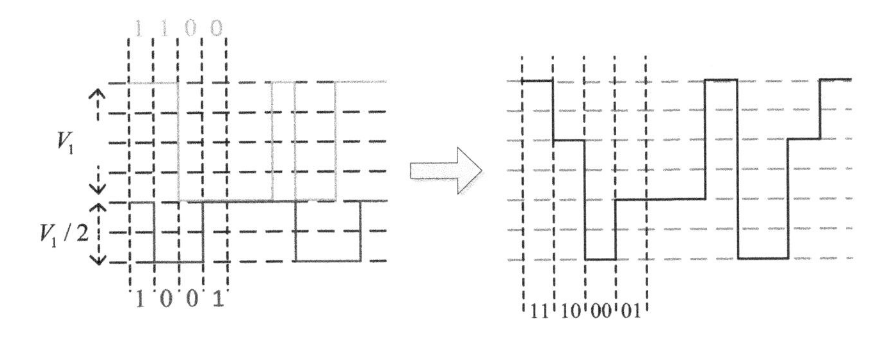

Figure 3.2 Illustration of 4-SPAM modulation scheme with two LEDs.

where $\mathcal{F}(\alpha, z)$ is defined as

$$\mathcal{F}(\alpha, z) \triangleq Q\left(\sqrt{\alpha\left(z^{-1}-1\right)}\right), z \in (0,1]. \tag{3.16}$$

Furthermore, since LED nonlinearity is also a challenge for the PAM-based VLC systems, several implementation methodologies have been proposed to mitigate its negative effects, such as superposed pulse amplitude modulation (SPAM) and grouped modulation [10, 11].

In SPAM, various groups of LED chips are placed together for simultaneous illumination, which have different intensities when suitable driving voltages are applied. For each single LED, it can only stay in either "on" or "off" status, i.e., two-level OOK. Therefore, there exist no LED nonlinearity issues. However, if they are controlled by the input bit vector as a whole, different intensities are combined together to support multilevel modulations. Specifically, N LEDs having intensity ranges of V, $V/2$, $V/4$, ..., $V/2^{N-1}$ are used in the 2^N-SPAM. An example of 4-SPAM is illustrated in Fig. 3.2, where two OOK signals of V_1 and $V_1/2$ drive two corresponding LEDs to produce four-level optical signals. Since only one PD is required to collect the optical signals at the SPAM receiver, it possesses much lower complexity.

The principle of SPAM can be extended by way of adopting several LEDs with the same intensity, which are divided into N groups for 2^N-level modulation. In each group, the number of illuminating LEDs is different to maintain the required intensity, which is referred to as grouped modulation [11]. An example of two-grouped modulation with a nine-LED squared array is shown in Fig. 3.3, which is installed in the center of the ceiling. Group 1 contains three LEDs (in red color), while Group 2 consists of the other six LEDs (in black color). When the same driving voltage is applied on each LED, the ratio of optical power emitted by Group 1 over Group 2 is $1:2$, which is similar to the previous 4-SPAM example. There exist $\binom{9}{3}$ possible choices for group partition of those 9 LEDs. In order to provide uniform illumination, it is preferred to choose 1st, 6th, and 8th LEDs for Group 1 as illustrated in

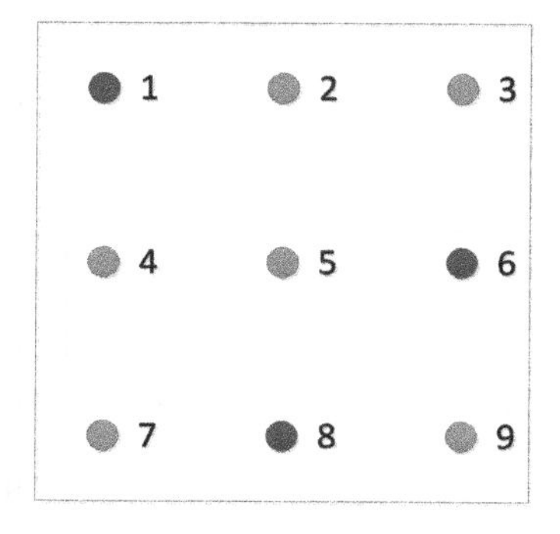

Figure 3.3 Illustration of two-grouped modulation with a nine-LED squared array.

Fig. 3.3 [11].

If the size of LED array is $(2^N - 1) \times (2^N - 1)$, it could be divided into N groups for 2^N-level modulation. In the mth group, where $m = 1, 2, ..., N$, $2^{m-1} \times (2^N - 1)$ LEDs with identical driving voltage are illuminated simultaneously. The distribution of N small-sized LED groups can be calculated carefully to get a uniform illumination. In each group, a two-level OOK optical signal is generated and corresponds to binary 0 and 1, respectively. A 2^N level constellation can be realized with the combination of binary modulation in different groups. Therefore, N bit/s/Hz can be achieved via grouped modulation. Similar to SPAM method [10], multiple-level modulation scheme realized by grouped modulation is simple since only two-level modulation is utilized in different groups. Compared with its SPAM counterpart, grouped modulation is easier to be implemented since all the LEDs have the same driving voltage. However, it requires more LEDs in order to meet the same throughput requirement.

3.2 Pulse position modulation

PPM is an orthogonal modulation scheme, which has been widely used in optical wireless communication systems [12]. In L-PPM, one symbol duration is equally divided into L time slots, and only one of the L slots is utilized for data transmission, which is occupied by a given pulse. The power of the activated pulse is constant, denoted as LP_t, while the other $L - 1$ time slots are set to zero. The transmitted data is represented by the position of the pulse instead of its intensity, and $\log_2 L$ bits can be conveyed. The transmitted waveforms for 2-PPM and 4-PPM are shown in Fig. 3.4. Compared with PAM, PPM requires much lower average-power since only $1/L$ of the symbol duration is activated, which is achieved at the expense of an increased bandwidth. For L-PPM, the required bandwidth is $L/\log_2 L$ times that of OOK with the same data throughput. For example, when 16-PPM is used, it

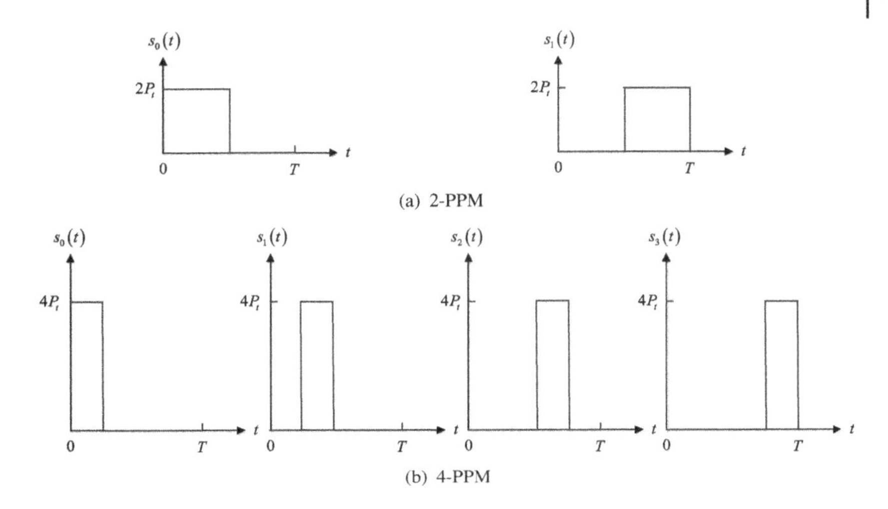

Figure 3.4 Transmitted waveforms for 2-PPM and 4-PPM.

requires four times frequency bandwidth. When L increases, the average power of L-PPM decreases by a factor of $1/L$ and the noise is only increased by $L/\log_2 L$, considering the bandwidth expansion. Therefore, the received SNR is dramatically improved. From another point of view, since the transmitted signal is zero in most of the time, the peak-to-average power ratio (PAPR) is significantly increased, which requires higher transmitted peak power.

In an LOS channel without multipath distortion, the optimum maximum likelihood (ML) receiver for L-PPM employs a continuous-time filter matched to one timeslot/one chip, whose output is sampled at the chip rate [12]. Afterwards, the L samples in each symbol duration are sent to a block decoder, which decodes the transmitted $\log_2 L$ information bits. When a soft-decision decoder is used, the largest sample is estimated as the activated pulse, and the actual samples are kept. When a hard-decision decoder is used, each sample is simply quantized to "on" or "off" with a threshold detector, and the transmitted bits are determined by the position of "on" pulses. Accidental occasions such as no sample or more than one sample is "on" have to be treated specially in the hard-decision decoder. Compared with soft-decision decoder, hard-decision decoder is easier for implementation with the sacrifice of around 1.5 dB optical power penalty [12].

In a multipath channel, the "on" pulse may induce interference in the other "off" chips, which is denoted as ISI. Due to its negative effect, the optimal soft-decision decoder proposed for LOS channel does not work well. Therefore, maximum likelihood sequence detection (MLSD) is proposed in [13], which could eliminate the ISI at the expense of high complexity and large processing delay, which is not favorable in practical applications. Decision feedback equalizers (DFEs) are proposed in [14], which consist of a symbol-rate DFE, a chip-rate DFE, and a correcting-chip-rate

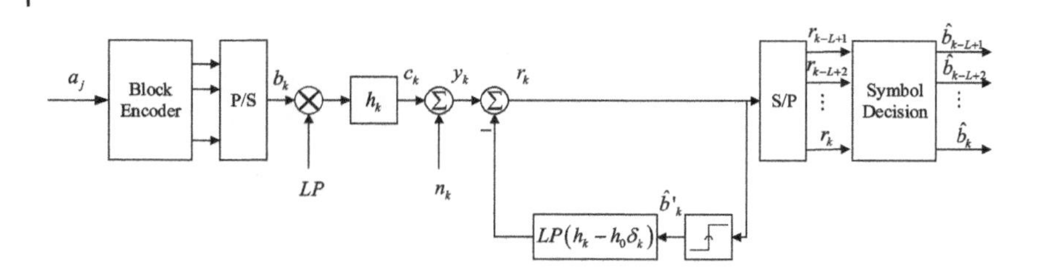

Figure 3.5 L-PPM system with chip-rate zero-forcing decision feedback equalizer.

DFE. A minimum mean squared error (MMSE) DFE is also derived in [15].

The multipath channel with ISI can be expressed as

$$y(t) = x(t) \otimes h(t) + n(t), \tag{3.17}$$

where $x(t)$ and $y(t)$ denote the transmitted and received signals, respectively. The symbol $\otimes$ indicates the convolution operation, and $n(t)$ is the additive white Gaussian noise. $h(t)$ is the channel response and its frequency response is given by $H(f)$. The optical path loss can be calculated as

$$H(0) = \int_{-\infty}^{\infty} h(t)\, dt, \tag{3.18}$$

and the received optical power is $P_r = H(0)\, P_t$.

The block diagram of an L-PPM system with the chip-rate zero-forcing DFE (ZF-DFE) is illustrated in Fig. 3.5 [16]. The input bit sequences $\{a_j\}$ are sent to a block encoder with the rate of $\log_2 L/L$, which generates length-L symbol vectors with only one unit chip value and $L-1$ zero chip values. The chip sequences $\{b_k\}$ are then scaled by LP for transmission. The effects of transmitter filter, multipath channel, and whitened-matched filter are combined as a causal, minimum-phase discrete-equivalent channel impulse response h_k, which is normalized as $\sum_k h_k = 1$. The samples $\{n_k\}$ are AWGN noise with zero mean and variance of N_0. In the chip-rate ZF-DFE, the first decision in the feedback loop in Fig. 3.5 makes chip-by-chip decisions $\hat{b}'_k$, where $\hat{b}'_k = 1$ if $r_k > \lambda$ and $\hat{b}'_k = 0$ otherwise, and λ is the predefined threshold. Afterwards, the chip-by-chip decisions are fed back through a reverse filter $LP\,(h_k - h_0\delta_k)$ containing the strictly causal portion of h_k. The second decision makes symbol-by-symbol decisions, based on which is the largest of the L sample values $r_k - L + 1$, ..., r_k, independent of the chip-by-chip decision.

The transmitted PPM codeword is denoted as $\mathbf{b} = [b_0\ b_1\ \cdots\ b_{L-1}]^{\mathrm{T}}$, and X indicates the event that all previous chip decisions and symbol decisions from prior PPM codewords are correct. When $\mathbf{e}_l$ is defined as an $L \times 1$ vector with all zeros but a single one at position $l \in \{0, 1, \ldots, L-1\}$, we have

$$
\begin{aligned}
&P\left(\text{symbol error}|X\right) \qquad (3.19)\\
=&\frac{1}{L}\sum_{l=0}^{L-1} P\left(\text{error}|X, \mathbf{b}=\mathbf{e}_l\right)\\
=&\frac{1}{L}\sum_{l=0}^{L-1}\{P\left(\text{error}|X, \hat{\mathbf{b}}'=\mathbf{b}, \mathbf{b}=\mathbf{e}_l\right) P\left(\hat{\mathbf{b}}'=\mathbf{b}|X, \mathbf{b}=\mathbf{e}_l\right)\\
&+\sum_{j=0}^{l-1} P\left(\text{error}|X, \hat{\mathbf{b}}'-\mathbf{b}=\mathbf{e}_j, \mathbf{b}=\mathbf{e}_l\right) P\left(\hat{\mathbf{b}}'-\mathbf{b}=\mathbf{e}_j|X, \mathbf{b}=\mathbf{e}_l\right)\\
&+P\left(\text{error}|X, \hat{\mathbf{b}}'=\mathbf{0}, \mathbf{b}=\mathbf{e}_l\right) P\left(\hat{\mathbf{b}}'=\mathbf{0}|X, \mathbf{b}=\mathbf{e}_l\right)\\
&+\sum_{j=l+1}^{L-1} P\left(\text{error}|X, \hat{\mathbf{b}}'-\mathbf{b}=\mathbf{e}_j, \mathbf{b}=\mathbf{e}_l\right) P\left(\hat{\mathbf{b}}'-\mathbf{b}=\mathbf{e}_j|X, \mathbf{b}=\mathbf{e}_l\right)\\
&+P\left(\text{error}|X, w_H\left(\hat{\mathbf{b}}'-\mathbf{b}\right)\geq 2, \mathbf{b}=\mathbf{e}_l\right)P\left(w_H\left(\hat{\mathbf{b}}'-\mathbf{b}\right)\geq 2|X, \mathbf{b}=\mathbf{e}_l\right)\},
\end{aligned}
$$

where w_H denotes the Hamming weight. For simplicity, we can rewrite (3.19) as

$$
\begin{aligned}
&P\left(\text{symbol error}|X\right) \qquad (3.20)\\
=&\frac{1}{L}\sum_{l=0}^{L-1}\left(P_0\pi_0+\sum_{j=0}^{l-1}P_{1,j}\pi_{1,j}+P_2\pi_2+\sum_{j=l+1}^{L-1}P_{3,j}\pi_{3,j}+P_4\pi_4\right).
\end{aligned}
$$

Since a symbol error occurs only when a chip is wrong, it is obvious that the symbol-decision error $P_0 = 0$. For the ease of derivation, let $\eta_k = LP_r h_k/\sqrt{N_0}$ denote the normalized impulse response, and the normalized threshold for the chip-rate slicer is set to the minimax value $\lambda = \eta_0/2$, which maximizes the probability that all chips are successfully detected. The symbol-decision error $P_{1,j}$ in (3.20) occurs when a chip j is detected above the threshold before the actual transmitted chip l. Therefore, one noise sample is above λ, which is denoted as z_1. The probability of the symbol error $P_{1,j}$ is then given by

$$
\begin{aligned}
P_{1,j} =&1-E\{(1-Q_{1,0}\left(z_1+\eta_0-\eta_{l-j}\right))^j \qquad (3.21)\\
&\times\left(1-Q_{1,j}\left(z_1+\eta_0-\eta_{l-j}\right)\right)\\
&\times\prod_{i=j+1\neq l}^{L-1}\left(1-Q\left(z_1+\eta_0-\eta_{l-j}+\eta_{i-j}\right)\right)\},
\end{aligned}
$$

where

$$
Q_{1,i}\left(x\right)=\max\left\{\left(Q\left(x\right)-Q\left(\lambda+\eta_{i-j}\right)\right)/\left(1-Q\left(\lambda+\eta_{i-j}\right)\right), 0\right\}, i\neq j, \qquad (3.22)
$$

and

$$Q_{1,j}(x) = \min\{Q(x)/Q(\lambda), 1\}. \tag{3.23}$$

Besides, the probability of the event that one chip error happens before the transmitted chip occurs is expressed as [16]

$$\begin{aligned}\pi_{1,j} =&(1-Q(\lambda))^j Q(\lambda) \prod_{i=j+1}^{l-1} (1-Q(\eta_{i-j}+\lambda)) \\ &\times Q(\eta_{i-j}-\eta_0+\lambda) \prod_{i=l+1}^{L-1} (1-Q(\eta_{i-j}-\eta_{i-l}+\lambda)).\end{aligned} \tag{3.24}$$

The symbol-decision error P_2 in (3.20) indicates that the transmitted chip is not detected, i.e., none of the L chips is greater than λ. Denoting z_2 as the noise sample in position l, which is smaller than λ, we have

$$P_2 = 1 - E\left\{(1-Q_{2,0}(z_2+\eta_0))^l \prod_{i=l+1}^{L-1} (1-Q_{2,l}(z_2\eta_0-\eta_{i-l}))\right\}, \tag{3.25}$$

where $Q_{2,i}$ has the similar expression as $Q_{1,i}$ and η_{i-j} is replaced by η_{i-l} in (3.22). The probability of the event that no chips are larger than λ is given by

$$\pi_2 = (1-Q(\lambda))^l \prod_{i=0}^{L-l-1} Q(\eta_i - \lambda). \tag{3.26}$$

The symbol-decision error $P_{3,j}$ in (3.20) presents that the chip j is detected above λ. Denoting z_3 as the noise sample that is above the threshold, we have

$$P_{3,l} = 1 - E\left\{(1-Q_{1,0}(z_3+\eta_0))^j \prod_{i=j+1}^{L-1} (1-Q_{1,i}(z_3+\eta_0+\eta_{i-j}))\right\}. \tag{3.27}$$

The symbol-decision error P_4 in (3.20) means that more than one chip is detected incorrectly, which is difficult to calculate. Therefore, we obtain a lower bound for $P(\text{error}|X)$ by simply setting $P_4 = 0$, and an upper bound for $P(\text{error}|X)$ by setting $P_4 = (L-1)/L$. The probability π_4 can be calculated as $\pi_4 = 1 - (\pi_0 + \pi_1 + \pi_2 + \pi_3)$, where $\pi_0 = 1 - LQ(\lambda)$ denotes the probability that all chips are detected correctly.

The block diagram of an L-PPM system with the symbol-rate ZF-DFE is illustrated in Fig. 3.6, which feedbacks symbol decisions instead of intermediate chip decisions. Different from its chip-rate counterpart, the feedback filter in the symbol-rate ZF-DFE feedbacks appropriate information to cancel all postcursor ISI, but it does not cancel intra-symbol interference. At each time $k = JL - 1$, where J is an integer,

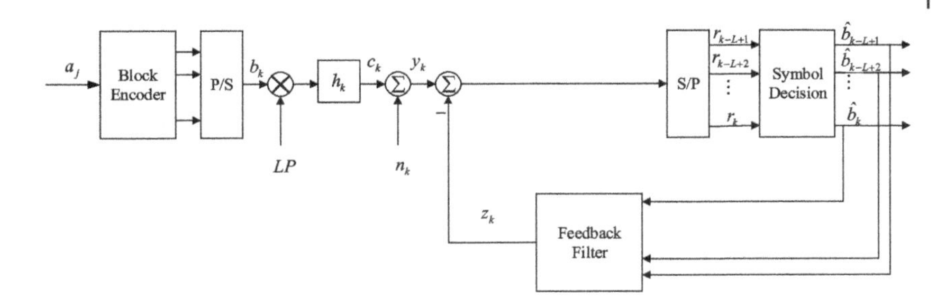

Figure 3.6 L-PPM system with symbol-rate zero-forcing decision feedback equalizer.

a symbol decision is made. The resultant length-L vector $\hat{b}_{k-L+1}$, ..., $\hat{b}_k$ is passed to the feedback filter and generates the output as

$$z_k = LP \sum_{i=1}^{N_s-1} (b_{k-iL+l_i}\delta_k \otimes h_{k+iL-l_i}u_k), \tag{3.28}$$

where $l_i \in \{0, 1, ..., L-1\}$, $i = 1, 2, ..., N_s - 1$ represent the symbols already detected previously, and N_s is the largest number of symbols that can be spanned by h_k. The received samples r_{k-L+1}, ..., r_k are passed through a minimum-Euclidean-distance detector, and the symbol decision module selects the detected symbol which minimizes $\sum_{k=0}^{L-1} \left| r_k - LP\left(\hat{b}_k \otimes h_k\right)\right|^2$.

Since only one pulse is "on" during one symbol duration, the spectral efficiency of PPM is limited. Several variants of modulation schemes have been proposed recently. Figure 3.7 provides several examples of the modified PPM schemes. Differential PPM (DPPM) is a simple modification of PPM, which omits the "off" chips when the "on" chip appears [17]. As a result, less time is required to convey the same information, and symbols have unequal durations. Since DPPM has variable symbol durations, the receiver does not know the symbol boundaries, MLSD is preferred to be used in the optimal soft-decision decoder, which has high complexity. If a hard-decision decoder is used, it might be easier to decode DPPM than PPM since the former does not require symbol-level timing recovery. For a fixed modulation order L, DPPM requires much less bandwidth and slightly more power than PPM because of the reduced duty cycle. It was indicated in [12] that 16-DPPM achieved 3.1 dB optical power gain compared with 4-PPM with only 6% increased bandwidth. Multipulse PPM (MPPM) achieves an increased spectral efficiency with more than one "on" chip in a single symbol duration [18]. Besides, MPPM has lower PAPR compared with PPM. Overlapping PPM (OPPM) is a generalization of conventional PPM, which also transmits more than one "on" consecutive chip in a symbol duration [19]. However, the "on" chips in different symbols can be overlapped. It was shown in [19] that OPPM could not only achieve a higher spectral efficiency compared to PPM, but also support dimming control. Expurgated PPM (EPPM) was

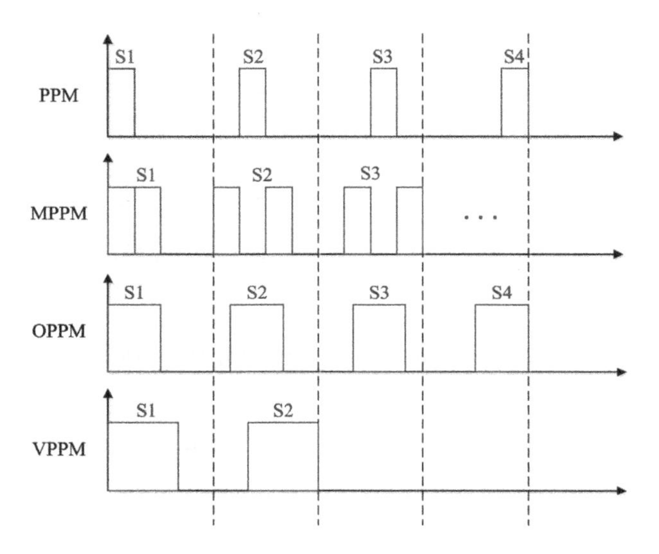

Figure 3.7 Examples of modified PPM schemes.

proposed in [20, 21], where the symbols in the MPPM are expurgated to maximize the inter-symbol distance. It can achieve the same spectral efficiency as PPM and provide dimming support in VLC systems at the same time since it can realize arbitrary levels of illumination by changing the number of pulses per symbol and the length of the symbol. Moreover, it can mitigate the flickering with multiple pulses in a single symbol duration. In the IEEE 802.15.7 standard, variable PPM (VPPM) is used, which is a hybrid of PPM and pulse width modulation (PWM) [3]. In VPPM, the bits are encoded by the position of pulse similar to PPM. Meanwhile, the pulse width could be adjusted as well. Therefore, dimming control can be achieved and the simplicity and robustness inherited from PPM can be maintained.

3.3 Carrierless amplitude phase modulation

CAP was firstly developed by Bell Labs as a variant of quadrature amplitude modulation (QAM) [22, 23]. Due to its high spectral efficiency and simple implementation, CAP has been successfully employed in asymmetric digital subscriber line (ADSL). Recently, it is also considered as a promising modulation format for optical fiber communications and visible light communications [24–30].

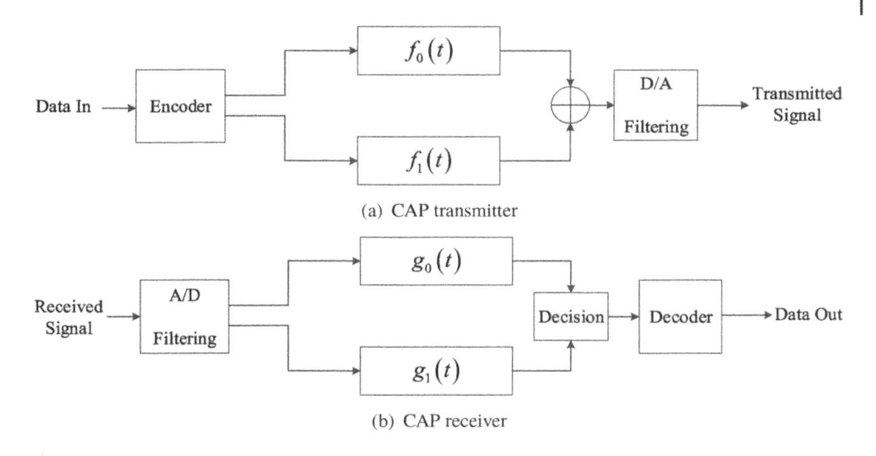

Figure 3.8 Diagram of CAP transceiver.

3.3.1 Principles of CAP

CAP is a bandwidth-efficient two-dimensional (2-D) passband transmission scheme. The basic principle of CAP is to choose two orthogonal filters to modulate two different data streams, which are combined for simultaneous transmission. Those two orthogonal filters can be obtained by multiplying a root raised cosine (RRC) filter with sine or cosine functions, which are given by [31]

$$\begin{aligned} f_0(t) &= g(t)\cos(2\pi f_c t), \\ f_1(t) &= g(t)\sin(2\pi f_c t), \end{aligned} \tag{3.29}$$

where $g(t)$ is the RRC filter with roll-off factor α, f_c is the symbol frequency, which is larger than the highest frequency in $g(t)$.

The diagram of CAP transceiver is shown in Fig. 3.8. The data stream is encoded into two independent symbol streams $\{s_{0,k}\}$ and $\{s_{1,k}\}$, which are modulated by their corresponding orthogonal filters. The combined signal is given by

$$x(t) = \sum_{k=-\infty}^{\infty} \left[s_{0,k} f_0\left(t - \frac{k}{f_c}\right) + s_{1,k} f_1\left(t - \frac{k}{f_c}\right)\right]. \tag{3.30}$$

At the receiver, the receiving filters $g_0(t)$ and $g_1(t)$ are used to invert the channel response and transmitting filters, which are then used for decoding. Several techniques have been proposed for data recovery, including the linear equalizer and the decision feedback equalizer [31]. In order to achieve perfect reconstruction performance, a fractionally spaced linear equalizer (FSLE) can be utilized together with a training sequence to update the filter taps adaptively via the least mean square (LMS) algorithm [31].

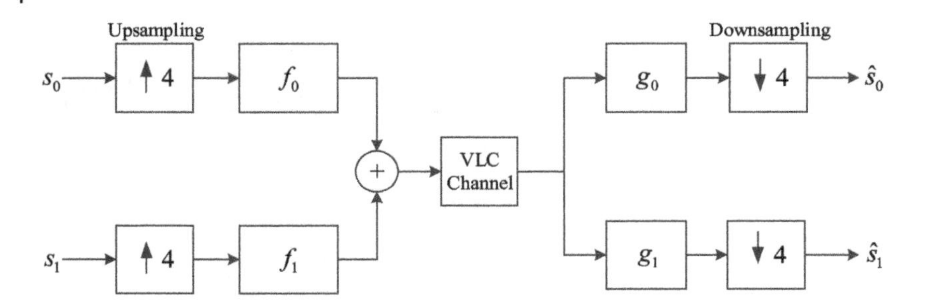

Figure 3.9 Diagram of CAP as a multirate transmultiplexer.

Alternatively, the CAP transceiver can be regarded as a transmultiplexer. The shaping filter at the transmitter is implemented by an upsampled finite impulse response (FIR) window from the original continuous-time noncausal signal, while the equalizer at the receiver is downsampled to match the original symbol rate. The upsampling and downsampling factor ζ is chosen carefully considering the tradeoff between performance and implementation complexity. In particular, $\zeta = 4$ would be a reasonable choice [31], and the corresponding diagram of the transmultiplexer is illustrated in Fig. 3.9. When perfect reconstruction is considered, the CAP transceiver topology is an obvious multirate transmultiplexer, which can be written as

$$\begin{pmatrix} \hat{s}_0 \\ \hat{s}_1 \end{pmatrix} = T \begin{pmatrix} s_0 \\ s_1 \end{pmatrix}, \tag{3.31}$$

where T is the overall system transfer matrix given by

$$T = G(z)\,\Gamma F(z), \tag{3.32}$$

where

$$G\left(e^{j\omega}\right) = \begin{pmatrix} G_0\left(\frac{\omega}{4}\right) & G_0\left(\frac{\omega}{4}+\frac{\pi}{2}\right) & G_0\left(\frac{\omega}{4}+\pi\right) & G_0\left(\frac{\omega}{4}+\frac{3\pi}{2}\right) \\ G_1\left(\frac{\omega}{4}\right) & G_1\left(\frac{\omega}{4}+\frac{\pi}{2}\right) & G_1\left(\frac{\omega}{4}+\pi\right) & G_1\left(\frac{\omega}{4}+\frac{3\pi}{2}\right) \end{pmatrix} \tag{3.33}$$

denotes the type I polyphase decomposition of the receiver filter bank, and

$$F\left(e^{j\omega}\right) = \begin{pmatrix} F_0\left(\frac{\omega}{4}\right) & F_1\left(\frac{\omega}{4}\right) \\ F_0\left(\frac{\omega}{4}+\frac{\pi}{2}\right) & F_1\left(\frac{\omega}{4}+\frac{\pi}{2}\right) \\ F_0\left(\frac{\omega}{4}+\pi\right) & F_1\left(\frac{\omega}{4}+\pi\right) \\ F_0\left(\frac{\omega}{4}+\frac{3\pi}{2}\right) & F_1\left(\frac{\omega}{4}+\frac{3\pi}{2}\right) \end{pmatrix} \tag{3.34}$$

is the type II polyphase decomposition of the transmitter filter bank [32]. Γ is a permutation matrix representing the fractional delay in the receiver downsampling. F_0, F_1, G_0, and G_1 denote the corresponding discrete-time Fourier transforms for the filters f_0, f_1, g_0, and g_1, respectively. In order to achieve perfect reconstruction performance, T should satisfy

$$T = z^{-n} I, \tag{3.35}$$

which only contains the identity matrix I and n delay elements z^{-n}. Since n does not have to be a multiple of the symbol period in (3.7), the fractional part of the delay should be adjusted accordingly. Moreover, if the transmitter polyphase decomposition matrix F is paraunitary, i.e.,

$$F^{-1} = z^{-m} F^{\mathrm{H}}, \tag{3.36}$$

the perfect reconstruction receiver would be the matched filter pair of the transmitter. Since the CAP system utilizes shaping waveforms which span over multiple symbol periods, the Hilbert pair constitutes a nonparaunitary system with the receiver filters. Therefore, the unique minimum norm solution can be derived for the underdetermined system, which is given by

$$\begin{aligned} (P(f_0))\,\mathbf{g}_0 &= \boldsymbol{\delta}, \\ (P(f_1))\,\mathbf{g}_0 &= \mathbf{O}, \\ (P(f_0))\,\mathbf{g}_1 &= \mathbf{O}, \\ (P(f_1))\,\mathbf{g}_1 &= \boldsymbol{\delta}, \end{aligned} \tag{3.37}$$

where $\mathbf{g}_0$ and $\mathbf{g}_1$ are the vectors of samples corresponding to the receiver filters g_0 and g_1. $P(\cdot)$ is a permutation matrix of the transmitter filter, $\mathbf{O}$ is a vector with all the elements being zero, and $\boldsymbol{\delta}$ is a vector with all zero elements but one '1' in the middle.

From (3.30), the transmitted signal $x(t)$ is a cyclostationary random process with the period $T = 1/f_c$ for both PAM and CAP, and we can restrict analysis to the time interval $(0, T]$ for simplicity. Furthermore, even when the roll-off factor is extremely small, i.e., $\alpha = 0.99$, 99% of the energy is still constrained in K = 2000 symbol periods [33]. When only K symbols are considered, the signals for PAM and CAP are given by

$$x^{\mathrm{PAM}}(t) = \sum_{k=1}^{K} s_k f_k(t) \tag{3.38}$$

and

$$x^{\mathrm{CAP}}(t) = \sum_{k=1}^{K} \left(s_{0,k} f_{0,k}(t) + s_{1,k} f_{1,k}(t)\right), \tag{3.39}$$

where $f_k(t) = f(t - (k - K/2)T)$, $f_{0,k}(t) = f_0(t - (k - K/2)T)$, and $f_{1,k}(t) = f_1(t - (k - K/2)T)$. Denote $x_k^{\mathrm{PAM}}(t)$ and $x_k^{\mathrm{CAP}}(t)$ as

$$x_k^{\mathrm{PAM}}(t) = s_k f_k(t), \tag{3.40}$$

and

$$x_k^{\text{CAP}}(t) = s_{0,k} f_{0,k}(t) + s_{0,k} f_{0,k}(t), \tag{3.41}$$

it is obvious that (3.38) and (3.39) are the sum of $x_k^{\text{PAM}}(t)$ and $x_k^{\text{CAP}}(t)$. The signal can be considered as an interference of past and future symbols multiplied by the filter response with different time shifts. To calculate the PAPR for both PAM and CAP, the characteristic functions of (3.40) and (3.41) are exploited. Specifically, for a random variable X, its characteristic function is given by the inverse Fourier transform of its probability density function (PDF) [33]

$$\varphi_X(f) = E\left\{e^{j2\pi X}\right\} = \int_{-\infty}^{\infty} p_X(x)\, e^{j2\pi x} dx, \tag{3.42}$$

where $p_X(x)$ is the PDF. Therefore, the characteristic function of M-ary PAM is expressed as

$$\varphi_{x_k^{\text{PAM}}(t)}(f) = \frac{2}{M} \sum_{m=1}^{M/2} \cos\left(2\pi f f_k(t)\, s_m c_{\text{PAM}}\right), \tag{3.43}$$

where $\{s_m\}$ is the constellation set for PAM and c_{PAM} is the corresponding normalization factor to maintain the average energy of the constellations as 1. Similarly the characteristic functions for two-dimensional CAP are given by

$$\varphi_{x_{0,k}^{\text{CAP}}(t)}(f) = \frac{2}{\sqrt{M}} \sum_{m=1}^{\sqrt{M}/2} \cos\left(2\pi f f_{0,k}(t)\, s_m c_{\text{CAP}}\right), \tag{3.44}$$

$$\varphi_{x_{1,k}^{\text{CAP}}(t)}(f) = \frac{2}{\sqrt{M}} \sum_{m=1}^{\sqrt{M}/2} \cos\left(2\pi f f_{1,k}(t)\, s_m c_{\text{CAP}}\right).$$

Therefore, the characteristic functions of both PAM and CAP signals in (3.38) and (3.39) are written as

$$\varphi_{x^{\text{PAM}}(t)}(f) = \prod_{k=1}^{K} \varphi_{x_k^{\text{PAM}}(t)}(f), \tag{3.45}$$

and

$$\varphi_{x^{\text{CAP}}(t)}(f) = \prod_{k=1}^{K} \varphi_{x_{0,k}^{\text{CAP}}(t)}(f)\, \varphi_{x_{1,k}^{\text{CAP}}(t)}(f). \tag{3.46}$$

After Fourier transform, the PDFs of $x^{\text{PAM}}(t)$ and $x^{\text{CAP}}(t)$ can be obtained. When $x(t)$ is used to replace either $x^{\text{PAM}}(t)$ or $x^{\text{CAP}}(t)$, and $p_{x(t)}(x)$ is defined as its PDF, we have

$$p_X(x) = \frac{1}{T} \int_0^{\infty} p_{x(t)}(x)\, dt, \tag{3.47}$$

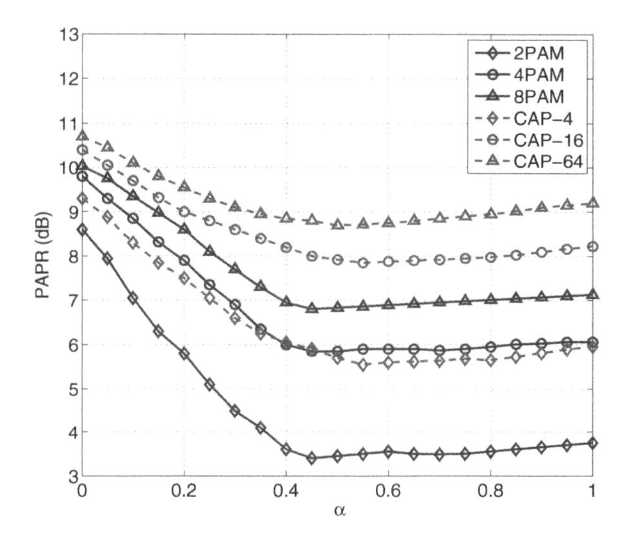

Figure 3.10 PAPR at CCDF = 10^{-5} for PAM and CAP.

and the complementary cumulative distribution function (CCDF) is given by

$$\Theta_X(z) = \int_z^{\infty} p_X(x)\,dx. \tag{3.48}$$

The calculation of the characteristic functions can be implemented by fast Fourier transform (FFT) and inverse fast Fourier transform (IFFT), which have low complexity. When the interval (0, T] is equally divided into 64 parts, the amplitude range is limited to (-32, 32) and the size of FFT/IFFT is set to 2^{16}, Fig. 3.10 illustrates the required PAPR at CCDF = 10^{-5} for CAP and PAM, where CAP-4, CAP-16, and CAP-64 employ 2PAM, 4PAM, and 8PAM in each dimension, respectively. It can be seen that the PAPR of CAP is always higher than PAM when the same modulation order is applied, and their PAPRs vary when the filter roll-off factor α changes.

3.3.2 Multidimensional CAP

In conventional 2-D CAP, two orthogonal filters are generated by multiplying a single RRC filter with sine or cosine functions, which are used to modulate two different data streams. If the third orthogonal filter is adopted, one additional data stream can be transmitted, which is referred to as 3-D CAP and can be seen as an extension of the conventional 2-D CAP. Denoting the 3-D symbol vector as $\mathbf{s}_k = [s_{0,k}\ s_{1,k}\ s_{2,k}]$, and the orthogonal filters as $f_0(t)$, $f_1(t)$, and $f_2(t)$, the combined 3-D CAP signal

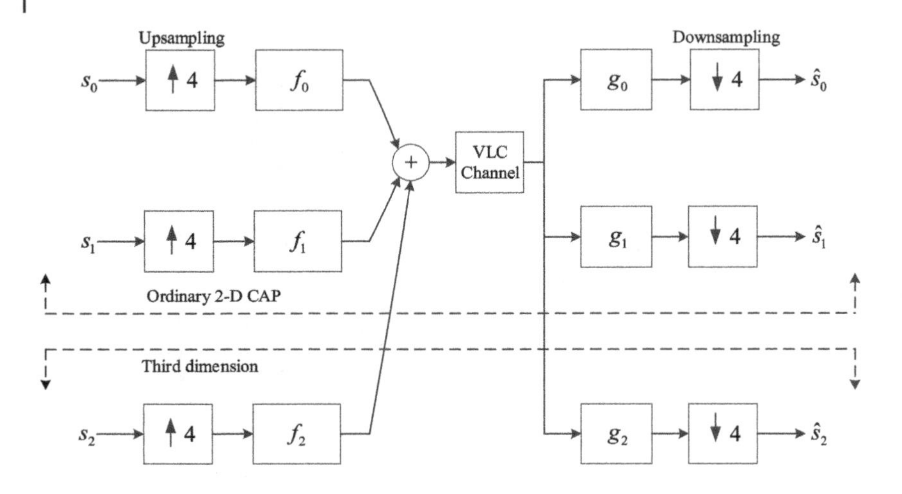

Figure 3.11 Diagram of 3-D CAP as a multirate transmultiplexer.

can be generated as

$$x(t) = \sum_{k=-\infty}^{\infty} \sum_{i=0}^{2} s_{i,k} f_i \left(t - \frac{k}{f_c} \right). \tag{3.49}$$

The diagram of 3-D CAP is illustrated in Fig. 3.11, which is obviously an extension of the 2-D CAP. The upsampling and downsampling factors are still $\zeta = 4$. Therefore, the symbol rate per dimension remains unchanged, leading to increased overall throughput. However, in order to allow for the extra dimension, the excess bandwidth of the shaping filter should be increased as well. Fortunately, it has no effect on the performance for additive white noise over a flat channel [34]. Besides, increasing excess bandwidth might achieve better performance than increasing either the symbol rate or the number of bits per symbol [31]. Since there exist more than two orthogonal shaping filters, the original Hilbert pair signals do not work well. It is illustrated in [35] that when the upsampling factor is 4, perfect reconstruction performance can be maintained for up to four orthogonal filters. However, it has to use the entire sampling range and surpass the allowed frequency bandwidth, which is unfavorable for actual VLC systems. Therefore, only three filters are designed in the 3-D CAP system.

The input data is firstly divided into three bit streams, which are listed in the vector form as $[s_{0,k}\ s_{1,k}\ s_{2,k}]$. The output $[\hat{s}_{0,k}\ \hat{s}_{1,k}\ \hat{s}_{2,k}]$ after the multirate transmultiplexer should be as close as possible to $[s_{0,k}\ s_{1,k}\ s_{2,k}]$. Afterwards, the receiver-side equalizer can be used to achieve the perfect reconstruction. Similar to (3.32), the transmultiplexer can be modeled as the multiple-input multiple-output transfer matrix, where the polyphase decomposition of the filters can be rewritten as

$$G\left(e^{j\omega}\right)=\begin{pmatrix} G_0\left(\frac{\omega}{4}\right) & G_0\left(\frac{\omega}{4}+\frac{\pi}{2}\right) & G_0\left(\frac{\omega}{4}+\pi\right) & G_0\left(\frac{\omega}{4}+\frac{3\pi}{2}\right) \\ G_1\left(\frac{\omega}{4}\right) & G_1\left(\frac{\omega}{4}+\frac{\pi}{2}\right) & G_1\left(\frac{\omega}{4}+\pi\right) & G_1\left(\frac{\omega}{4}+\frac{3\pi}{2}\right) \\ G_2\left(\frac{\omega}{4}\right) & G_2\left(\frac{\omega}{4}+\frac{\pi}{2}\right) & G_2\left(\frac{\omega}{4}+\pi\right) & G_2\left(\frac{\omega}{4}+\frac{3\pi}{2}\right) \end{pmatrix}, \quad (3.50)$$

and

$$F\left(e^{j\omega}\right)=\begin{pmatrix} F_0\left(\frac{\omega}{4}\right) & F_1\left(\frac{\omega}{4}\right) & F_2\left(\frac{\omega}{4}\right) \\ F_0\left(\frac{\omega}{4}+\frac{\pi}{2}\right) & F_1\left(\frac{\omega}{4}+\frac{\pi}{2}\right) & F_2\left(\frac{\omega}{4}+\frac{\pi}{2}\right) \\ F_0\left(\frac{\omega}{4}+\pi\right) & F_1\left(\frac{\omega}{4}+\pi\right) & F_2\left(\frac{\omega}{4}+\pi\right) \\ F_0\left(\frac{\omega}{4}+\frac{3\pi}{2}\right) & F_1\left(\frac{\omega}{4}+\frac{3\pi}{2}\right) & F_2\left(\frac{\omega}{4}+\frac{3\pi}{2}\right) \end{pmatrix}. \quad (3.51)$$

Usually, the adaptive equalizer at the receiver side is preferred to be implemented by an FIR filter instead of an infinite impulse response (IIR) filter. The problem to find three input signals can be modeled as an optimization problem traversing the entire space and several constraints have to be included in the optimization problem. The perfect reconstruction condition should be satisfied, and the frequency-domain characteristics should be as close as possible to the raised-cosine frequency characteristics since the latter has relatively reliable performance in the VLC link. The minimax optimization was performed in [31] to find three signals based on the sequential quadratic programming method [36]. In the optimization problem, three signals $\{f_0, f_1, f_2\}$ are solved at the same time, aiming to minimize the ∞-norm of the difference in frequency domain [31]

$$\min_{\{f_0,f_1,f_2\}} \{\max\left(|H-R|\right)\} \text{ subject to } \quad G\left(z\right)\Gamma F\left(z\right)=z^{-n}I, \quad (3.52)$$

where H denotes the frequency magnitude characteristic for a given signal $f_i \in \{f_0, f_1, f_2\}$, R represents the required passband frequency magnitude response, and F is the polyphase decomposition of f_i.

The calculation of matrix G is completed by computing the polyphase decomposition of the receiver corresponding to three filters $\{f_0, f_1, f_2\}$, while filters at the receiver are obtained by directly computing the unique minimum norm solution of the underdetermined problem, which maximizes the signal-to-noise ratio (SNR) when channel distortion is not considered. Figure 3.12(a) illustrates the time-domain waveform of three signals obtained by the minimax optimization algorithm, and their corresponding frequency responses are demonstrated in Fig. 3.12(b).

From Fig. 3.12(b), the frequency responses of three solved signals are nonzero at high frequency. Nevertheless, only negligible transmitted power is allocated. Compared with 2-D CAP with the same bandwidth and modulation order per dimension, there exists some kind of SNR penalty in 3-D CAP when near-end cross-talk is considered [37]. However, to achieve the same throughput, the modulation order of the conventional 2-D CAP has to be enlarged and its required SNR becomes higher. For example, if 4PAM is adopted in each dimension of 3-D CAP, 2-D CAP needs to employ 8PAM in each dimension to maintain the same throughput, which requires around 6 dB receiver sensitivity improvement. Another merit of 3-D CAP is to serve three users simultaneously in a multiple access manner.

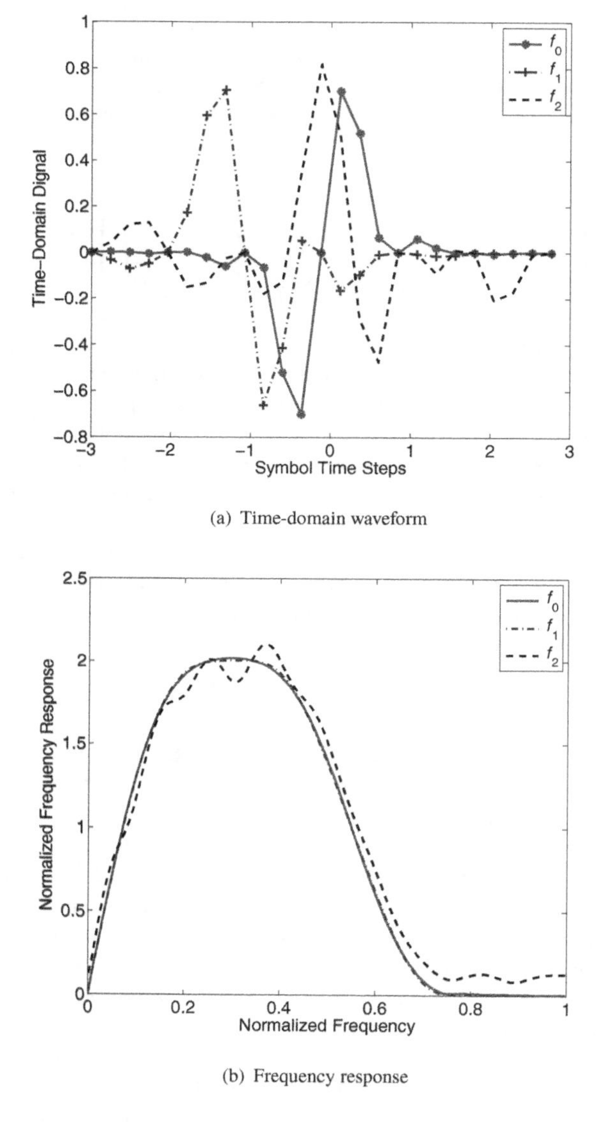

(a) Time-domain waveform

(b) Frequency response

Figure 3.12 The three signals solved using the minimax optimization algorithm.

3.4 Modulation and coding schemes for dimmable VLC

In indoor VLC systems, communication and illumination should be maintained at the same time, where the light can be dimmed to satisfy various illumination and power requirements [38]. Besides, dimming technology is preferred to be energy efficient. As illustrated in Fig. 3.13, the LEDs can be dimmed at daytime to save energy, while high optical power is required at night to provide sufficient illumination. However, the dimming operation will inevitably interfere with the communication function of VLC systems since it will alter the received optical power and SNR. The conventional modulation schemes should be modified to support dimming control, and coding schemes should be combined to achieve reliable performance.

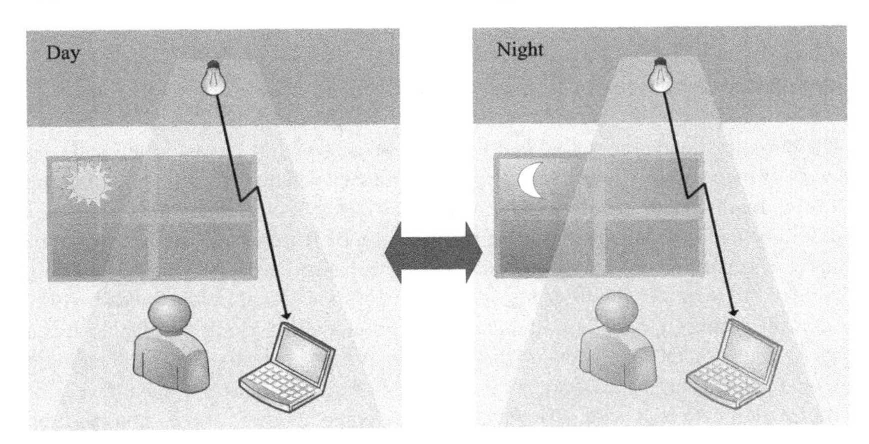

Figure 3.13 Visible light communications with dimming control.

The inherent nonlinearity of LEDs is a challenge for orthogonal frequency division multiplexing (OFDM) methodology since OFDM has high PAPR. The input of an LED has a minimum threshold value that can generate current, which is referred to as turn-on voltage (TOV). When the input voltage is above the TOV, the voltage-current and current-power characteristics are nonlinear. Several methods have been proposed to mitigate the effect of LED nonlinearity, where the transfer characteristic of the LED is considered to be quasi-linear in a limited range after predistortion. That is, if $v_{\min}$ and $v_{\max}$ are denoted as the minimum and maximum allowed signals according to the voltage levels permitted by LED, the transfer characteristic of the LED between $[v_{\min}, v_{\max}]$ is assumed to be linear. Specifically, the relationship between the emitted optical power and the input voltage is given by

$$P_{opt}(t) = \begin{cases} 0, & v(t) < v_{\min}, \\ \eta\left(v(t) - v_{\min}\right), & v_{\min} \leq v(t) \leq v_{\max}, \\ \eta\left(v_{\max} - v_{\min}\right), & v(t) > v_{\max}, \end{cases} \tag{3.53}$$

where η and $v(t)$ are the voltage-power transfer coefficient and instantaneous input voltage, respectively.

Since the illumination level is proportional to the average emitted optical power, dimming control can be implemented by adjusting the average optical power of LEDs. The measured dimming level d is defined as

$$d = E\left(P_{opt}(t)\right) / \left(\eta\left(v_{\max} - v_{\min}\right)\right), \tag{3.54}$$

which obviously falls in the interval $[0, 1]$. When the required dimming level is adjusted, the received optical power and effective SNR are changed, which will inevitably alter the achievable data rate for a given BER requirement. The modulation schemes preferred for VLC systems should support relatively high data rate under different dimming targets.

3.4.1 Modulation schemes for dimmable VLC

The dimming control for OOK and PAM can be achieved by changing the intensity levels of the symbols. It can keep the symbol rate constant when different dimming level is required, the required SNR is however changed, and the modulation order should be decreased at some time to maintain the BER performance. Alternatively, one can keep the levels of information symbols unchanged and adjust the average duty cycle of the waveform by inserting some compensation symbols, which can be fully on or off during the required time interval to provide dimming performance. It can be regarded as a DC bias added to the waveform, which is positive when the required optical power is high and negative otherwise. For example, if the average intensity of information data is A with period T_1 and the average intensity of the compensation symbols is B with period T_2, the resulting average intensity is expressed as [39]

$$N = \frac{AT_1 + BT_2}{T_1 + T_2}. \tag{3.55}$$

Apparently, this compensation scheme results in different symbol rate when the dimming level changes. An example of OOK is illustrated in Fig. 3.14, where Fig. 3.14(a) is the original OOK signal, Figs. 3.14(b) and 3.14(c) show the dimmed signal with $d = 30\%$ and $d = 70\%$, respectively.

Another modulation scheme is variable OOK (VOOK), which inserts zero to the "on" chip when the dimming level is below 50% and one to the "off" chip when the dimming level is above 50% [40]. In this way, the average emitted optical power could be altered. The dimming control of PPM can be achieved via VPPM introduced in Section 3.2, which is similar to VOOK. For MPPM, the symbol duration T is divided into n chips, and w $(1 \leq w \leq n)$ optical pulses are emitted during one symbol duration. For adjusting dimming level, the number of optical pulses w should be modified and the dimming factor of MPPM can be written as

$$d = \frac{w}{n}. \tag{3.56}$$

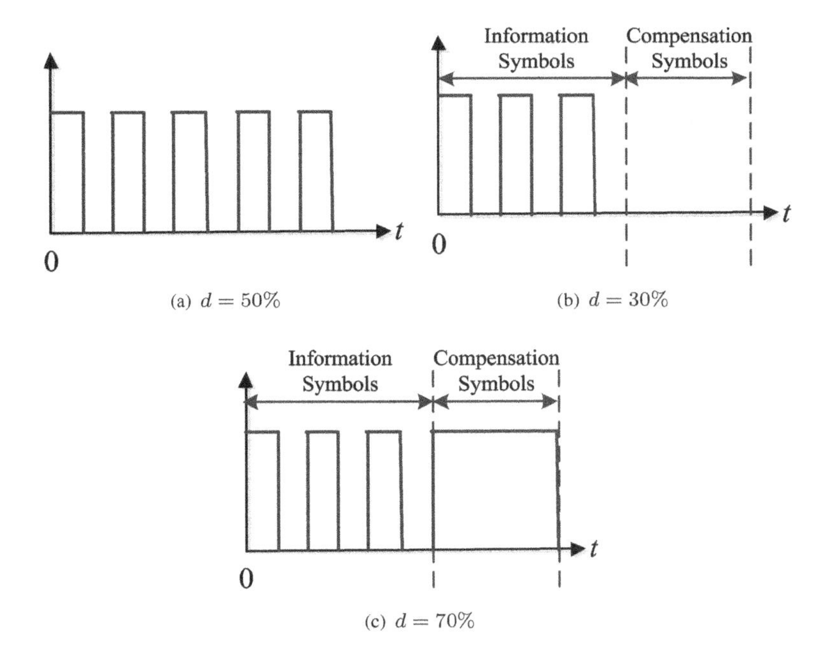

Figure 3.14 An example of the compensation scheme for OOK.

Specifically, the system is under full brightness when $w = n$, and $d = 100\%$. n different dimming levels could be supported, which are $1, (n-1)/n, ..., 1/n$, respectively. The MPPM symbol is given by $x(t) = \sum_{i=0}^{n-1} c_i \text{rect}(nt/T - i)$, where $\mathbf{c} = [c_0\ c_1\ \cdots\ c_{n-1}]$ is a binary vector with length n and weight w, $rect(t)$ is defined as one for $0 \leq t \leq 1$ and zero for otherwise.

The power requirement and spectral efficiency for these modulation schemes under dimming control are analyzed. When high SNR is considered, the BER is dominated by the nearest two different symbols, which can be well approximated as $Q(D_{\min}/2\sigma)$, where $D_{\min}$ is the minimum Euclidean distance between arbitrary two symbols. OOK is used as a benchmark for comparison. The power required to achieve the same BER as OOK is defined as $P \approx (D_{\text{OOK}}/D_{\min}) P_{\text{OOK}}$ and $P_{\text{OOK}} = Q^{-1}(\text{BER})\sqrt{N_0 R_b}$, and R_b is the bit rate. The Euclidean distance of VOOK is given by

$$D_{\text{VOOK}} = \begin{cases} P\sqrt{\frac{2d}{R_b}}, & 0 \leq d \leq 0.5, \\ P\sqrt{\frac{2(1-d)}{R_b}}, & 0.5 < d \leq 1. \end{cases} \tag{3.57}$$

The ratio of D_{OOK} to D_{VOOK} indicates that there is a power increase in VOOK compared with OOK. The Euclidean distance of VPPM is given by

$$D_{\text{VPPM}} = \begin{cases} P\sqrt{\frac{2d}{R_b}}, & 0 \le d \le 0.5, \\ P\sqrt{\frac{2(1-d)}{R_b}}, & 0.5 < d \le 1, \end{cases} \tag{3.58}$$

which is the same as VOOK. Therefore, the required power for VPPM is the same as that for VOOK.

For MPPM, the minimum distance is expressed as

$$D_{\text{MPPM}} = (P/n)\sqrt{2n\log_2\binom{n}{w}/R_b}, \tag{3.59}$$

which is altered according to different dimming requirements. The spectral efficiencies of VOOK and VPPM are given by

$$\nu_{\text{VOOK}} = \begin{cases} 2d, & 0 \le d \le 0.5, \\ 2(1-d), & 0.5 < d \le 1, \end{cases} \tag{3.60}$$

and

$$\nu_{\text{VPPM}} = \begin{cases} d, & 0 \le d \le 0.5, \\ (1-d), & 0.5 < d \le 1, \end{cases} \tag{3.61}$$

respectively. It can be seen that VOOK could realize twice the spectral efficiency of VPPM, while the spectral efficiency of MPPM can be expressed as

$$\nu_{\text{MPPM}} = \binom{n}{w}/n. \tag{3.62}$$

3.4.2
Coding schemes for dimmable VLC

In addition to the modulation schemes, coding is also considered in the design of dimming support. In the IEEE 802.15.7 standard, convolutional code and Reed-Solomon code are employed for forward error correction (FEC). After FEC encoding, run length limited (RLL) line encoding is adopted to provide DC balance and flicker mitigation, which includes Manchester, 4B6B, and 8B10B codes [41].

Apart from the RLL line coding, the source coding can be redesigned to increase the data rate. For a given codeword length, the construction of the codebook should contain as many codewords as possible. The inverse source coding (ISC) employs inverse mapping of lossless compression for OOK/PAM signals, and has been implemented with a reversal of practical Huffman encoding [42]. Since the Huffman code can achieve optimal lossless compression, ISC could achieve the maximal data rate for a given dimming target. It can also be extended to M-PAM signals [43].

For a given dimming target d, the ratio of transmitting "1" and "0" is $d : 1 - d$. Therefore, the maximum efficiency is determined by the entropy as

$$E_p = -d\log_2 d - (1-d)\log_2(1-d). \tag{3.63}$$

In order to achieve the maximum data rate in (3.63), the distribution of data bits has to be adjusted, where the probability of "1" is d while the probability of "0" is $1 - d$. It is the inverse operation of source coding, which is referred to as ISC.

The ratio of the efficiency of ISC and its compensation scheme is given by [39]

$$\frac{E_P}{E_O} = \begin{cases} \frac{-d\log_2 d-(1-d)\log_2(1-d)}{2d}, & 0 < d \leq 0.5, \\ \frac{-d\log_2 d-(1-d)\log_2(1-d)}{2(1-d)}, & 0.5 < d < 1, \end{cases} \tag{3.64}$$

where E_O is the efficiency of the compensation scheme. It can be seen that the upper bound of ISC is better than the compensation scheme, except for the cases that $d = 0$, 0.5, or 1 when both schemes have the same performance. When the dimming target stays far from 50%, large performance gain can be achieved by ISC. Specifically, for the dimming target of 29% or 71%, the performance gain is 50% and it is increased to 100% when the dimming target becomes 16% or 84%. Even though the performance improvement is huge when the dimming target is close to 0 or 100%, the absolute value becomes very small, which is not preferred for data transmission. When the required data rate is set to at least 50% of the maximum transmission rate, the dimming target should be within the range [11%, 89%]. The dimming range could be extended to [3.1%, 96.9%] when the required data rate is set to only 20% of the maximum transmission rate.

An example of ISC was proposed in [42], which employed Huffman codes to transform the data with equal distribution of "0" and "1" to the data with the probability of d for "1" and the probability of 1 - d for "0". For example, if the dimming target is set to $d = 70\%$, the probabilities of "1" and "0" should be 70% and 30%, respectively. When Huffman encoding is used to compress the data, since the probability of "1" is high, the following bit after "1" has to be jointly considered. An example of Huffman encoding is listed in Table 3.1, where the probability of "1" is 70%. The average lengths of the codewords before and after encoding are 1.7 and 1.51, respectively, where the compression ratio is 0.888. The entropy for 70% dimming is 0.881 according to (3.63). As a result, more than 94% compression is achieved compared to its optimum solution. Table 3.2 is the inverse operation of Table 3.1, which transforms the evenly distributed binary data to the data with 70% of "1" by inverse Huffman coding. The average lengths before and after inverse Huffman encoding are 1.5 and 1.75, respectively, and the decompression ratio is 1.17. The resultant dimming level is around 71.4%, which is close to the target value of 70%. If more bits are included in Huffman and inverse Huffman encoding, the achievable dimming level will be more close to the required dimming target.

To facilitate arbitrary dimming targets in noisy environments, the codebook is designed to select binary sequences with nonuniform probability, which equals the dimming target. However, only a subset of codewords should be chosen so that their elements are separated far enough from each other in Hamming distance to provide error correcting capability. To facilitate practical implementation, the rules of both encoder and decoder should abide by linear property. Therefore, feasible dimming targets are discrete rather than arbitrary. The codebook generated from a linear codebook where all codewords have constant Hamming weight is a useful choice for

Table 3.1 An example of Huffman encoding.

Symbol/Length	Probability	Codeword/Length
0/1	0.3	00/2
10/2	0.21	01/2
11/2	0.49	1/1

Table 3.2 An example of inverse Huffman encoding.

Symbol/Length	Probability	Codeword/Length
00/2	0.25	0/1
01/2	0.25	10/2
1/1	0.5	11/1

dimming support, and was applied to Reed–Muller (RM) codes in [44]. Since only dimming target in the form of $1/2^b$, $b = 1, 2, \ldots$, could be bolstered, it needs to be combined with compensation symbol insertion to support an arbitrary dimming level.

Although a codebook with the constant-weight property has a sub-exponential number of codewords, its code rate might be less than actual achievable rate required by high throughput transmission [41]. Therefore, it is not preferred for high-rate transmission and the constraint of constant-weight has to be relaxed to furnish more choices of code rates, which is realistic since the dimming requirement is imposed on the average optical power instead of the instantaneous optical power. For example, by scrambling a linear codeword with a random sequence, the dimming target of $d = 0.5$ can be easily realized. The weight of a codeword can be viewed as a binomial random variable, and dimming level is represented by the weight divided by the codeword length, which is also treated as a random variable. Its mean value is the dimming target and its variance is inversely proportional to the codeword length. In order to facilitate arbitrary dimming target, puncturing technique can be applied to the convolutional code and turbo code [45, 46]. However, puncturing does not perform well in the low dimming region since lots of useful bits might be pruned away.

3.5 Conclusion

In this chapter, single carrier and carrierless modulation schemes have been introduced for VLC systems. Specifically, PAM with frequency-domain equalization and PPM with decision feedback equalization are presented to combat ISI caused by multipath distortion and their BER performances are analyzed. In order to overcome the effect of LED nonlinearity, PAM can be implemented with multiple LEDs and each

LED is modulated by OOK independently. Besides, 2-D CAP is also addressed due to its high spectral efficiency and simple implementation, which is then extended to multidimensional CAP in order to improve the spectral efficiency. Finally, various modulation and coding schemes have been illustrated for dimming control in single carrier VLC systems to support communication and illumination simultaneously.

References

1 G. Stepniak, M. Schuppert, and C. A. Bunge, "Advanced modulation formats in phosphorous LED VLC links and the impact of blue filtering," *J. Lightw. Technol.*, vol. 33, no. 21, pp. 4413–4423, Nov. 2015.

2 D. J. F. Barros, S. K. Wilson, and J. M. Kahn, "Comparison of orthogonal frequency-division multiplexing and pulse-amplitude modulation in indoor optical wireless links," *IEEE Trans. Commun.*, vol. 60, no. 1, pp. 153–163, Jan. 2012.

3 IEEE Std. 802.15.7-2011, *Part 15.7: Short-Range Wireless Optical Communication Using Visible Light*, Sep. 2011.

4 N. Fujimoto and H. Mochizuki, "477 Mbit/s visible light transmission based on OOK-NRZ modulation using a single commercially available visible LED and a practical LED driver with a pre-emphasis circuit," in *Proc. Optical Fiber Communication Conference and Exposition and the National Fiber Optic Engineers Conference (OFC/NFOEC) 2013* (Anaheim, CA), Mar. 17–21, 2013, JTh2A. 73.

5 B. Fahs, A. J. Chowdhury, and M. M. Hella, "A 12-m 2.5-Gb/s Lighting Compatible Integrated Receiver for OOK Visible Light Communication Links," *J. Lightw. Technol.*, vol. 34, no. 16, pp. 3768–3775, Aug. 2016.

6 X. Li, N. Bamiedakis, X. Guo, J. J. D. McKendry, E. Xie, R. Ferreira, E. Gu, M. D. Dawson, R. V. Penty, and I. H. White, "Wireless visible light communications employing feed-forward pre-equalization and PAM-4 modulation," *J. Lightw. Technol.*, vol. 34, no. 8, pp. 2049–2055, Apr. 2016.

7 A. Nuwanpriya, S. W. Ho, J. A. Zhang, A. J. Grant, and L. Luo, "PAM-SCFDE for optical wireless communications," *J. Lightw. Technol.*, vol. 33, no. 14, pp. 2938–2949, Jul. 2015.

8 K. Lee, H. Park, and J. Barry, "Indoor channel characteristics for visible light communications," *IEEE Commun. Lett.*, vol. 15, no. 2, pp. 217–219, Feb. 2011.

9 Y. P. Lin and S. M. Phoong, "BER minimized OFDM systems with channel independent precoders," *IEEE Trans. Signal Process.*, vol. 51, no. 9, pp. 2369–2380, Sep. 2003.

10 J. F. Li, Z. T. Huang, R. Q. Zhang, F. X. Zeng, M. Jiang, and Y. F. Ji, "Superposed pulse amplitude modulation for visible light communication," *Opt. Exp.*, vol. 21, no. 25, pp. 31006–31011, Dec. 2013.

11 A. Yang, Y. Wu, M. Kavehrad, and G. Ni, "Grouped modulation scheme for led array module in a visible light communication system," *IEEE Wirel. Commun.*, vol. 22, no. 2, pp. 24–28, Apr. 2015.

12 J. M. Kahn and J. R. Barry, "Wireless infrared communications," *Proc. IEEE*, vol. 85, no. 2, pp. 265–298, Feb. 1997.

13 M. D. Audeh, J. M. Kahn, and J. R. Barry, "Performance of pulseposition modulation on measured nondirected indoor infrared channels," *IEEE Trans. Commun.*, vol. 44, no. 6, pp. 654–659, Jun. 1996.

14 J. R. Barry, "Sequence detection and equalization for pulse-position modulation," in *Proc. IEEE International Conference on Communications (ICC) 1994* (New Orleans, LA), May 1–5, 1994, pp. 1561–1565.

15 A. G. Klein and P. Duhamel, "Decision-feedback equalization for

pulse-position modulation," *IEEE Trans. Signal Process.*, vol. 55, no. 11, pp. 5361–5369, Nov. 2007.

16 M. D. Audeh, J. M. Kahn, and J. R. Barry, "Decision-feedback equalization of pulse-position modulation on measured nondirected indoor infrared channels," in *IEEE Trans. Commun.*, vol. 47, no. 4, pp. 500–503, Apr. 1999.

17 D. S. Shiu and J. M. Kahn, "Differential pulse-position modulation for power-efficient optical communication," *IEEE Trans. Commun.*, vol. 47, no. 8, pp. 1201–1210, Aug. 1999.

18 H. Sugiyama and K. Nosu, "MPPM: A method for improving the band-utilization efficiency in optical PPM," *J. Lightw. Technol.*, vol. 7, no. 3, pp. 465–472, Mar. 1989.

19 B. Bai, Z. Xu, and Y. Fan, "Joint LED dimming and high capacity visible light communication by overlapping PPM," in *Proc. Wireless and Optical Communications Conference (WOCC) 2010* (Shanghai, China), May 14–15, 2010, pp. 1–5.

20 M. Noshad and M. Brandt-Pearce, "Expurgated PPM using symmetric balanced incomplete block designs," *IEEE Commun. Lett.*, vol. 16, no. 7, pp. 968–971, Jul. 2012.

21 M. Noshad and M. Brandt-Pearce, "Application of expurgated PPM to indoor visible light communications-Part I: Single-user systems," *J. Lightw. Technol.*, vol. 32, no. 5, pp. 875–882, Mar. 2014.

22 G. H. Im and J. J. Werner, "Bandwidth-efficient digital transmission up to 155 Mb/s over unshielded twisted pair wiring," in *Proc. IEEE International Conference on Communications (ICC) 1993* (Geneva, Switzerland), May 23–26, 1993, vol. 3, pp. 1797–1803.

23 G. H. Im and J. J. Werner, "Bandwidth-efficient digital transmission over unshielded twisted-pair wiring," *IEEE J. Sel. Areas Commun.*, vol. 13, no. 9, pp. 1643–1655, Dec. 1995.

24 J. L. Wei, J. D. Ingham, D. G. Cunningham, R. V. Penty, and I. H. White, "Performance and power dissipation comparisons between 28 Gb/s NRZ, PAM, CAP and optical OFDM systems for data communication applications," *J. Lightw. Technol.*, vol. 30, no. 20, pp. 3273–3280, Oct. 2012.

25 F. M. Wu, C. T. Lin, C. C. Wei, C. W. Chen, H. T. Huang, and C. H. Ho, "1.1 Gb/s white-LED-based communication employing carrier-less amplitude and phase modulation," *IEEE Photon. Technol. Lett.*, vol. 24, no. 19, pp. 1730–1732, Oct. 2012.

26 J. L. Wei, D. G. Cunningham, R. V. Penty, and I. H. White, "Study of 100 gigabit ethernet using carrierless amplitude/phase modulation and optical OFDM," *J. Lightw. Technol.*, vol. 31, no. 9, pp. 1367–1373, May 2013.

27 L. Tao, Y. Ji, J. Liu, A. P. T. Lau, N. Chi, and C. Lu, "Advanced modulation formats for short reach optical communication systems," *IEEE Netw.*, vol. 27, no. 6, pp. 6–13, Nov./Dec. 2013.

28 F. M. Wu, C. T. Lin, C. C. Wei, C. W. Chen, Z. Y. Chen, and K. Huang, "3.22-Gb/s WDM visible light communication of a single RGB LED employing carrier-less amplitude and phase modulation," in *Proc. Optical Fiber Communication Conference and Exposition and the National Fiber Optic Engineers Conference (OFC/NFOEC) 2013* (Anaheim, CA), Mar. 17–21, 2013, OTh1G.4.

29 M. Sharif, J. K. Perin, and J. M. Kahn, "Modulation schemes for single-laser 100Gb/s links: Single-carrier," *J. Lightw. Technol.*, vol. 33, no. 20, pp. 4268–4277, Oct. 2015.

30 P. A. Haigh, A. Burton, K. Werfli, H. L. Minh, E. Bentley, P. Chvojka, W. O. Popoola, I. Papakonstantinou, and S. Zvanovec, "A multi-CAP visible-light communications system with 4.85-b/s/Hz spectral efficiency," *IEEE J. Sel. Areas Commun.*, vol. 33, no. 9, pp. 1771–1779, Sep. 2015.

31 A. F. Shalash and K. K. Parhi, "Multidimensional carrierless AM/PM systems for digital subscriber loops," *IEEE Trans. Commun.*, vol. 47, no. 11, pp. 1655–1667, Nov. 1999.

32 P. P. Vaidyanathan, *Multirate Systems and Filter Banks*. Englewood Cliffs, NJ: Prentice-Hall, 1993.

33 S. Stern and R. Fischer, "Efficient assessment of the instantaneous power distributions of pulse-shaped single-and multi-carrier signals," in *Proc. International Black Sea Conference on Communications and Networking (BlackSeaCom) 2013*

(Batumi, Georgia), Jul. 3–5, 2013, pp. 12–17.
34 J. J. Werner, *Tutorial on carrierless AM/PM-Part II: Performance of bandwidth efficient line codes*, T1E1 Contribution, vol. T1E1.4/93-058, 1992.
35 M. Vetterli and J. Kovacevic, *Wavelets and Subband Coding*. Englewood Cliffs, NJ: Prentice-Hall, 1993.
36 R. K. Brayton, S. W. Director, G. D. Hachtel, and L. M. Vidigal, "A new algorithm for statistical circuit design based on quasinewton methods and function splitting," *IEEE Trans. Circuits Syst.*, vol. 26, no. 9, pp. 784–794, Sep. 1979.
37 A. F. Shalash and K. K. Parhi, "Three-dimensional carrierless AM/PM line code for the UTP cables," in *Proc. IEEE International Symposium on Circuits and Systems (ISCAS) 1997* (Hong Kong, China), Jun. 09–12, 1997, pp. 2136–2139.
38 A. Tsiatmas, C. P. M. J. Baggen, F. M. J. Willems, J. M. G. Linnartz, and J. W. M. Bergmans, "An illumination perspective on visible light communications," *IEEE Commun. Mag.*, vol. 52, no. 7, pp. 64–71, Jul. 2014.
39 S. Rajagopal, R. D. Roberts, and S. K. Lim, "IEEE 802.15.7 visible light communication: Modulation schemes and dimming support," *IEEE Commun. Mag.*, vol. 50, no. 3, pp. 72–82, Mar. 2012.
40 K. Lee and H. Park, "Modulations for visible light communications with dimming control," *IEEE Photon. Technol. Lett.*, vol. 23, no. 16, pp. 1136–1138, Aug. 2011.
41 S. H. Lee, D. Y. Jung, and J. K. Kwon, "Modulations and coding for dimmable visible light communication," *IEEE Commun. Mag.*, vol. 53, no. 2, pp. 136–143, Feb. 2015.
42 J. K. Kwon, "Inverse source coding for dimming in visible vight communications using NRZ-OOK on reliable links," *IEEE Photonics Tech. Lett.*, vol. 22, no. 19, pp. 1455–57, Oct. 2010.
43 K. I. Ahn and J. K. Kwon, "Capacity analysis of M-PAM inverse source coding in visible light communications," *J. Lightw. Tech.*, vol. 30, no. 10, pp. 1399–1404, May. 2012.
44 S. Kim and S. Jung, "Novel FEC coding scheme for dimmable visible light communication based on the modified Reed-Muller codes," *IEEE Photon. Technol. Lett.*, vol. 23, no. 20, pp. 1514–1516, Oct. 2011.
45 S. Lee and J. Kwon, "Turbo code-based error correction scheme for dimmable visible light communication systems," *IEEE Photon. Technol. Lett.*, vol. 24, no. 17, pp. 1463–1465, Sep. 2012.
46 J. Kim and H. Park, "A coding scheme for visible light communication with wide dimming range," *IEEE Photon. Technol. Lett.*, vol. 26, no. 5, pp. 465–468, Mar. 2014.

4
Multicarrier Modulation

In this chapter, we present a review of optical orthogonal frequency division multiplexing (OFDM) schemes for broadband and high-data-rate visible light communications (VLCs). We also introduce the recent development on optical OFDM with respect to performance enhancement, power- and spectral-efficiency promotion and discuss modified optical OFDM schemes under lighting constraints. A comprehensive comparison of existing and proposed optical OFDM techniques is provided as well.

Unlike radio frequency (RF) transmission, VLC usually adopts intensity-modulation and direct-detection (IM/DD), where the optical OFDM signals are directly modulated onto the luminance of the emitted visible light. In the IM/DD scheme, the amplitude of the optical OFDM signals is constrained to be real-valued and non-negative. Since the conventional RF OFDM scheme is not feasible for intensity modulation, several optical OFDM schemes have been proposed to satisfy the specific signal constraints in VLC, such as DC-biased optical OFDM (DCO-OFDM), asymmetrically clipped optical OFDM (ACO-OFDM), pulse-amplitude-modulated discrete multitone (PAM-DMT), and unipolar OFDM (U-OFDM). The recapitulative architecture and mechanism of these schemes are presented in Section 4.1. The performance comparison is investigated as well.

The time-domain signal of optical OFDM is the sum of a large number of orthogonal harmonic components. The peak power of optical OFDM signals in time domain is much higher than the average, which increases the probability that the amplitude of optical OFDM signals exceeds the dynamic region of light emitting diode (LED). This inherent high peak-to-average power ratio (PAPR) issue is prone to induce severe nonlinear distortion and impair the transmission performance of VLC systems. In Section 4.2, we introduce the recent inspiring researches on the performance enhancement for optical OFDM. This topic involves three major perspectives including optimization of direct current (DC) bias and scaling factor, mitigating the nonlinear effect of LED, and PAPR reduction.

The power and spectral efficiency is another important issue for optical OFDM schemes in VLC applications. The existing optical OFDM schemes, depicted in Section 4.1, either require an additional DC bias or sacrifice part of the subcarriers to satisfy the real-valued and non-negative constraints. Hence, considerable spectral

Visible Light Communications: Modulation and Signal Processing. First edition. Zhaocheng Wang, Qi Wang, Wei Huang, and Zhengyuan Xu.
Published 2017 by John Wiley & Sons, Inc.

or power efficiency loss is inevitable in these conventional optical OFDM schemes. Recent research exhibits great potential to enhance the spectral and power efficiency of optical OFDM. State-of-the-art power- and spectral-efficient optical OFDM, exemplified by hybrid optical OFDM, enhanced U-OFDM (eU-OFDM), and Layered ACO-OFDM (LACO-OFDM), are discussed in Section 4.3.

In VLC applications, the illumination function should be considered besides information transmission. Dimming control is used to adjust brightness of the illumination and to satisfy different illumination requirements in daily scenarios. To support dimming control, optical OFDM must be capable of operating under lighting constraints on user's demand. In Section 4.4, optical OFDM techniques under lighting constraints, including pulse width modulation (PWM), reverse polarity optical OFDM (RPO-OFDM), and asymmetrical hybrid optical OFDM (AHO-OFDM), are introduced comprehensively.

4.1
Optical OFDM for visible light communications

There has been a long history of OFDM technology since its embryonic form first appeared in a patent by R. W. Chang from Bell Labs in 1966 [1]. From then on, several critical techniques including fast Fourier transform (FFT) and cyclic prefix (CP) were proposed to complement OFDM schemes [2]. Nowadays, practical communication systems, for example, the Long Term Evolution-Advanced (LTE-A) standard [3], have successfully applied OFDM to provide high-rate transmission and multiple access. Since OFDM exhibits a strong resistance to inter-symbol interference (ISI) in dispersive channel, OFDM technology has become a promising multicarrier modulation format for VLC applications. In this section, we introduce four mainstream optical OFDM schemes for VLC systems.

4.1.1
DC-biased optical OFDM

In conventional RF OFDM schemes, the data are transmitted in parallel on multicarriers. The orthogonality of subcarriers ensures that the symbols in the same OFDM block do not interfere with each other. Because OFDM is capable of mitigating ISI effectively, it is an ideal modulation scheme for VLC for high-rate transmission. The optical OFDM scheme inherits the basic attributes from its RF counterparts but also exhibits several differences. Most of all, optical OFDM is directly modulated on the intensity of emitted light and constrained to be real-valued and nonnegative. Modification is embedded in optical OFDM to satisfy this constraint. In this subsection, a representative multicarrier scheme, referred to as DCO-OFDM [4], is introduced.

Figure 4.1 depicts the architecture of DCO-OFDM transceiver. Assuming that total N subcarriers are allocated in a single OFDM block, where N is typically a large even number. At the transmitter, the serial bit stream is first converted to a paral-

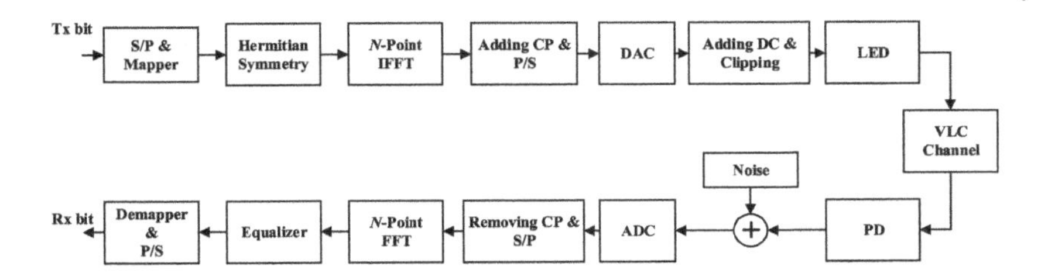

Figure 4.1 The block diagram of DCO-OFDM transceiver for visible light communications.

lel sequence and then mapped to the $N/2-1$ complex-valued symbols according to the specific modulation constellation $\mathcal{X}$ such as quadrature amplitude modulation (QAM) [5]. The modulated OFDM block $\mathbf{X} = [X_0\ X_1\ \cdots\ X_{N-1}]$, where subscript number denotes the associated subcarrier index, is constructed as follows: $X_0 = 0$ and X_1 to $X_{N/2-1}$ carry the $N/2-1$ information symbols, while $X_{N/2}$ to X_{N-1} satisfy the Hermitian symmetry as

$$X_k = X_{N-k}^*, k = N/2,\ ...,\ N-1, \tag{4.1}$$

where the superscript mark "*" represents the conjugate operation. Setting $X_0 = X_{N/2} = 0$ is to avoid the DC and complex-valued harmonic components, while the Hermitian symmetry of $\mathbf{X}$ enables the transmitter to generate real-valued time-domain signals.

The OFDM symbol vector $\mathbf{X}$ is fed to the processor of inverse fast Fourier transform (IFFT) and converted to discrete time-domain samples efficiently as

$$x_n = \frac{1}{\sqrt{N}} \sum\nolimits_{k=0}^{N-1} X_k \exp\left(j\frac{2\pi kn}{N}\right), n = 0, 1,\ ...,\ N-1, \tag{4.2}$$

where x_n represents the nth discrete time-domain sample. Considering the imposed Hermitian symmetry, the IFFT operation (4.2) is further expressed as

$$\begin{aligned} x_n &= \frac{1}{\sqrt{N}} \sum\nolimits_{k=1}^{N/2-1} \left(X_k \exp\left(j\frac{2\pi kn}{N}\right) + X_{N-k} \exp\left(j\frac{2\pi (N-k) n}{N}\right)\right) \\ &= \frac{1}{\sqrt{N}} \sum\nolimits_{k=1}^{N/2-1} \left(X_k \exp\left(j\frac{2\pi kn}{N}\right) + X_k^* \exp\left(-j\frac{2\pi kn}{N}\right)\right) \\ &= \frac{2}{\sqrt{N}} \sum\nolimits_{k=1}^{N/2-1} \mathrm{Re}\left(X_k \exp\left(j\frac{2\pi kn}{N}\right)\right), n = 0, 1,\ ...,\ N-1, \end{aligned} \tag{4.3}$$

where $\mathrm{Re}\,(Z)$ denotes the real part of Z. The imaginary parts of the time-domain signal samples are forced to zero. It is worth emphasizing again that the total number of used subcarriers is $N/2-1$ and the rest of the subcarriers are exploited to impose the Hermitian symmetry. After IFFT, a CP of length L_{CP}, which is the copy of the last L_{CP} samples of each time-domain DCO-OFDM block, is appended in its front. CP provides guard interval without destroying the orthogonality of subcarriers. When L_{CP} exceeds the maximum delay of the dispersive channel, ISI is totally discarded.

Then the discrete samples are converted into a serial sequence and fed to a digital-to-analog converter (DAC). The converted electrical signal $x(t)$ is still bipolar and is not feasible for intensity modulation. A DC bias B_{DC} should be added to $x(t)$. In DCO-OFDM, the DC bias B_{DC} is set to

$$B_{\mathrm{DC}} = \mu \sqrt{E\left\{x^2(t)\right\}}, \tag{4.4}$$

where $E\{\cdot\}$ denotes the expectation operation and μ is a constant coefficient. Considering the power gain induced by B_{DC}, the DC bias level is evaluated as $10 \log\left(\mu^2 + 1\right)$. Besides that, the negative amplitudes of the biased signals are clipped as zero for intensity modulation. After biasing and clipping, the electrical DCO-OFDM signal $x_{\mathrm{DCO}}(t)$ is used to drive the LED and to be modulated on the intensity of illumination. According to the central limit theorem, $x(t)$ approximates a Gaussian distribution with zero mean when $N \geq 64$ [4]. Therefore, the optical power of DCO-OFDM is B_{DC}, while its electric power is $\left(\mu^2 + 1\right) E\left\{x^2(t)\right\}$.

At the receiver, the photodiode (PD) component captures the optical signal from the VLC channel and transforms it into the electrical signal $y(t)$. A lens can be placed in front of the PD to filter the background light. In the PD, the thermal noise and shot noise interfere with the received signal. These two types of noise can be both modeled as additive white Gaussian noise (AWGN) [6]. Thus, for the dispersive VLC channel with the impulse response $h(t)$, the received signal $y(t)$ is expressed as

$$y(t) = h(t) \star x_{\mathrm{DCO}}(t) + w(t), \tag{4.5}$$

where the notation "$\star$" denotes the convolution operation and $w(t)$ is the AWGN with zero mean. After analog-to-digital converter (ADC), the received DCO-OFDM discrete sample block is acquired with the CP removed and then is reshaped as the parallel sequence $\{y_n, n = 0, 1, \ldots, N-1\}$.

To recover the transmitted data symbols, the N-point FFT component converts the time-domain samples into the frequency-domain symbols $\left\{Y_k, k = 0, 1, \ldots, N-1\right\}$. For each subcarrier, the associated channel is flat. Hence, simple one-tap equalization method is imposed. The equalizer divides the received symbols $\{Y_k, k = 1, 2, \ldots, N/2-1\}$ on the used subcarriers by the associated channel state information (CSI) H_k, which can be estimated according to the pilot embedded in DCO-OFDM signals. Based on these equalized symbols, a maximum likelihood (ML) detection is invoked as

$$X_k^r = \arg \min_{\overline{X}_k \in \mathcal{X}} \left\|Y_k - H_k \overline{X}_k\right\|^2, \tag{4.6}$$

where $\overline{X}_k$ and X_k^r represent the candidate and recovered symbols on the kth subcarrier, respectively, while $\|\cdot\|$ indicates the Euclidean distance. The demapper then maps the complex symbol X_k^r into the corresponding bits to recover the transmitted data.

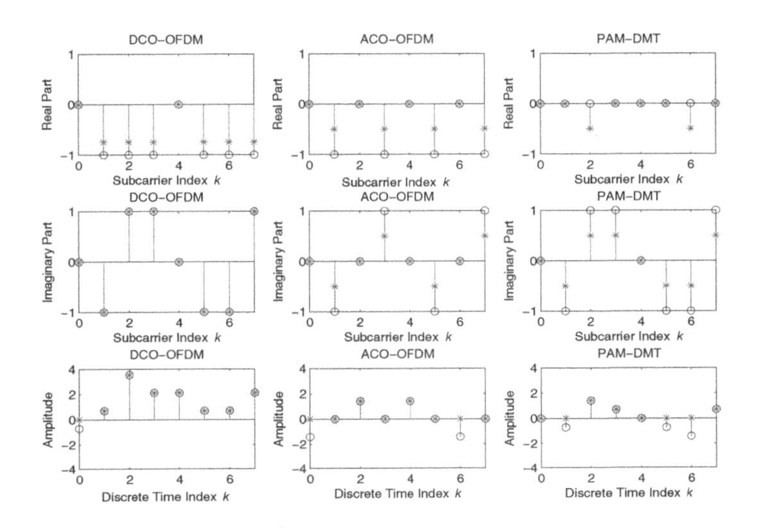

Figure 4.2 Three different optical OFDM signals: DCO-OFDM (left-column plots), ACO-OFDM (center-column plots), and PAM-DMT (right-column plots).

A simple DCO-OFDM block with eight subcarriers is depicted in the left-column plots of Fig. 4.2, where circles and crosses denote non-clipped and clipped signals, respectively. The three data symbols are modulated according to 4QAM constellation and allocated to the first to third subcarriers, while the fifth to seventh subcarriers are occupied by X_3^*, X_2^*, and X_1^*, respectively. In this case, a 3 dB DC bias, denoted by the black dash line, is added to the DCO-OFDM signals. It is observed that clipping operation results in slight distortions on all the subcarriers. The clipping distortions reduce the amplitudes of the received symbols and impair the performance of ML detection. Since the DC part of the DCO-OFDM signal carries no data information, 3 dB power cost is just for mitigating the distortions.

DCO-OFDM is a simple and spectral-efficient multicarrier scheme for VLC transmission. However, it has the inherent issue of low power efficiency. This is because a high DC bias level is required to raise the negative peak over zero. This induces a severe power-efficiency loss, which restricts the achievable performance of DCO-OFDM systems.

4.1.2 ACO-OFDM and PAM-DMT

In order to enhance the power efficiency of VLC multicarrier systems, several modified optical OFDM schemes were proposed in [7, 8]. Among them, ACO-OFDM is the most typical and inspiring scheme for power-efficient transmission. The basic idea of ACO-OFDM is sacrificing even subcarriers to accommodate the clipping distortions. This strategy avoids DC bias and reduces the power cost significantly, at the considerable cost of reduced spectrum efficiency. PAM-DMT is another alternative asymmetric optical OFDM scheme [8].

4.1.2.1 ACO-OFDM

The block diagram of ACO-OFDM is similar to its DCO-OFDM counterpart. The major difference is the allocation of data symbols. In the ACO-OFDM scheme, the data symbols are only placed on the odd subcarriers of the first $N/2$ subcarriers and hence, an ACO-OFDM block of N subcarriers can only accommodate $N/4$ information symbols. Considering the constraint of real-valued amplitude, the Hermitian symmetry constraint of (4.1) is also imposed on the $(N/2)$th to $(N-1)$th subcarriers. At the transmitter, after the IFFT operation, the discrete time-domain samples have the asymmetry as

$$\begin{aligned} x_n &= \frac{1}{\sqrt{N}} \sum_{k=0}^{N-1} X_k \exp\left(j\frac{2\pi kn}{N}\right) \\ &= \frac{1}{\sqrt{N}} \sum_{m=0}^{N/2-1} X_{2m+1} \exp\left(j\frac{2\pi(2m+1)n}{N}\right) \\ &= -\frac{1}{\sqrt{N}} \sum_{m=0}^{N/2-1} X_{2m+1} \exp\left(j\frac{2\pi(2m+1)(n+N/2)}{N}\right) \\ &= -x_{n+N/2}, n = 0, 1, \ldots, N/2-1. \end{aligned} \tag{4.7}$$

The asymmetry in (4.7) manifests the fact that the amplitudes of the first half samples are identical to those of the second half samples but with opposite signs. The negative-valued sample $x_n < 0$ can be simply recovered by observing its asymmetry sample $x_{n+N/2}$ or $x_{n-N/2}$. Hence, the ACO-OFDM signal $x_{\text{ACO}}(t)$ is directly clipped at zero without adding any DC bias, and then transformed into an optical signal. Assuming that x_n^c is the clipped sample, where

$$x_n^c = \begin{cases} x_n, & x_n > 0, \\ 0, & x_n \leq 0, \end{cases} \tag{4.8}$$

the distorted data symbols X^c_{2m+1} on the odd subcarriers $m = 0, 1, \ldots, N/2 - 1$ can be derived as

$$\begin{aligned} X^c_{2m+1} =& \frac{1}{\sqrt{N}} \sum_{n=0}^{N-1} x^c_n \exp\left(-j\frac{2\pi(2m+1)n}{N}\right) \\ =& \frac{1}{\sqrt{N}} \sum_{n=0}^{N/2-1} \left(x^c_n - x^c_{n+N/2}\right) \exp\left(-j\frac{2\pi(2m+1)n}{N}\right) \\ =& \frac{1}{2\sqrt{N}} \sum_{n=0}^{N/2-1} \left(x_n - x_{n+N/2}\right) \exp\left(-j\frac{2\pi(2m+1)n}{N}\right) \\ =& \frac{1}{2\sqrt{N}} \sum_{n=0}^{N/2-1} x_n \exp\left(-j\frac{2\pi(2m+1)n}{N}\right) \\ &+ \frac{1}{2\sqrt{N}} \sum_{n=0}^{N/2-1} x_{n+N/2} \exp\left(-j\frac{2\pi(2m+1)\left(n+N/2\right)}{N}\right) \\ =& \frac{1}{2} X_{2m+1}. \end{aligned} \tag{4.9}$$

It can be seen that clipping operation reduces the power of an ACO-OFDM signal by half.

Since x_n approximates a Gaussian distribution when $N \geq 64$, the distribution of x^c_n could be written as [4]

$$p_{\text{ACO}}\left(x^c_n\right) = \begin{cases} \frac{1}{2}, & x^c_n = 0, \\ \frac{1}{\sqrt{2\pi}\sigma_{\text{ACO}}} \exp\left(-\frac{\left(x^c_n\right)^2}{2\sigma^2_{\text{ACO}}}\right), & x^c_n > 0, \end{cases} \tag{4.10}$$

where σ_{ACO} denotes the root mean square (RMS) of the unclipped ACO-OFDM signal x_n. The optical power of x^c_n is then calculated by $E\left\{x^c_n\right\} = \sigma_{\text{ACO}}/\sqrt{2\pi}$, while the electric power of x^c_n is given by $E\left\{\left(x^c_n\right)^2\right\} = {\sigma_{\text{ACO}}}^2/2$ [9].

At the receiver of an ACO-OFDM system, which shares a similar procedure as the DCO-OFDM scheme, only the symbols Y_{2m+1} on the odd subcarriers $m = 0, 1, \ldots, N/4 - 1$ are extracted after the FFT operation and fed to the equalizer and demapper. Taking into account the clipping distortion in (4.8), the ML detection for the ACO-OFDM demapper is modified as

$$X^r_k = \arg\min_{\overline{X}_k \in \mathcal{X}} \left\|2Y_k - H_k\overline{X}_k\right\|^2. \tag{4.11}$$

Notice that one element of the transmitted sample pair $\left\{x^c_n, x^c_{n+N/2}\right\}$, where $n = 0, 1, \ldots, N/2 - 1$, is zero-valued. This feature of ACO-OFDM signal can be used to enhance the receiver performance. Assuming that y^e_n represents the equalized time sample, which is simply acquired by computing the IFFT of the equalized frequency-domain sequence, a pairwise ML detection [10] can be carried out to mitigate the noise as

$$\left\{y^r_n, y^r_{n+N/2}\right\} = \begin{cases} \left\{y^e_n, 0\right\}, & y^e_n > y^e_{n+N/2}, \\ \left\{0, y^e_{n+N/2}\right\}, & y^e_n < y^e_{n+N/2}, \end{cases} \quad n = 0, 1, \ldots, N/2 - 1.$$

(4.12)

By performing the FFT and the ML detection of (4.11) according to the detected sample sequence $\{y_n^r\}$, the data symbols can be recovered more accurately. The pairwise ML detection removes about half of the noise and it achieves performance gains of 1.3 dB and 1 dB over the flat and dispersive channels, respectively [10].

4.1.2.2 PAM-DMT

In the PAM-DMT scheme, the data symbol is only allocated on the imaginary part of a subcarrier, whereas the real part of the subcarrier is always set to zero. The frequency-domain PAM-DMT block of N subcarriers can be expressed as

$$\mathbf{X} = \left[0\, jX_1^{\text{PAM}}\, jX_2^{\text{PAM}} \cdots jX_{N/2-1}^{\text{PAM}}\, 0\, -jX_{N/2-1}^{PAM} \cdots -jX_1^{\text{PAM}}\right], \quad (4.13)$$

where X_k^{PAM} is real-valued mapped symbol taking value from the pulse amplitude modulation (PAM) constellation $\mathcal{X}_{\text{Re}}$. After the IFFT operation, the time-domain PAM-DMT signal follows the asymmetry as

$$\begin{aligned} x_n &= -\frac{2}{\sqrt{N}} \sum_{k=0}^{N-1} X_k^{\text{PAM}} \sin\left(\frac{2\pi kn}{N}\right) \\ &= \frac{2}{\sqrt{N}} \sum_{k=0}^{N-1} X_k^{\text{PAM}} \sin\left(\frac{2\pi k(N-n)}{N}\right) \\ &= -x_{N-n},\ n = 1, 2, \ldots, N/2-1. \end{aligned} \quad (4.14)$$

Since the two elements of the pair $\{x_n, x_{N-n}\}$ are opposite to each other, the asymmetric clipping can be executed without any DC bias, just as in the case of ACO-OFDM. Similarly, the optical power and electric power of PAM-DMT are $\sigma_{\text{PAM}}/\sqrt{2\pi}$ and $\sigma_{\text{PAM}}{}^2/2$, where σ_{PAM} denotes the RMS of the unclipped PAM-DMT signal [9]. The distortion on the imaginary part is derived as

$$\begin{aligned} \text{Im}\left(X_k^c\right) &= -\frac{1}{\sqrt{N}} \sum_{n=0}^{N-1} x_n^c \sin\left(\frac{2\pi kn}{N}\right) \\ &= -\frac{1}{\sqrt{N}} \sum_{n=0}^{N/2-1} \left(x_n^c - x_{N-n}^c\right) \sin\left(\frac{2\pi kn}{N}\right) \\ &= -\frac{1}{2\sqrt{N}} \sum_{n=0}^{N/2-1} \left(x_n - x_{N-n}\right) \sin\left(\frac{2\pi kn}{N}\right) \\ &= \frac{1}{2} X_k^{\text{PAM}},\ k = 1, 2, \ldots, N/2-1, \end{aligned} \quad (4.15)$$

where $\text{Im}\,(Z)$ denotes the imaginary part of Z. Hence, the amplitude of the distorted symbol in the imaginary part is half of the original data symbol. In order to recover the data symbol $X_k^{\text{PAM},r}$ at the receiver, the ML detection for the PAM-DMT signal is imposed as

$$X_k^{\text{PAM},r} = \arg \min_{\overline{X}_k \in \mathcal{X}_{\text{Re}}} \left\| \text{Im}\left(2Y_k - jH_k \overline{X}_k\right) \right\|^2. \quad (4.16)$$

The pairwise ML detection [10] can also be applied to mitigate the noise effect.

Both the ACO-OFDM and PAM-DMT signals are also exemplified in Fig. 4.2. For the ACO-OFDM scheme depicted in the center-column plots of Fig. 4.2, two 4QAM symbols are modulated on the two odd subcarriers. As for the PAM-DMT signal shown in the right-column plots of Fig. 4.2, the real parts of all the subcarriers are set to zero, and three 2PAM symbols are allocated on the imaginary parts of the subcarriers 1, 2, and 3. It is observed that both schemes exhibit asymmetry in discrete samples. For ACO-OFDM, the 0th, 1st, 2nd, and 3rd samples are opposite to the $4, 5, 6, 7$th samples, whereas the 1st, 2nd, and 3rd samples are opposite to the $7, 6, 5$th samples in PAM-DMT. The distorted data symbols are all reduced to half in both schemes. The leftover non-regular distortions fall on the even subcarrier and the real part in ACO-OFDM and PAM-DMT, respectively.

Compared to DCO-OFDM, ACO-OFDM and PAM-DMT achieve significant power efficiency enhancement. However, since half of the subcarriers in ACO-OFDM signals are sacrificed to accommodate the asymmetry clipping distortions, the spectral efficiency is half of its DCO-OFDM counterpart when the same modulation constellation is applied. The same disadvantage also exists in PAM-DMT systems, as only a real-valued constellation is employed. Hence, a careful tradeoff between power- and spectral- efficiency is required to select a suitable optical OFDM to meet practical requirements.

4.1.3 Unipolar OFDM

For both ACO-OFDM and PAM-DMT schemes, specific part of the spectrum is exploited to accommodate the clipping distortions. By contrast, U-OFDM [11], which

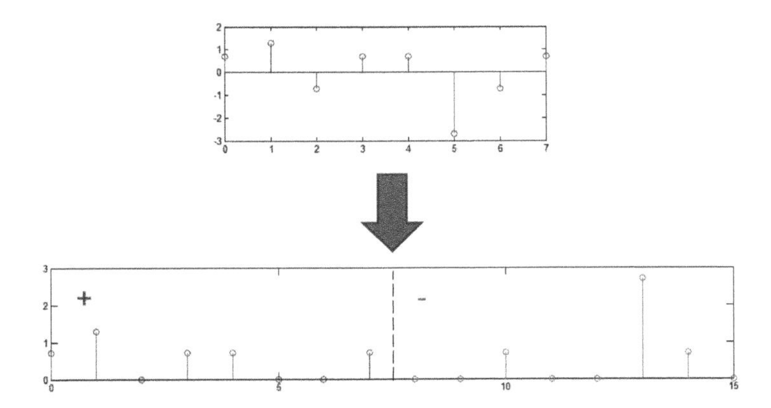

Figure 4.3 Unipolar optical OFDM signal encoding.

was also referred to as Flip-OFDM in [12, 13], doubles the OFDM block length to transmit the negative amplitudes without requiring clipping operation. Hence, the block length of U-OFDM is $2N$, and each block carries $N/2-1$ information symbols with the first subcarrier set to zero while the Hermitian symmetry of (4.1) is deployed to generate X_k for $0 \leq k \leq N-1$. The bipolar discrete time samples $\{x_n, n=0,1,\ ...,N-1\}$ are acquired after the IFFT processor. In order to generate the unipolar signals for intensity modulation, the U-OFDM block is extended to $\{x_n^u, n=0,1,\ ...,2N-1\}$. Each sample x_n is encoded into the new pair $\{x_n^u, x_{n+N}^u\}$. When the amplitude of the original OFDM sample x_n is positive, the first sample of the new pair is labeled as "active" and the second one as "inactive". Otherwise, the first sample is set as "inactive" and the second sample is set as "active". The active sample is equal to the absolute value of x_n and the inactive sample is zero. This encoding in U-OFDM is equivalent to the bijection as

$$x_n \to \{x_n^u, x_{n+N}^u\} = \begin{cases} \{x_n, 0\}, & x_n > 0, \\ \{0, -x_n\}, & x_n \leq 0, \end{cases} \quad n = 0,1,\ ...,N-1. \tag{4.17}$$

Figure 4.3 exemplifies the encoding procedure in U-OFDM. The eight-point bipolar sequence is the original OFDM signal. After adopting the bijection of (4.17), this sample sequence is extended into a 16-point sample signal. The first half of the U-OFDM signal, denoted by the blue circle, allocates the positive part of the original OFDM signal and sets the negative sample as zero. In the second half with the legend of red circle, the absolute values of the original negative samples are placed

The transmitter modulates the unipolar signals $\{x_n^u\}$ on the emitting light. The receiver captures the signals $\{y_n^u, n=0,1,\ ...,2N-1\}$ and recovers the bipolar OFDM signals $\{y_n, n=0,1,\ ...,N-1\}$ as $y_n = y_n^u - y_{n+N}^u$. The pairwise ML decoder [10] can be applied to enhance the detection performance. The power-efficient U-OFDM scheme does not require DC bias. However, since the length of the OFDM block is doubled, the spectral efficiency is reduced to half of the DCO-OFDM scheme.

4.1.4 Performance comparison

In this section, four mainstream optical OFDM schemes have been introduced. DCO-OFDM enjoys an advantage of high spectral efficiency, whereas ACO-OFDM, PAM-DMT, and U-OFDM enhance the power efficiency at the expense of spectral efficiency. A simulation is carried out to compare these optical OFDM schemes. In the simulated VLC system, the total subcarrier number is set to 512. QAM constellation is adopted for DCO-OFDM, ACO-OFDM, and U-OFDM signals, while PAM is applied for PAM-DMT signal. Figure 4.4 depicts the simulation results. The spectral efficiencies, constellations, modulation orders, and DC bias levels employed are attached near the corresponding curves. The vertical axis of Fig. 4.4 is $E_{\mathrm{b,elec}}/N_0$ which denotes the ratio between the average electrical energy per bit $E_{\mathrm{b,elec}}$ and the noise power spectral density N_0. For fair comparison, the power of DC bias is in-

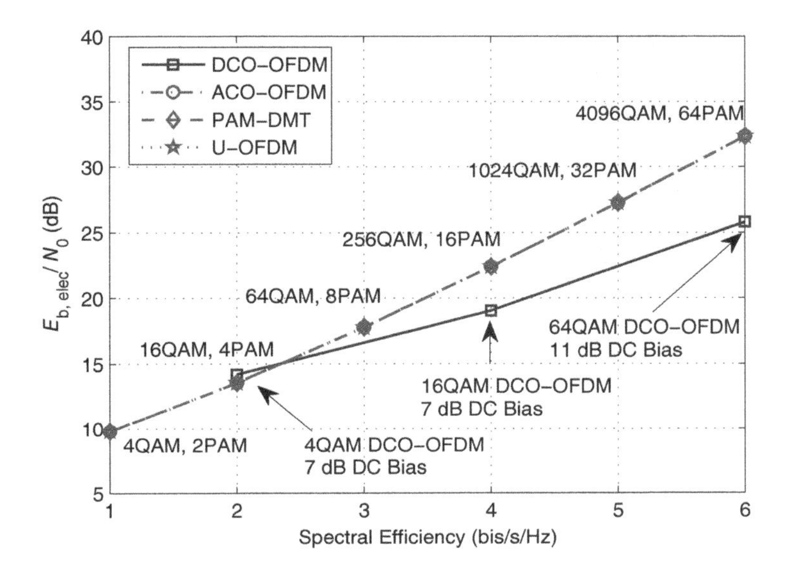

Figure 4.4 Performance comparison of DCO-OFDM, ACO-OFDM, PAM-DMT, and U-OFDM schemes.

cluded in $E_{\mathrm{b,elec}}$. Since bit errors at the bit error rate (BER) level of 10^{-3} can be corrected by using forward error correction (FEC) code, the demodulation threshold for $E_{\mathrm{b,elec}}/N_0$ at BER $= 10^{-3}$ is selected to evaluate the power cost of each optical OFDM scheme. As expected, the ACO-OFDM, PAM-DMT, and U-OFDM schemes share the same performance, and the demodulation threshold curves of these three optical OFDM schemes coincide with each other. When the required spectral efficiency is below 2 bit/s/Hz, the ACO-OFDM scheme achieves a slight performance gain of 0.7 dB compared to DCO-OFDM. However, when the required spectral efficiency exceeds 4 bit/s/Hz, the DCO-OFDM scheme enjoys lower power cost and exhibits higher power-efficiency than the other three schemes. In summary, DCO-OFDM is a preferred multicarrier scheme for high spectral-efficiency VLC transmission.

4.2 Performance enhancement for optical OFDM

Although the above-mentioned optical OFDM schemes provide high-rate transmission for VLC multicarrier systems, their achievable performance are restricted by the inherent high PAPR issue. In practical VLC systems, the deployed LEDs exhibit a strong nonlinearity and a narrow dynamic region. Due to the high PAPR of an OFDM signal, its peak amplitude is prone to exceed the dynamic region and the signal is considerably distorted when driving LEDs. The nonlinear distortion spreads to all the subcarriers after FFT operation and induces a severe performance loss. Hence, several major performance enhancement methods are presented in this section.

4.2.1 DC bias and scaling optimization

Due to the p-n junction barrier and saturation effect, the relationship between the forward voltage across an LED component and the forward current is nonlinear [14]. This relationship may be modeled by the nonlinear transfer characteristic between the input electrical power and the output optical power. Thus, the LED nonlinearity is generally defined by the relationship between the input current amplitude z, which is linearly proportional to the square root of the input electrical power, and the output current $F(z)$, which is linearly proportional to the output optical power. Figure 4.5 exemplifies a typical LED nonlinear transfer characteristic between z and $F(z)$. In Fig. 4.5, the dynamic region, denoted by $[\lambda_{\text{lower}}, \lambda_{\text{upper}}]$, can be regarded as the linear range of the LED transfer curve. In an optical OFDM scheme, a suitable DC bias is exploited to allocate the OFDM signal $z = x + B_{\text{DC}}$ to within the dynamic region. The amplitude outside the dynamic region should be clipped to mitigate the nonlinear effect. Hence, the LED nonlinear transfer function can be modeled as double side

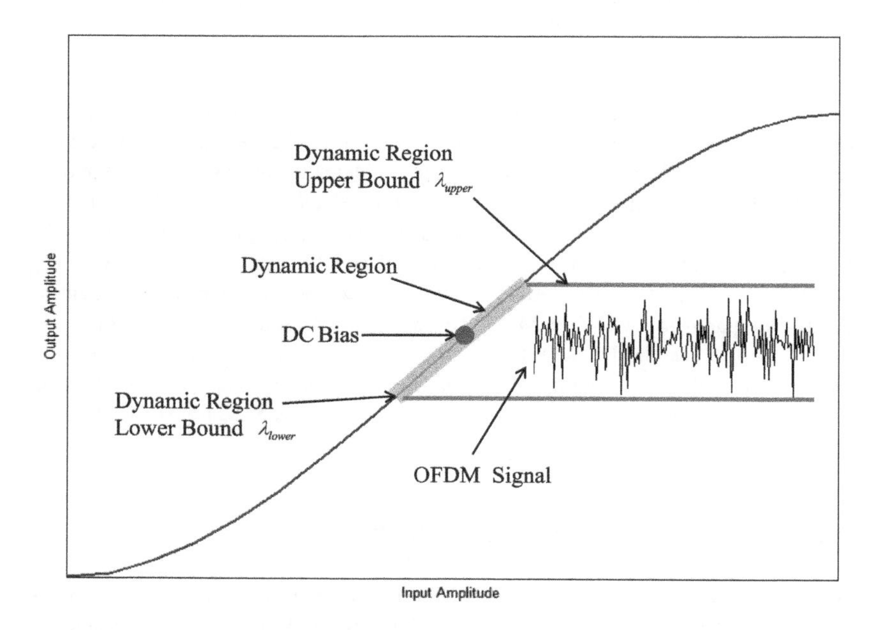

Figure 4.5 LED nonlinear transfer characteristic and DC bias setting for DCO-OFDM signal.

clipping [14], which is expressed as

$$F\left(x+B_{\mathrm{DC}}\right)=\begin{cases}\lambda_{\mathrm{lower}}, & x+B_{\mathrm{DC}}<\lambda_{\mathrm{lower}},\\ x+B_{\mathrm{DC}}, & \lambda_{\mathrm{lower}}\leq x+B_{\mathrm{DC}}<\lambda_{\mathrm{upper}},\\ \lambda_{\mathrm{upper}}, & x+B_{\mathrm{DC}}\geq\lambda_{\mathrm{upper}}.\end{cases} \tag{4.18}$$

Obviously, the clipping level is related to the LED dynamic region and the average power of input signal x. In order to evaluate the effect of clipping distortion, clipping level CL is defined as

$$CL=10\log\frac{\left\|\lambda_{\mathrm{upper}}-\lambda_{\mathrm{lower}}\right\|^{2}}{E\left\{x^{2}\right\}}\ \mathrm{dB}. \tag{4.19}$$

The DCO-OFDM time-domain signal x satisfies a Gaussian distribution [14]. By applying Bussgang theorem [15], the clipped signal $F\left(z\right)$ can be modeled as

$$F\left(z\right)-B_{\mathrm{DC}}=Kx+w_{\mathrm{clip}}, \tag{4.20}$$

where K denotes the attenuation factor and w_{clip} the non-correlative clipping noise. At the receiver, the ML detection regards the clipping distortion as part of the Gaussian noise and equalizes the data symbol with the attenuation factor K. Hence, the effective signal-to-noise ratio (SNR) $\Gamma_{\mathrm{b,elec}}$ on the kth subcarrier is equal to

$$\Gamma_{\mathrm{b,elec}}=\frac{K^{2}\sigma_{x}^{2}}{\sigma_{\mathrm{clip}}^{2}+\sigma_{\mathrm{AWGN}}^{2}/\left|H_{k}\right|^{2}}, \tag{4.21}$$

where σ_{x}^{2} and $\sigma_{\mathrm{AWGN}}^{2}$ denote the powers of the transmitted DCO-OFDM signal and the AWGN noise, respectively [16]. The DC bias B_{DC} is not included in the effective SNR since it does not convey any information. According to [14], $\sigma_{\mathrm{clip}}^{2}$ and K are the functions of B_{DC} and σ_{x}^{2}. When the dynamic region $\left[\lambda_{\mathrm{lower}},\ \lambda_{\mathrm{upper}}\right]$ is known, DC bias and signal power can be optimized to maximize the effective SNR and thus enhance the performance. In order to support flexible power adjustment for DCO-OFDM, scaling factor $\alpha>0$ is introduced. Assuming that the OFDM signal x^{normal} converted by IFFT components is normalized in electrical power, the transmitted DCO-OFDM signal x is amplified as $\alpha x^{\mathrm{normal}}$ and the signal power σ_{x}^{2} is equal to α^{2}. The optimization for mitigating the clipping noise can then be expressed as

$$\left\{B_{\mathrm{DC}}^{\mathrm{op}},\alpha^{\mathrm{op}}\right\}=\arg\max_{B_{\mathrm{DC}}>0,\alpha>0}\Gamma_{\mathrm{b,elec}}\left(B_{\mathrm{DC}},\alpha\right). \tag{4.22}$$

The DC bias and scaling factor optimization methods are classified as static and dynamic methods. In the static method, the DC bias and scaling factor of each DCO-OFDM block are fixed. The optimum DC bias can be formulated according to Bussgang theorem and the Gaussian distribution characteristics of DCO-OFDM signals, which is reported in [17]. The optimum scaling factor is acquired by minimizing the required average normalized optical signal-to-noise ratio (OSNR) for the target BER level of 10^{-3} [16]. However, the peak amplitudes of different DCO-OFDM blocks vary in a large range and the fixed scaling factor cannot make full use of dynamic region. Hence, a dynamic DC bias and scaling factor optimization method was proposed in [16], whereby the DC bias and scaling factor are adaptively adjusted for each DCO-OFDM block. Since all the samples in the normalized sample sequence $\{x_n^{\text{normal}}, n = 0, 1, \ldots, N-1\}$ generated by the IFFT converter are multiplied by the scaling factor, a large scaling factor increases the power of clipping noise, whereas a small scaling factor reduces the effective signal power. Considering the tradeoff between the powers of clipping noise and signal, the scaling factor is recommended to set according to [16]

$$\alpha^{\text{op}} = \frac{2\left(\lambda_{\text{upper}} - \lambda_{\text{lower}}\right)}{x_{\max}^{\text{normal}} + x_{\text{s}\max}^{\text{normal}} - x_{\min}^{\text{normal}} - x_{\text{s}\max}^{\text{normal}}}, \tag{4.23}$$

where $x_{\max}^{\text{normal}}$ and $x_{\min}^{\text{normal}}$ denote the maximum and minimum elements of $\{x_n^{\text{normal}}\}$, respectively, while $x_{\text{s}\max}^{\text{normal}}$ and $x_{\text{s}\min}^{\text{normal}}$ represent the second maximum and second minimum samples, respectively. By adopting this scaling factor, only the positive and negative peak amplitudes of OFDM signal may exceed the dynamic region, and this limits the clipping distortion level. Although α^{op} given in (4.23) is not the true optimum scaling factor that maximizes the effective SNR, it achieves a desired balance between the powers of clipping noise and signal effectively without high calculation complexity. After α^{op} is determined, the DC bias is acquired by minimizing the power of clipping noise according to [16]

$$B_{\text{DC}}^{\text{op}} = \arg \min_{B_{\text{DC}}>0} \sum_{n=0}^{N-1} \left(F\left(\alpha^{\text{op}} x_n^{\text{normal}} + B_{\text{DC}}\right) - \alpha^{\text{op}} x_n^{\text{normal}} - B_{\text{DC}}\right)^2. \tag{4.24}$$

The optimization of (4.24) still imposes considerable computation especially when the subcarrier number is large. Since α^{op} acquired in (4.23) guarantees that the range of the signal amplitudes approaches the dynamic region of the LED, an approximate optimum $B_{\text{DC}}^{\text{op}}$ is recommended as [16]

$$B_{\text{DC}}^{\text{op}} = \left(\lambda_{\text{upper}} + \lambda_{\text{lower}}\right)/2 - \alpha^{\text{op}}\left(x_{\max}^{\text{normal}} + x_{\min}^{\text{normal}}\right)/2. \tag{4.25}$$

After adopting this approximation, the middle point of the amplitude range in each DCO-OFDM block is raised to coincide with the center of the LED dynamic region. Since the optimum scaling factor is unknown to the receiver, an additional pilot should be embedded in each DCO-OFDM block for equalization. The DC bias is converted onto the zeroth subcarrier after the FFT operation and it does not affect the data symbols. Only simple addition and multiplication are required in the solution of Eqs. (4.23) and (4.25). The peak and second peak amplitudes are obtained by simply traversing the sample sequence once. Thus, the DC bias and scaling factor optimization proposed in [16] is of extremely low complexity and suitable for practical implementation.

In Fig. 4.6, the simulated BER curves of both the static and dynamic optimization methods are presented. In the simulation system, each DCO-OFDM block contains

256 subcarriers and adopts 64QAM constellation. The results show that the dynamic optimization method enjoys a performance gain of around 1 dB compared to the static counterpart. Furthermore, the approximate optimum DC bias of (4.25) is observed to achieve almost the same BER performance as the true optimal solution. This confirms that this low-complexity near-optimal DC bias is preferred for practical implementation.

4.2.2 LED nonlinearity mitigation

As introduced in the previous subsection, the nonlinear distortion is often modeled as part of the non-correlative noise at the conventional receiver. Although the negative effect of the clipping noise is restricted by the DC bias and scaling factor optimization, the nonlinear distortion still impairs the performance of optical OFDM systems. In this subsection, a state-of-the-art algorithm is introduced to significantly mitigate the nonlinear effect of LED.

According to [18], the achievable capacity of OFDM transmission over nonlinear channel only suffers a slight performance loss even when narrow dynamic region is imposed. Hence, the correlation between the clipping distortion and received signal

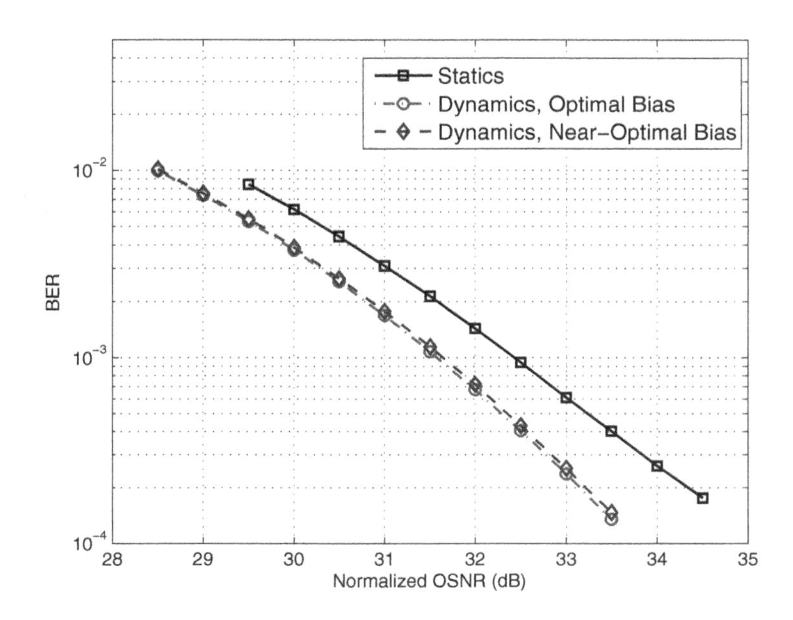

Figure 4.6 Performance comparison of static and dynamic DC bias and scaling factor optimization.

can be exploited to enhance the performance of the clipped DCO-OFDM. A novel maximum likelihood sequence detection (MLSD) was proposed to mitigate the LED nonlinear effect in [19]. Unlike the conventional receiver which detects the data symbol on each subcarrier by using (4.6), the DCO-OFDM data symbol sequence is jointly detected in the proposed method.

For clear clarification, denote the Ns-length candidate symbol sequence of each DCO-OFDM block by $\overline{\mathbf{X}} = \left[\overline{X}_1\, \overline{X}_2\, \cdots\, \overline{X}_{Ns}\right] \in \mathcal{X}^{Ns}$, where $0 < Ns < N/2$, and the associated candidate DCO-OFDM sample sequence after the operations of the Hermitian symmetry in (4.1), IFFT in (4.2) and LED double side clipping in (4.18) by $\mathbf{r}\left(\overline{\mathbf{X}}\right)$. Since the noise vector obeys the N-dimensional independent and identical Gaussian distribution with each element having zero mean and variance of σ^2, the conditional probability density function (PDF) or likelihood of the candidate symbol sequence $\overline{\mathbf{X}}$ is given by

$$p\left(\mathbf{r}\left|\mathbf{X} = \overline{\mathbf{X}}\right.\right) = \frac{1}{\left(2\pi\sigma^2\right)^{N/2}} \exp\left(-\frac{1}{2\sigma^2}\left\|\mathbf{y} - \mathbf{r}\left(\overline{\mathbf{X}}\right)\right\|^2\right), \tag{4.26}$$

where $\mathbf{y}$ is the received sample sequence and a flat channel is assumed. Hence, the MLSD of [19] is expressed as

$$\mathbf{X}^{\mathrm{r}}_{\mathrm{MLSD}} = \arg \min_{\overline{\mathbf{X}} \in \mathcal{X}^{Ns}} \left\|\mathbf{y} - \mathbf{r}\left(\overline{\mathbf{X}}\right)\right\|^2. \tag{4.27}$$

However, the minimization in (4.27) needs to be solved by traversing all the possible candidate symbol sequences. For M-ary constellation, the complexity of the MLSD is on the order of $O(M^{Ns})$, which makes this detection infeasible.

A low-complexity suboptimal algorithm was further proposed in [19]. Since the clipping only distorts a symbol around its transmitted constellation point, the symbol sequence $\mathbf{X}^{\mathrm{r}}_{\mathrm{MLD}}$ recovered by the conventional subcarrier based maximum likelihood detection (MLD) of (4.6) is close to $\mathbf{X}^{\mathrm{r}}_{\mathrm{MLSD}}$. In the proposed suboptimal algorithm, $\mathbf{X}^{\mathrm{r}}_{\mathrm{MLD}}$ is set as the initial candidate symbol sequence as $\overline{\mathbf{X}} = \mathbf{X}^{\mathrm{r}}_{\mathrm{MLD}}$. Starting on the first subcarrier, $\overline{X}_1$ of $\overline{\mathbf{X}}$ is substituted by its 4 nearest neighbor constellation points to generate the four candidate sequences $\overline{\mathbf{X}}^{(i)}$, $i = 1, 2, 3, 4$. Compute the Euclidean distances $\left\|\mathbf{y} - \mathbf{r}\left(\overline{\mathbf{X}}^{(i)}\right)\right\|^2$, $i = 1, 2, 3, 4$, and find the sequence $\overline{\mathbf{X}}^{(i_{\min})}$ with the minimum Euclidean distance. If $\left\|\mathbf{y} - \mathbf{r}\left(\overline{\mathbf{X}}^{(i_{\min})}\right)\right\|^2 < \left\|\mathbf{y} - \mathbf{r}\left(\overline{\mathbf{X}}\right)\right\|^2$, update $\overline{\mathbf{X}}$ with $\overline{\mathbf{X}}^{(i_{\min})}$. Then move to the next subcarrier, and continue the same update until all the subcarriers are updated. Then return to the first subcarrier and repeat the same procedure until the iteration number reaches its maximum number L. The detailed algorithm is summarized in Algorithm 4.1.

The final output $\widehat{\mathbf{X}}^{\mathrm{r}}_{\mathrm{MLSD}}$ is a suboptimal solution of (4.27). With this algorithm, the complexity of the sequence detection is reduced to $O(LMNs)$. According to the simulation results presented in [19], the approximate MLSD algorithm can significantly mitigate the clipping distortions of DCO-OFDM signals in comparison to the conventional subcarrier-based MLD. When CL is set as 11 dB and 16QAM is

Algorithm 4.1 Low Complexity Near MLSD Algorithm.

Input: Received signal sequence $\mathbf{y}$; Constellation set $\mathcal{X}$;
Output: Near optimal MLSD solution $\widehat{\mathbf{X}}^{\mathrm{r}}_{\mathrm{MLSD}}$;
1: **Pre-processing**: For each $X \in \mathcal{X}$, find its four nearest constellation points $X_{\text{near}}(X,0)$, $X_{\text{near}}(X,1)$, $X_{\text{near}}(X,2)$, and $X_{\text{near}}(X,3)$; Store all the resulting subsets by look-up table;
2: **Initialization**: Find the conventional MLD solution $\mathbf{X}^{\mathrm{r}}_{\mathrm{MLD}}$ with the ML detection (4.6); Set $\overline{\mathbf{X}} = \mathbf{X}^{\mathrm{r}}_{\mathrm{MLD}}$, reconstruct the clipped DCO-OFDM signal $\mathbf{r}\left(\overline{\mathbf{X}}\right)$ and calculate the MLSD metric $\left\|\mathbf{y} - \mathbf{r}\left(\overline{\mathbf{X}}\right)\right\|^2$;
3: **for** $l = 1 : L$ **do**
4: **for** $k = 1 : N/2 - 1$ **do**
5: **for** $i = 0 : 3$ **do**
6: Find $\overline{X}_{k,i} = X_{\text{near}}\left(\overline{X}_k, i\right)$ for $\overline{X}_k$ from the look-up table;
7: Generate $\overline{\mathbf{X}}^{(i)}$ by substituting $\overline{X}_k$ in $\overline{\mathbf{X}}$ with $\overline{X}_{k,i}$;
8: Reconstruct $\mathbf{r}\left(\overline{\mathbf{X}}^{(i)}\right)$, and compute $\left\|\mathbf{y} - \mathbf{r}\left(\overline{\mathbf{X}}^{(i)}\right)\right\|^2$;
9: **end for**
10: Find $i_{\min} = \arg\min_{0 \le i \le 3} \left\|\mathbf{y} - \mathbf{r}\left(\overline{\mathbf{X}}^{(i)}\right)\right\|^2$;
11: **if** $\left\|\mathbf{y} - \mathbf{r}\left(\overline{\mathbf{X}}^{(i_{\min})}\right)\right\|^2 < \left\|\mathbf{y} - \mathbf{r}\left(\overline{\mathbf{X}}\right)\right\|^2$ **then**
12: Set $\overline{\mathbf{X}} = \overline{\mathbf{X}}^{(i_{\min})}$ with $\left\|\mathbf{y} - \mathbf{r}\left(\overline{\mathbf{X}}\right)\right\|^2 = \left\|\mathbf{y} - \mathbf{r}\left(\overline{\mathbf{X}}^{(i_{\min})}\right)\right\|^2$;
13: **end if**
14: **end for**
15: **end for**
16: **return** $\widehat{\mathbf{X}}^{\mathrm{r}}_{\mathrm{MLSD}} = \overline{\mathbf{X}}$;

adopted, the BER performance of this approximate MLSD algorithm converges to almost the same level of the ideal case, where the DCO-OFDM signal is not clipped at the transmitter, with only $L = 1$ iteration. As for higher-order constellations, such as 64QAM and 256QAM, the approximate MLSD algorithm achieves a performance enhancement of around 2 dB at BER $= 10^{-3}$ compared to the conventional subcarrier-based MLD. Moreover, the BER curves of the approximate MLSD almost coincide with those of the ideal-case counterparts. The results of [19] hence demonstrate the effectiveness of the approximate MLSD.

A similar method for mitigating the LED nonlinearity is applied to encoded DCO-OFDM [20]. More specifically, a bit-interleaved coded modulation with iterative demapping and decoding (BICM-ID) scheme is proposed for clipped DCO-OFDM to further mitigate the LED nonlinearity. The architecture of this proposed DCO-OFDM BICM-ID system is depicted in Fig. 4.7. At the transmitter, the source bit vector $\mathbf{u}$ is firstly encoded as $\mathbf{c}$, interleaved as $\mathbf{c}^{\pi}$ and then mapped into symbol vector $\mathbf{X}$. After imposing Hermitian symmetry and IFFT operation, the time-domain signal $\mathbf{x}$ is generated, which is then added with DC bias and clipped as $\mathbf{x}_{\text{DCO}}$ to

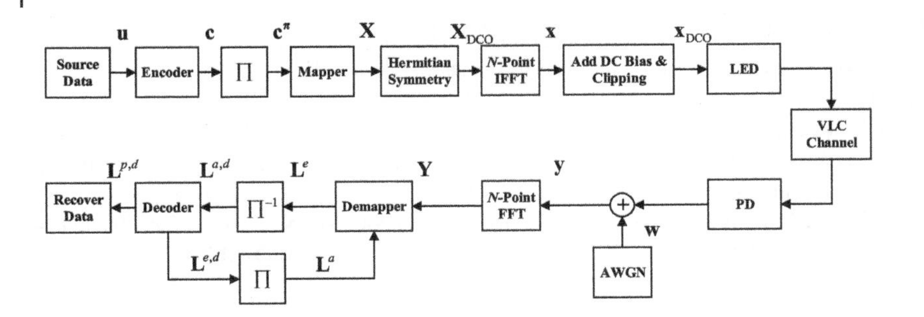

Figure 4.7 The block diagram of BICM-ID scheme for clipped DCO-OFDM in visible light communications.

accommodate the dynamic region of LED. At the receiver, the demapper calculates the extrinsic log-likelihood ratio (LLR) $\mathbf{L}^e$ as soft information and feeds it to the channel decoder after de-interleaving as *a priori* information $\mathbf{L}^{a,d}$. Then, the decoder outputs the *a posteriori* LLR $\mathbf{L}^{p,d}$ to recover the received bits and the extrinsic LLR $\mathbf{L}^{e,d}$ to provide the feedback *a priori* information for the next iterative demapping. For an arbitrary encoded bit b, its LLR is defined as

$$L_b = \log \frac{P\left(b=0\right)}{P\left(b=1\right)}, \tag{4.28}$$

where $P\left(A\right)$ represents the probability of event A. For the M-ary constellation, where $M = 2^m$, let $L^e_{b_{k,i}}$ and $L^a_{b_{k,i}}$ denote the demapping extrinsic LLR and decoding feedback LLR of the ith labeled bit $b_{k,i}$ on the kth subcarrier, respectively, the conventional soft demapping criterion is expressed as [20]

$$\begin{aligned} L^e_{b_{k,i}} = & \max_{\overline{X}_k \in \mathcal{X}_i^{(0)}} \left(-\frac{1}{\sigma^2} \left\| \overline{X}_k - Y_k \right\|^2 + \frac{1}{2} \mathbf{s}\left(\overline{X}_k\right) \left(\mathbf{L}^a_{b_k}\right)^{\mathrm{T}} \right) \\ & - \max_{\overline{X}_k \in \mathcal{X}_i^{(1)}} \left(-\frac{1}{\sigma^2} \left\| \overline{X}_k - Y_k \right\|^2 + \frac{1}{2} \mathbf{s}\left(\overline{X}_k\right) \left(\mathbf{L}^a_{b_k}\right)^{\mathrm{T}} \right) - L^a_{b_{k,i}}, \end{aligned} \tag{4.29}$$

where $(\cdot)^{\mathrm{T}}$ stands for the transpose operator and $\mathcal{X}_i^{(b)}$, $b = 0, 1$, denotes the constellation set with the ith labeled bit $b_i = b$. In (4.29), the vector $\mathbf{s}\left(\overline{X}_k\right)$ is equal to $\mathbf{1} - 2\mathbf{b}\left(\overline{X}_k\right)$, where $\mathbf{b}\left(\overline{X}_k\right)$ is the labeled bit vector of candidate symbol $\overline{X}_k$ and $\mathbf{1}$ is the vector with all the elements being 1, while $\mathbf{L}^a_{b_k}$ is the associated vector of the feedback LLRs on the kth subcarrier, denoted as $\left[L^a_{b_{k,0}}\ L^a_{b_{k,1}}\ \cdots\ L^a_{b_{k,m}}\right]$.

The above-mentioned soft demapper still models the nonlinear distortion as non-correlative noise. In order to mitigate the LED nonlinearity, a novel demapping criterion was further presented in [20]. The feedback LLR vector $\mathbf{L}^a = \left[\mathbf{L}^a_{b_1}\ \mathbf{L}^a_{b_2}\ \cdots\ \mathbf{L}^a_{b_{N_s}}\right]$ from the channel decoder is an accurate estimation of the encoded bits in a DCO-OFDM block. The proposed demapping method exploits this fact to recover the encoded bits. According to the definition of LLR, the encoded bit vector $\mathbf{b}^a\left(\mathbf{L}_{b_k}^a\right)$ is calculated by the following hard decision

$$b_i^k\left(\mathbf{L}_{b_k}^a\right)=\begin{cases}0, & L_{b_k,i}^a\geq 0,\\ 1, & L_{b_k,i}^a<0,\end{cases} \tag{4.30}$$

where $b_i^k\left(\mathbf{L}_{b_k}^a\right)$ is the ith element of $\mathbf{b}^a\left(\mathbf{L}_{b_k}^a\right)$. Then, $b_i^k\left(\mathbf{L}_{b_k}^a\right)$ is mapped onto the estimated data symbol $X^a\left(\mathbf{L}_{b_k}^a\right)$ based on the constellation $\mathcal{X}$, whereas the data symbol sequence $\mathbf{X}\left(\mathbf{L}^a\right)$ is simply generated as

$$\mathbf{X}\left(\mathbf{L}^a\right)=\left[X^a\left(\mathbf{L}_{b_1}^a\right)\ X^a\left(\mathbf{L}_{b_2}^a\right)\ \cdots\ X^a\left(\mathbf{L}_{b_{N_s}}^a\right)\right]. \tag{4.31}$$

By replacing the kth element $X^a\left(\mathbf{L}_{b_k}^a\right)$ of $\mathbf{X}\left(\mathbf{L}^a\right)$ with $\overline{X}_k$, a candidate symbol sequence $\overline{\mathbf{X}}\left(\mathbf{L}^a,\overline{X}_k\right)$ is acquired. Transfer $\overline{\mathbf{X}}\left(\mathbf{L}^a,\overline{X}_k\right)$ into the clipped DCO-OFDM signal $\mathbf{r}\left(\overline{\mathbf{X}}\left(\mathbf{L}^a,\overline{X}_k\right)\right)$ and extract the kth distorted candidate symbol $R_k^d\left(\mathbf{L}^a,\overline{X}_k\right)$ from the frequency-domain sequence of $\mathbf{r}\left(\overline{\mathbf{X}}\left(\mathbf{L}^a,\overline{X}_k\right)\right)$ after FFT operation, which can be simply calculated as

$$R_k^d\left(\mathbf{L}^a,\overline{X}_k\right)=\frac{1}{\sqrt{N}}\mathbf{e}_k\left(\mathbf{r}\left(\overline{\mathbf{X}}\left(\mathbf{L}^a,\overline{X}_k\right)\right)\right)^{\mathrm{T}}, \tag{4.32}$$

where $\mathbf{e}_k=\left[\exp\left(j\frac{2\pi}{N}0k\right)\ \exp\left(j\frac{2\pi}{N}1k\right)\ \cdots\ \exp\left(j\frac{2\pi}{N}\left(N-1\right)k\right)\right]$ is the discrete Fourier transform (DFT) harmonic vector. The novel demapping criterion is then presented as [20]

$$\begin{aligned}L_{b_k,i}^e=&\max_{\overline{X}_k\in\mathcal{X}_i^{(0)}}\left(-\frac{1}{\sigma^2}\left\|R_k^d\left(\mathbf{L}^a,\overline{X}_k\right)-Y_k\right\|^2+\frac{1}{2}\mathbf{s}\left(\overline{X}_k\right)\left(\mathbf{L}_{b_k}^a\right)^{\mathrm{T}}\right)\\&-\max_{\overline{X}_k\in\mathcal{X}_i^{(1)}}\left(-\frac{1}{\sigma^2}\left\|R_k^d\left(\mathbf{L}^a,\overline{X}_k\right)-Y_k\right\|^2+\frac{1}{2}\mathbf{s}\left(\overline{X}_k\right)\left(\mathbf{L}_{b_k}^a\right)^{\mathrm{T}}\right)-L_{b_k,i}^a.\end{aligned} \tag{4.33}$$

With the aid of the decoding feedback LLR $\mathbf{L}^a$, the associated LED nonlinearity distortion on the candidate symbol $\overline{X}_k$ can be effectively mitigated, which enhances the accuracy of the output demapping LLR. The convergence performance of the proposed BICM-ID receiver with the soft demapping criterion of (4.33) was reported in [20], which also confirms that a BER performance enhancement of around 1 dB is achieved by the proposed iterative demapper in comparison to the conventional counterpart when a low density parity check (LDPC) code is adopted as the channel code.

4.2.3 PAPR reduction

Besides the narrow dynamic region of LEDs, the high PAPR of optical OFDM signal is another major cause of severe nonlinear effects. PAPR reduction techniques can decrease the probability that the peak amplitude is nonlinearly distorted and enhance the performance over the nonlinear channel. Many PAPR reduction algorithms have been presented for RF multicarrier systems, as were reported in the overview [21]. Since VLC transmission systems adopt intensity modulation, these PAPR reduction techniques for RF OFDM cannot be directly applied in optical OFDM systems. In this subsection, a comprehensive introduction of PAPR reduction techniques for optical OFDM is presented.

In optical OFDM, the PAPR of an OFDM block is defined as

$$\text{PAPR} = \frac{\left\| \max_{0 \le n < N} \{x_n\} \right\|^2}{E\{x_n^2\}}. \tag{4.34}$$

Since the PAPR varies in a large range for different OFDM blocks, the complementary cumulative distribution function (CCDF) is used to depict the distribution of PAPR, which is defined as the probability that the PAPR exceeds the specific threshold PAPR_0 as

$$CCDF\,(\text{PAPR}_0) = Pr\,(\text{PAPR} > \text{PAPR}_0)\,. \tag{4.35}$$

The CCDF curve is an effective metric to evaluate the PAPR reduction performance.

In the work [22], a universal PAPR reduction method for optical OFDM schemes is proposed based on precoding technique. At the transmitter, the data symbol vector $\mathbf{X}^s = [X_1\ X_2\ \cdots\ X_{Ns}]$ is multiplied by a precoding matrix as

$$(\mathbf{X}^p)^{\mathrm{T}} = \mathbf{P}\,(\mathbf{X}^s)^{\mathrm{T}}\,, \tag{4.36}$$

where $\mathbf{P}$ is an $Ns \times Ns$ matrix which can be expressed as

$$\mathbf{P} = \begin{bmatrix} p_{1,1} & p_{1,2} & \cdots & p_{1,Ns} \\ p_{2,1} & p_{2,2} & \cdots & p_{2,Ns} \\ \vdots & \vdots & \ddots & \vdots \\ p_{Ns,1} & p_{Ns,2} & \cdots & p_{Ns,Ns} \end{bmatrix}. \tag{4.37}$$

The precoded symbol vector $\mathbf{X}^p$ is allocated on the associated subcarriers with the Hermitian symmetry (4.1) imposed for DCO-OFDM and ACO-OFDM or (4.13) imposed for PAM-DMT. Then, the frequency-domain symbol sequence is fed to the IFFT converter to generate the time-domain sample sequence. The research [22] recommended DFT, Zadoff–Chu (ZC) sequence or discrete cosine transform (DCT) based precoding matrix for DCO-OFDM and ACO-OFDM signals. Denoting the elements of $\mathbf{P}$ as $p^{\text{DFT}}_{n+1,k+1}$, $p^{\text{ZC}}_{n+1,k+1}$, and $p^{\text{DCT}}_{n+1,k+1}$, $0 \le n, k < N_s$, for the DFT, ZC sequence, and DCT precoding methods, respectively, one has

$$p^{\text{DFT}}_{n+1,k+1} = \frac{1}{\sqrt{Ns}} \exp\left(j\frac{2\pi}{Ns}nk\right), \tag{4.38}$$

$$p^{\text{ZC}}_{n+1,k+1} = \begin{cases} \frac{1}{\sqrt{Ns}} \exp\left(j\frac{2\pi r}{Ns^2}\left(\frac{(nNs+k)^2}{2} + q\,(nNs+k)\right)\right), & Ns \text{ even}, \\ \frac{1}{\sqrt{Ns}} \exp\left(j\frac{2\pi r}{Ns^2}\left(\frac{(nNs+k)^2}{2} + \left(q+\frac{1}{2}\right)(nNs+k)\right)\right), & Ns \text{ odd}, \end{cases}$$

(4.39)

and

$$p^{\mathrm{DCT}}_{n+1,k+1} = \begin{cases} \sqrt{\frac{1}{Ns}}, & n = 0, \\ \sqrt{\frac{2}{Ns}} \cos\left(\pi n \left(\frac{2k+1}{2N_s}\right)\right), & n = 1, \ldots, Ns - 1. \end{cases} \tag{4.40}$$

In (4.39), r is the code index relatively prime to Ns and q is any integer. Since the precoded symbols are restricted to be real-valued in PAM-DMT, only DCT precoding can be used. The precoding method flattens the distribution of the time-domain sample amplitude and reduces the PAPR effectively. The simulation results presented in [22] confirm that the precoding methods with DCT and ZC precoding matrices reduce the PAPR threshold at the CCDF level of 10^{-4} by over 3 dB for both ACO-OFDM and PAM-DMT signals.

Another PAPR reduction technique called selective mapping (SLM) is also introduced for DCO-OFDM [23]. The transmitter randomly generates R groups of Ns-length phase factors $\mathbf{p}^r_{\mathrm{SLM}} = [p^r_1 \, p^r_2 \, \cdots \, p^r_{Ns}]$ for $1 \leq r \leq R$, where $p^r_k = \exp\left(j \frac{w}{W} 2\pi\right)$ with w randomly taking the value from the set $\{0, 1, \ldots, W - 1\}$. The DCO-OFDM block $\mathbf{X}\left(\mathbf{p}^r_{\mathrm{SLM}}\right)$ after applying SLM is

$$\mathbf{X}\left(\mathbf{p}^r_{\mathrm{SLM}}\right) = \left[p^r_1 X_1 \, p^r_2 X_2 \, \cdots \, p^r_{Ns} X_{Ns}\right]. \tag{4.41}$$

By imposing the Hermitian symmetry and applying the IFFT operation, the time-domain sample sequence $\{x_n\left(\mathbf{p}^r_{\mathrm{SLM}}\right)\}$ is generated. The transmitter compares all the PAPR values of the R groups of $\{x_n\left(\mathbf{p}^r_{\mathrm{SLM}}\right)\}$ and selects the sequence with the minimum PAPR for modulation into DCO-OFDM signal. Since the receiver requires the information of the selected phase factor $\mathbf{p}^r_{\mathrm{SLM}}$ for demodulation, an additional pilot block is embedded in the DCO-OFDM signal. In order to further reduce the PAPR, the pilot-assisted SLM method [23] applies the same $\mathbf{p}^r_{\mathrm{SLM}}$ for consecutive U DCO-OFDM blocks to minimize the PAPR of the $(U+1)N$-length sample sequence including the additional pilot block. This pilot-assisted SLM technique reduces the PAPR threshold at the CCDF level of 10^{-4} by around 2 dB.

Partial transmit sequence (PTS) is another PAPR reduction technique that is capable of enhancing the performance of ACO-OFDM over nonlinear channel [24]. In this technique, the symbol block is divided into the V disjoint clusters as $\mathbf{X}^s = [\mathbf{X}_1 \, \mathbf{X}_2 \, \cdots \, \mathbf{X}_V]$. By multiplying a V-length phase factor vector $\mathbf{p}^r_{\mathrm{PTS}} = [p^r_1 \, p^r_2 \, \cdots \, p^r_V]$ with $\mathbf{X}^s$, it is transformed into

$$\mathbf{X}\left(\mathbf{p}^r_{\mathrm{PTS}}\right) = \left[p^r_1 \mathbf{X}_1 \, p^r_2 \mathbf{X}_2 \, \cdots \, p^r_V \mathbf{X}_V\right], \tag{4.42}$$

which are allocated on the odd subcarriers and converted to the ACO-OFDM sample sequence $\{x_n\left(\mathbf{p}^r_{\mathrm{PTS}}\right)\}$. The transmitter is required to compute the optimal $\mathbf{p}^r_{\mathrm{PTS}}$ to minimize the PAPR of $\{x_n\left(\mathbf{p}^r_{\mathrm{PTS}}\right)\}$. In order to transmit the side information of phase factor, the proposed scheme in [24] modulates the optimal $\mathbf{p}^r_{\mathrm{PTS}}$ into a low-power PAM-DMT signal, whereby $\mathbf{p}^r_{\mathrm{PTS}}$ is converted to a burst pulse shift key (B-PSK) symbol and allocated on the imaginary parts of the even subcarriers. The transmitter adds up the ACO-OFDM and PAM-DMT signals to drive the LED. Since the clipping distortion of PAM-DMT falls on the real parts of even subcarriers and does not affect the data symbol on odd subcarriers, the receiver can recover the two signals. Hence, the modified ACO-OFDM scheme based on PTS proposed in [24] does not occupy extra spectral resource to transmit the information of phase factor. According to the simulation results of [24], this modified PTS technique achieves a PAPR reduction of around 5 dB at the CCDF level of 10^{-4} and a performance gain of 1 dB under the LED nonlinearity, compared to the conventional ACO-OFDM.

The PAPR reduction techniques for optical OFDM discussed so far impose high computational cost. In [25], a low-complexity recoverable upper-clipping (RoC) was developed to reduce the PAPR of ACO-OFDM signals. Assuming that $\left\{x_n^c, x_{n+N/2}^c\right\}$ for $n = 0, 1, \ldots, N/2 - 1$ is the ACO-OFDM sample pair after imposing the constraint from (4.8). The RoC scheme is expressed as [25]

$$\left\{x_n^{\mathrm{RoC}}, x_{n+N/2}^{\mathrm{RoC}}\right\} = \begin{cases} \{x_n^c, 0\}, & 0 < x_n^c \le A, \\ \left\{0, x_{n+N/2}^c\right\}, & 0 < x_{n+N/2}^c \le A, \\ \{A, x_n^c - A\}, & A < x_n^c \le (1+\alpha)A, \\ \left\{x_{n+N/2}^c - A, A\right\}, & A < x_{n+N/2}^c \le (1+\alpha)A, \\ \{A, \alpha A\}, & x_n^c > (1+\alpha)A, \\ \{\alpha A, A\}, & x_{n+N/2}^c > (1+\alpha)A, \end{cases} \tag{4.43}$$

where A is the clipping threshold and $0 < \alpha < 1$ is a fixed factor. Figure 4.8 illustrates three different cases of upper-clipping. If the positive sample x_n^c ranges from A to $(1+\alpha)A$, x_n^c is clipped at A and the clipped error $x_n^c - A$ is placed at the $(n+N/2)$th sample, which is used to be zero. If x_n^c exceeds $(1+\alpha)A$, it is still clipped at A and the clipped error placed on the $(n+N/2)$th sample is restricted to αA. Since $\alpha < 1$, the clipped error is always lower than the associated recoverable upper-clipped sample. This property helps the receiver to distinguish which sample of the received sample pair $\left\{y_n^{\mathrm{RoC}}, y_{n+N/2}^{\mathrm{RoC}}\right\}$ is allocated with the positive amplitude of the ACO-OFDM signal sample. The original ACO-OFDM pair $\left\{y_n, y_{n+N/2}\right\}$ is

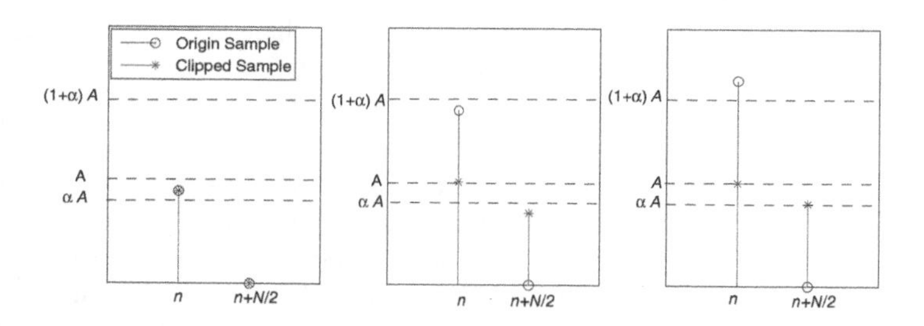

Figure 4.8 Three cases of recoverable upper-clipping in RoC aided ACO-OFDM scheme.

simply recovered as [25]

$$\{y_n, y_{n+N/2}\} = \begin{cases} \{y_n^{\mathrm{RoC}}, 0\}, & y_n^{\mathrm{RoC}} > y_{n+N/2}^{\mathrm{RoC}} \,\&\, y_n^{\mathrm{RoC}} + y_{n+N/2}^{\mathrm{RoC}} \leq A, \\ \left\{0, y_{n+N/2}^{\mathrm{RoC}}\right\}, & y_n^{\mathrm{RoC}} \leq y_{n+N/2}^{\mathrm{RoC}} \,\&\, y_n^{\mathrm{RoC}} + y_{n+N/2}^{\mathrm{RoC}} \leq A, \\ \{y_n^{\mathrm{RoC}} + A, 0\}, & y_n^{\mathrm{RoC}} > y_{n+N/2}^{\mathrm{RoC}} \,\&\, y_n^{\mathrm{RoC}} + y_{n+N/2}^{\mathrm{RoC}} > A, \\ \left\{0, y_{n+N/2}^{\mathrm{RoC}} + A\right\}, & y_n^{\mathrm{RoC}} \leq y_{n+N/2}^{\mathrm{RoC}} \,\&\, y_n^{\mathrm{RoC}} + y_{n+N/2}^{\mathrm{RoC}} > A. \end{cases} \tag{4.44}$$

A low clipping threshold A can reduce the peak amplitude significantly but induce severe clipping distortion and performance loss. The tradeoff between PAPR reduction and BER performance needs to be carefully considered to set an appropriate value of A. On the condition that the demodulation threshold SNR at the BER level of 10^{-3} almost coincides with the conventional ACO-OFDM, the RoC aided ACO-OFDM scheme achieves a maximum PAPR reduction of around 5 dB at the CCDF level of 10^{-4}. Furthermore, over the optical channel which includes the LED nonlinearity, a significant BER performance gain was observed with the proposed RoC aided ACO-OFDM scheme.

4.3 Spectrum- and power-efficient optical OFDM

As illustrated before, DCO-OFDM suffers from low power efficiency since a high DC bias is required to maintain the non-negativity of the transmitted signals. Although several power-efficient optical OFDM schemes, such as ACO-OFDM, PAM-DMT, and U-OFDM, have been proposed to avoid the DC bias, their spectral efficiencies are reduced considerably. Consequently, these power efficient schemes perform even worse than DCO-OFDM when a high spectral efficiency is required. This is because these schemes must employ higher constellation orders to meet the required spectral efficiency. Recently, several spectrum- and power-efficient optical OFDM schemes have been proposed to overcome this difficulty.

4.3.1 Hybrid optical OFDM

Since ACO-OFDM only occupies the odd subcarriers for data transmission, the even subcarriers can be employed to improve the spectral efficiency, and three hybrid optical OFDM have been proposed recently. Asymmetrically clipped DC biased optical OFDM (ADO-OFDM) is a combination of DCO-OFDM and ACO-OFDM, where the ACO-OFDM signal is transmitted on the odd subcarriers and the DCO-OFDM signal is transmitted on the even subcarriers [9, 26]. The symbols in a data block $\mathbf{X}$ are divided into two parts $\mathbf{X}_{\mathrm{odd}}$ and $\mathbf{X}_{\mathrm{even}}$, where $\mathbf{X}_{\mathrm{odd}} = [0\ X_1\ 0, X_3\ 0\ \cdots\ 0\ X_{N-1}]$ contains the odd-subcarrier symbols and $\mathbf{X}_{\mathrm{even}} = [X_0\ 0\ X_2\ 0\ \cdots\ X_{N-2}\ 0]$ includes the even-subcarrier symbols. Since $X_0 =$

$X_{N/2} = 0$ and the Hermitian symmetry of (4.1) is imposed on $\mathbf{X}$, both $\mathbf{X}_{\text{odd}}$ and $\mathbf{X}_{\text{even}}$ also satisfy the Hermitian symmetry. After the IFFT operation, the two time-domain signal vectors $\mathbf{x}_{\text{odd}} = [x_{\text{odd},0}\ x_{\text{odd},1}\ \cdots\ x_{\text{odd},N-1}]$ and $\mathbf{x}_{\text{even}} = [x_{\text{even},0}\ x_{\text{even},1}\ \cdots\ x_{\text{even},N-1}]$ are obtained. The elements in both $\mathbf{x}_{\text{odd}}$ and $\mathbf{x}_{\text{even}}$ are real and bipolar.

However, the structure of $\mathbf{x}_{\text{odd}}$ is the same as that of ACO-OFDM, and an ACO-OFDM signal vector can be obtained by clipping the negative signals in $\mathbf{x}_{\text{odd}}$, leading to

$$\mathbf{x}_{\text{ACO}} = \frac{1}{2}\mathbf{x}_{\text{odd}} + \mathbf{i}_{\text{ACO}}, \tag{4.45}$$

where $\mathbf{i}_{\text{ACO}}$ denotes the clipping distortion vector. It has been proven in [7] that the clipping distortion only falls on the even subcarriers after FFT, which will not interfere with the useful information on the odd subcarriers. As for $\mathbf{x}_{\text{even}}$, a DC bias B_{DC} is added and the remaining negative signals are clipped. The resultant DCO-OFDM signal is written by

$$\mathbf{x}_{\text{DCO}} = \mathbf{x}_{\text{even}} + B_{\text{DC}}\mathbf{1} + \mathbf{i}_{\text{DCO}}, \tag{4.46}$$

where $\mathbf{i}_{\text{DCO}}$ denotes the clipping distortion vector. Since $\mathbf{X}_{\text{even}}$ only occupies the even subcarriers, the elements of $\mathbf{x}_{\text{even}}$ follow the even symmetry

$$x_{\text{even},n} = x_{\text{even},n+N/2},\ n = 0, 1,\ ...,\ N/2 - 1. \tag{4.47}$$

After adding DC bias and clipping, the DCO-OFDM signal vector also follows the same even symmetry

$$x_{\text{DCO},n} = x_{\text{DCO},n+N/2},\ n = 0, 1,\ ...,\ N/2 - 1. \tag{4.48}$$

Therefore, the clipping distortion vector of this DCO-OFDM signal also follows the even symmetry

$$i_{\text{DCO},n} = i_{\text{DCO},n+N/2},\ n = 0, 1,\ ...,\ N/2 - 1, \tag{4.49}$$

and it will only interfere with information on the even subcarriers when transformed to the frequency domain, which means that both the useful information and clipping distortion are on the even subcarriers, and they will not affect the useful information in the ACO-OFDM signal. Hence, the two signal streams can be combined into the single ADO-OFDM signal vector

$$\mathbf{x}_{\text{ADO}} = \mathbf{x}_{\text{ACO}} + \mathbf{x}_{\text{DCO}} \tag{4.50}$$

for simultaneous transmission.

Assuming an AWGN channel, the time-domain received signal vector can be written as

$$\begin{aligned}\mathbf{y} &= \mathbf{x}_{\text{ACO}} + \mathbf{x}_{\text{DCO}} + \mathbf{w}_{\text{AWGN}} \\ &= \frac{1}{2}\mathbf{x}_{\text{odd}} + \mathbf{i}_{\text{ACO}} + \mathbf{x}_{\text{even}} + B_{\text{DC}}\mathbf{1} + \mathbf{i}_{\text{DCO}} + \mathbf{w}_{\text{AWGN,odd}} + \mathbf{w}_{\text{AWGN,even}},\end{aligned} \tag{4.51}$$

where the channel AWGN vector $\mathbf{w}_{\text{AWGN}}$ is decomposed into two parts $\mathbf{w}_{\text{AWGN,odd}}$ and $\mathbf{w}_{\text{AWGN,even}}$, whose DFTs $\mathbf{W}_{\text{AWGN,odd}}$ and $\mathbf{W}_{\text{AWGN,even}}$ correspond to the frequency-domain AWGN samples on the odd and even subcarriers, respectively. Furthermore, $\mathbf{W}_{\text{AWGN,odd}}$ only corrupts the odd subcarriers, while $\mathbf{W}_{\text{AWGN,even}}$ only distorts the even subcarriers. Therefore, after FFT, the symbols on the odd subcarriers can be written as

$$\mathbf{Y}_{\text{odd}} = \frac{1}{2}\mathbf{X}_{\text{odd}} + \mathbf{W}_{\text{AWGM,odd}}, \tag{4.52}$$

which can be used to detect the ACO-OFDM symbols directly. In order to demodulate the DCO-OFDM symbols, however, the interference on the even subcarriers induced by the ACO-OFDM signal should be removed first. An estimation of the transmitted ACO-OFDM signal can be made by performing the IFFT on $\mathbf{Y}_{\text{odd}}$ and clipping, yielding

$$\mathbf{y}_{\text{ACO}} = \frac{1}{2}\mathbf{x}_{\text{ACO}} + \mathbf{i}_{\text{ACO,est}} + \mathbf{w}_{\text{AWGN,odd}}, \tag{4.53}$$

where $\mathbf{i}_{\text{ACO,est}}$ is the estimated $\mathbf{i}_{\text{ACO}}$ since a noisy signal is used to reconstruct the ACO-OFDM signal. The estimated ACO-OFDM signal is then subtracted from the received signal $\mathbf{y}$ to obtain the DCO-OFDM signal by FFT

$$\mathbf{Y}'_{\text{even}} = \mathbf{X}_{\text{even}} + \mathbf{I}_{\text{DCO}} + B'_{\text{DCO}}\mathbf{1} + \mathbf{I}_{\text{ACO,est}} + \mathbf{W}_{\text{AWGN,even}}, \tag{4.54}$$

where B'_{DCO} denotes the DC component in the frequency domain, while $\mathbf{I}_{\text{DCO}}$ and $\mathbf{I}_{\text{ACO,est}}$ are the DFTs of $\mathbf{i}_{\text{DCO}}$ and $\mathbf{i}_{\text{ACO,est}}$, respectively. The DCO-OFDM symbols can be detected with $\mathbf{Y}'_{\text{even}}$, which contains not only the AWGN but also the estimated distortion of ACO-OFDM.

ADO-OFDM utilizes all the subcarriers for modulation, which improves the spectral efficiency compared to ACO-OFDM. However, DC bias is still required at the branch of DCO-OFDM, which is not power efficient. Therefore, a hybrid asymmetrical clipped optical OFDM (HACO-OFDM) is proposed, which replaces DCO-OFDM with unipolar PAM-DMT signals to avoid DC bias [27]. In the HACO-OFDM scheme, the ACO-OFDM signal generated on the odd subcarriers is identical to (4.45). As for the even subcarriers, however, only their imaginary parts are modulated by PAM, so that the corresponding time-domain signals follow the asymmetry as

$$x_{\text{even},n} = -x_{\text{even},N-n},\ n = 1,\ \ldots,\ N/2 - 1. \tag{4.55}$$

The PAM-DMT signal vector is obtained by clipping the negative signals in $\mathbf{x}_{\text{even}}$, resulting in

$$\mathbf{x}_{\text{PAM}} = \frac{1}{2}\mathbf{x}_{\text{even}} + \mathbf{i}_{\text{PAM}}, \tag{4.56}$$

where $\mathbf{i}_{\text{PAM}}$ denotes the clipping distortion of PAM-DMT. The clipping distortion only falls on the real parts of even subcarriers after FFT, which will not interfere

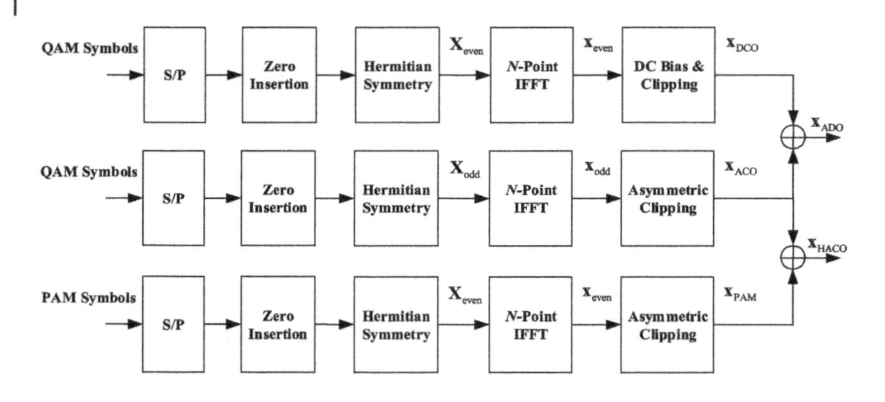

Figure 4.9 Block diagram of the transmitter for ADO-OFDM (top and middle branches) and HACO-OFDM (bottom and middle branches).

with the useful information on the imaginary parts. Since both the useful symbols and clipping distortion of $\mathbf{x}_{\text{PAM}}$ are on the even subcarriers, they will not interfere with the symbols of ACO-OFDM on the odd subcarriers. Therefore, PAM-DMT and ACO-OFDM can be combined to generate the HACO-OFDM signal as

$$\mathbf{x}_{\text{HACO}} = \mathbf{x}_{\text{ACO}} + \mathbf{x}_{\text{PAM}}. \tag{4.57}$$

For a convenient comparison, the transmitters of both ADO-OFDM and HACO-OFDM are illustrated in Fig. 4.9.

In HACO-OFDM, different modulation schemes are utilized for its ACO-OFDM and PAM-DMT components, which have different performance at the same SNR level. Specifically, PAM-DMT only employs one signal dimension of the even subcarriers and, therefore, the SNR value required for PAM-DMT to achieve the same BER performance is larger than that for ACO-OFDM with the same modulation orders. If different modulation orders are used for ACO-OFDM and PAM-DMT, the required SNRs are also different. For practical HACO-OFDM systems, it is highly desired that the information transmitted from both ACO-OFDM and PAM-DMT has similar performance. Therefore, the unequal power allocation of ACO-OFDM and PAM-DMT was proposed in [28].

For ACO-OFDM with a QAM constellation of size M_{ACO} and PAM-DMT with a PAM constellation of size M_{PAM}, the BER performance of QAM and PAM can be formulated as [29]

$$P_{\text{b,QAM}} \approx \frac{4(\sqrt{M_{\text{ACO}}}-1)}{\sqrt{M_{\text{ACO}}}\log_2(M_{\text{ACO}})} Q\left(\sqrt{\frac{3}{M_{\text{ACO}}-1}\frac{E_s}{N_0}}\right), \tag{4.58}$$

$$P_{\text{b,PAM}} \approx \frac{2(M_{\text{PAM}}-1)}{M_{\text{PAM}}\log_2(M_{\text{PAM}})} Q\left(\sqrt{\frac{6}{M_{\text{PAM}}^2-1}\frac{E_s}{N_0}}\right), \tag{4.59}$$

where E_s denotes the electrical energy per symbol and N_0 represents the power spectral density of the noise. Given a required BER value P_b, we can calculate numerically, the required E_s/N_0 values for the M_{ACO}-QAM-based ACO-OFDM and M_{PAM}-PAM-based PAM-DMT to achieve the same BER P_b, which are denoted by $E_{s,\text{ACO}}$ and $E_{s,\text{PAM}}$, respectively. Then, the power allocation of ACO-OFDM and PAM-DMT is determined by the power allocation factor

$$\eta_{\text{HACO}} = \frac{\sqrt{E_{s,\text{ACO}}}}{\sqrt{E_{s,\text{ACO}}} + \sqrt{E_{s,\text{PAM}}}}, \tag{4.60}$$

which specifies the proportion of the total optical power allocated to ACO-OFDM.

At the receiver, the detection of HACO-OFDM symbols is also divided into two steps, similar to ADO-OFDM [27]. The ACO-OFDM symbols on odd subcarriers can be detected first by simple FFT. Afterwards, the clipping distortion of ACO-OFDM is removed for the demodulation of the PAM-DMT symbols on even subcarriers. The entire detection procedure is the same as that given in Eqs. (4.51)–(4.54) but with the notation "DCO" replaced by "PAM-DMT" and without the DC bias term. Although the receiver of [27] is simple and straightforward, it does not eliminate the interference between ACO-OFDM and PAM-DMT signals thoroughly, which limits its performance. To further improve the performance of HACO-OFDM, an iterative receiver was proposed in [28], which reduces the interference between ACO-OFDM and PAM-DMT signals in the time domain by pairwise clipping.

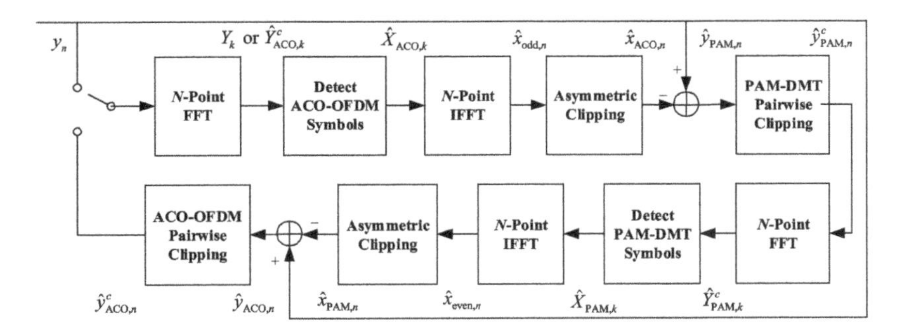

Figure 4.10 Block diagram of the iterative receiver for HACO-OFDM.

The block diagram of the iterative receiver for HACO-OFDM is illustrated in Fig. 4.10. In each iteration, the ACO-OFDM symbols on odd subcarriers are first detected using FFT. Then, the ACO-OFDM time-domain signal is regenerated and subtracted from the received signal. The residual signal is considered as the interfered PAM-DMT signal, and a pairwise clipping is applied to reduce the effect of noise and estimation error. The pairwise-clipped PAM-DMT signal is fed to the FFT module to detect the PAM-DMT symbols, and the resulting time-domain PAM-DMT signal is estimated. Afterwards, the estimated time-domain PAM-DMT signal is subtracted from the received signal and the pairwise clipping is employed again to acquire a more accurate ACO-OFDM signal. The receiver employs the received signal to detect the ACO-OFDM symbols in the first iteration but the pairwise-clipped ACO-OFDM signal is used in the subsequent iterations. We now detail the design of this iterative receiver.

First iteration. FFT is performed on the received signal directly and the detection of ACO-OFDM symbols is the same as (4.52). Let the resulting ML estimation be denoted by $\widehat{X}_{\text{ACO},k}$ for $k = 1, 3, \ldots, N/2 - 1$. The estimated time-domain ACO-OFDM samples $\widehat{x}_{\text{ACO},n}$, $n = 0, 1, \ldots, N-1$, are regenerated with the aid of IFFT and (4.45).

Unlike the conventional receiver where the subtraction takes place in the frequency domain with the estimated ACO-OFDM symbols, the estimated time-domain ACO-OFDM sample is subtracted from the received time-domain signal y_n, resulting in the interfered time-domain PAM-DMT signal as

$$\widehat{y}_{\text{PAM},n} = y_n - \widehat{x}_{\text{ACO},n},\ n = 0, 1, \ldots, N-1. \tag{4.61}$$

By considering the ACO-OFDM estimation error $i_{\text{ACO,est},n} = x_{\text{ACO},n} - \widehat{x}_{\text{ACO},n}$, which follows a Gaussian distribution according to the central limit theorem, and the receiver noise w_n, the above equation can be rewritten as

$$\widehat{y}_{\text{PAM},n} = x_{\text{PAM},n} + w_n + i_{\text{ACO,est},n},\ n = 0, 1, \ldots, N-1. \tag{4.62}$$

For PAM-DMT, half of the time-domain samples are set to zero by asymmetrical clipping, and the remaining samples are non-negative. Therefore, for the pair of transmitted samples $\{x_{\text{PAM},n}, x_{\text{PAM},N-n}\}$, where $n = 0, 1, \ldots, N/2 - 1$, one of them has to be zero. Similar to the algorithm proposed for ACO-OFDM in [10], the pairwise ML detector can be used for PAM-DMT. Specifically, for $n = 1, \ldots, N/2 - 1$, the following pairwise clipping is used

$$\widehat{y}^{c}_{\text{PAM},n} = \widehat{y}_{\text{PAM},n} I_{\{\widehat{y}_{\text{PAM},N-n} \le \widehat{y}_{\text{PAM},n}\}}, \tag{4.63}$$

$$\widehat{y}^{c}_{\text{PAM},N-n} = \widehat{y}_{\text{PAM},N-n} I_{\{\widehat{y}_{\text{PAM},N-n} > \widehat{y}_{\text{PAM},n}\}}, \tag{4.64}$$

where $I_{\{A\}}$ is the indicator function defined by: $I_{\{A\}} = 1$ if the event A is true, and $I_{\{A\}} = 0$ otherwise. If the estimation of zero-valued samples is sufficient accurate, which is the case under high SNR scenarios, half of the noise and ACO-OFDM estimation error are eliminated, and this achieves a considerable performance gain. Besides that, as $x_{\text{PAM},0} = x_{\text{PAM},N/2} = 0$, $\widehat{y}^{c}_{\text{PAM},0}$ and $\widehat{y}^{c}_{\text{PAM},N/2}$ are also set to zero to further reduce the noise and estimation error.

The pairwise-clipped PAM-DMT samples $\widehat{y}^{c}_{\text{PAM},n}$ are fed to the FFT block to obtain the frequency-domain symbols $\widehat{Y}^{c}_{\text{PAM},k}$, and the PAM-DMT symbols on even subcarriers are estimated according to

$$\widehat{X}_{\text{PAM},k} = \arg \min_{X \in \mathcal{X}_{\text{Re}}} \left\| X - 2\text{Im}\left(\widehat{Y}^{c}_{\text{PAM},k}\right) \right\|,\ k = 2, 4, \ldots, N/2 - 2. \tag{4.65}$$

$\widehat{X}_{\text{PAM},k}$ of (4.65) is more accurate since part of the noise and estimation error are eliminated. It is used to further improve the estimation accuracy of the ACO-OFDM symbols in an iterative way, where the time-domain PAM-DMT samples are also regenerated and subtracted from the received samples to yield

$$\widehat{y}_{\text{ACO},n} = y_n - \widehat{x}_{\text{PAM},n},\ n = 0, 1,\ ...,\ N-1, \tag{4.66}$$

where $\widehat{y}_{\text{ACO},n}$ contains both the receiver noise w_n and the PAM-DMT estimation error $i_{\text{PAM},n} = x_{\text{PAM},n} - \widehat{x}_{\text{PAM},n}$. Thus, the above equation can be rewritten as

$$\widehat{y}_{\text{ACO},n} = x_{\text{ACO},n} + w_n + i_{\text{PAM},n},\ n = 0, 1,\ ...,\ N-1. \tag{4.67}$$

As with the ACO-OFDM scheme, $i_{\text{PAM},n}$ also follows a Gaussian distribution. Similar to the PAM-DMT signal, a pairwise clipping could also be employed to the ACO-OFDM samples to reduce the noise and estimation error. Due to the symmetry structure of ACO-OFDM, for $n = 0, 1,\ ...,\ N/2-1$, one has

$$\widehat{y}^{c}_{\text{ACO},n} = \widehat{y}_{\text{ACO},n} I_{\{\widehat{y}_{\text{ACO},n+N/2} \leq \widehat{y}_{\text{ACO},n}\}}, \tag{4.68}$$

$$\widehat{y}^{c}_{\text{ACO},n+N/2} = \widehat{y}_{\text{ACO},n+N/2} I_{\{\widehat{y}_{\text{ACO},n+N/2} > \widehat{y}_{\text{ACO},n}\}}. \tag{4.69}$$

Second and subsequent iterations. The pairwise-clipped ACO-OFDM samples $\widehat{y}^{c}_{\text{ACO},n}$ are used for detection. After obtaining the frequency-domain symbols $\widehat{Y}^{c}_{\text{ACO},k}$ by FFT, the ACO-OFDM symbols on odd subcarriers are estimated according to

$$\widehat{X}_{\text{ACO},k} = \arg\min_{X \in \mathcal{X}} \left\| X - 2\widehat{Y}^{c}_{\text{ACO},k} \right\|,\ k = 1, 3,\ ...,\ N/2-1. \tag{4.70}$$

The iteration continues by conducting the estimation of PAM-DMT symbols, until the preset maximum number of iterations is reached.

Compared to the conventional counterpart given in [27], this iterative receiver requires one more IFFT and two pairwise-clipping modules. The IFFT module has low complexity since it is usually integrated in hardware, while the pairwise clipping can be realized by simple comparison operations. Therefore, this iterative receiver only increases the complexity slightly.

The BER performance of the iterative receiver is evaluated by simulations, in comparison to the conventional receiver of [27]. Two different HACO-OFDM systems are used: (Case 1) ACO-OFDM-4QAM and PAM-DMT-4PAM, and (Case 2) ACO-OFDM-16QAM and PAM-DMT-16PAM. The size of FFT/IFFT in the transmitter is set to 512 for both ACO-OFDM and PAM-DMT. The iteration number of our proposed iterative receiver is set to 2.

To guarantee that ACO-OFDM and PAM-DMT signals have similar BER performance at the same $E_{\text{b,elec}}/N_0$, unequal power allocation is considered. By setting the target BER to $P_b = 10^{-3}$, the corresponding power allocation factors are found to be 0.3942 and 0.2650 for Case 1 and Case 2, respectively. The BER performances are shown in Figs. 4.11 and 4.12 for those two cases, respectively. It can be seen that the required $E_{\text{b,elec}}/N_0$ values for ACO-OFDM and PAM-DMT are very close at the

BER level of 10^{-3} for both cases, which is consistent with the theoretical analysis. It can also be seen that the iterative receiver outperforms its conventional counterpart. Specifically, at the BER of 10^{-3}, the iterative receiver achieves 1.56 dB and 1.91 dB gains for the ACO-OFDM and PAM-DMT components in Case 1, while the performance gains become 2.05 dB and 2.62 dB in Case 2. At the BER level of 10^{-4}, the iterative receiver achieves 1.78 dB and 2.00 dB gains for the ACO-OFDM and PAM-DMT signals in Case 1, while the performance gains are 2.25 dB and 2.66 dB, respectively, in Case 2.

4.3.2 Enhanced U-OFDM

Hybrid optical OFDM schemes combine different optical OFDM schemes to utilize all the available subcarriers, which improves the spectral efficiency of ACO-OFDM. However, DC bias is still required in ADO-OFDM, which degrades the power efficiency of its ACO-OFDM component. In HACO-OFDM, only the imaginary parts of even subcarriers are employed for PAM-DMT modulation, which wastes a quarter

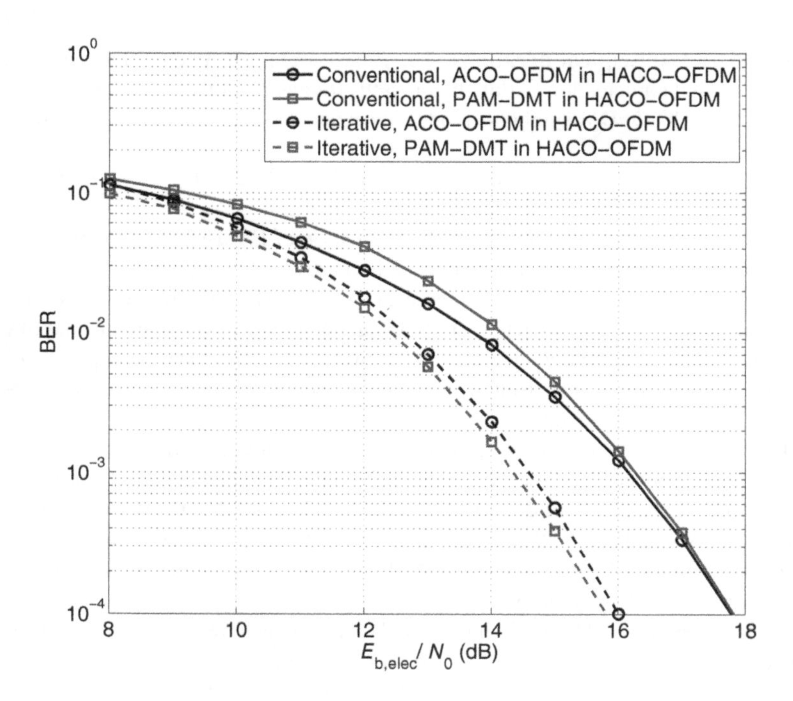

Figure 4.11 BER performance comparison of the conventional and iterative receivers for HACO-OFDM (Case 1 with power allocation factor $\eta = 0.3942$).

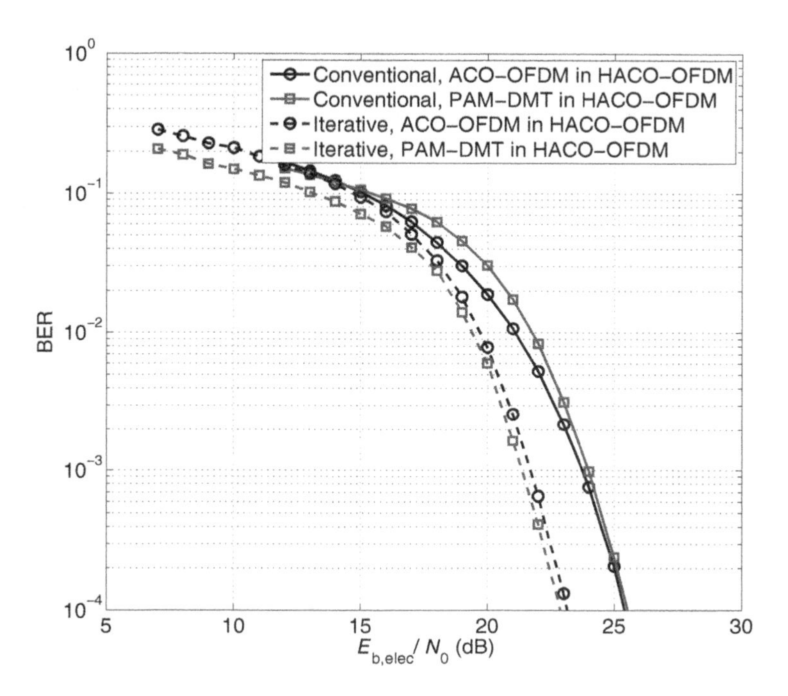

Figure 4.12 BER performance comparison of the conventional and iterative receivers for HACO-OFDM (Case 2 with power allocation factor $\eta = 0.2650$).

of the spectral resource. To avoid spectral efficiency loss, an enhanced U-OFDM (eU-OFDM) was proposed in [30, 31].

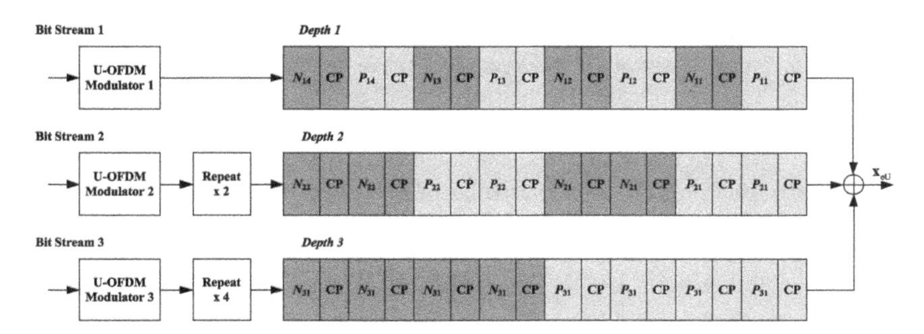

Figure 4.13 eU-OFDM transmitter with three modulation depths.

The transmitter of eU-OFDM is illustrated in Fig. 4.13, which combines different depths of U-OFDM streams for simultaneous transmission. Each depth employs the conventional U-OFDM with different repetitions, where the sub-frame corresponding to the positive signals is labeled with P and the reversed negative sub-frame is labeled with N. In the subscript, the first number denotes the depth index, while the second number represents the frame index. In *Depth 1*, the transmitted signal is exactly the same as the conventional U-OFDM, as described in Section 4.1.3. CP is added before each negative and positive frames to avoid ISI. In *Depth 2*, since the signals are superimposed on those of *Depth 1*, they should not interfere with the recovery of signals in *Depth 1*. In order to achieve this goal, the U-OFDM stream is repeated twice for both the positive and negative frames, as shown in Fig. 4.13. Since the demodulation of *Depth 1* U-OFDM is carried out by subtracting the negative frame from the positive frame, the *Depth 2* signals will not interfere with this demodulation as long as they are periodic within the U-OFDM frame in *Depth 1*. Therefore, in *Depth 2*, each U-OFDM frame is divided into two sub-frames for repetition. First, the positive sub-frame is repeated twice, which corresponds to the first frame of U-OFDM in *Depth 1*. Subsequently, the negative sub-frame is repeated twice corresponding to the second frame of U-OFDM in *Depth 1*. Since each U-OFDM signal is transmitted twice in *Depth 2*, the signals should be scaled by $\sqrt{1/2}$ to maintain the same energy per bit as in *Depth 1*.

Similarly, another stream can be added for simultaneous transmission, referred to as *Depth 3*. In *Depth 3*, the U-OFDM signals should be periodic within the U-OFDM frame in *Depth 2*, so that they can be eliminated at the demodulation of *Depth 2* signals. Specifically, the positive and negative sub-frames of U-OFDM are transmitted four times in *Depth 3*, respectively. In this way, the *Depth 3* signals are also periodic within the U-OFDM frame in *Depth 1*, which will not interfere with the demodulation of *Depth 1* either. In order to keep the energy per bit constant at all streams, the signals in *Depth 3* should be scaled by $\sqrt{1/4}$ since all the signals are transmitted for four times. Analogously, additional streams can be added for simultaneous transmission. For *Depth d*, the positive and negative sub-frames of U-OFDM are transmitted 2^{d-1} times and the signals are scaled by $\sqrt{2^{1-d}}$, so that they will not interfere with the demodulation of *Depth 1* to *Depth d−1*.

At the receiver, the demodulation of different depths is conducted in a serial manner. *Depth 1* can be simply demodulated as conventional U-OFDM as described in Section 4.1.3. Specifically, the negative sub-frame is subtracted from the positive sub-frame. For example, the demodulation of the first frame utilizes the FFT of $P_{11} - N_{11}$. Due to the special design of eU-OFDM, the interference from all the other depths for P_{11} is exactly the same as that for N_{11}, which can be completely eliminated by the subtraction operation. Therefore, the information bits in *Depth 1* can be successfully demodulated by the conventional U-OFDM receiver. After the demodulation of *Depth 1*, the information bits are used to regenerate the U-OFDM signals as an estimation of time-domain *Depth 1* signals, which are then subtracted from the received signals for the demodulation of other depths. The resultant signals contain the signals of *Depth 2* and subsequent depths, as well as the receiver noise and the estimation error of *Depth 1*. For the demodulation of *Depth 2*, since all the signals are transmitted twice, they need to be summed to obtain diversity gain. For example, the first two positive sub-frames corresponding to the time slot of P_{11} and

N_{11} are summed to get P_{21}, and the third and the fourth sub-frames corresponding to the time slot of P_{12} and N_{12} are summed to obtain N_{21}. Afterwards, $P_{21} - N_{21}$ is used for the demodulation of *Depth 2* as conventional U-OFDM demodulation. After demodulation of *Depth 2*, the information bits are then used for the estimation of the time-domain *Depth 2* signals, which are removed from the subtracted received signals of *Depth 1* to demodulate other depths in a similar way. It can be seen that in the demodulation of each depth, the interference from the subsequent depths will not interfere with the recovery due to the special structure of eU-OFDM. After the information bits are recovered at each depth, they are used to regenerate the U-OFDM signals as an estimation of the corresponding time-domain signals, which are then subtracted from the residual received signals for the demodulation of the subsequent depths.

The eU-OFDM scheme achieves higher spectral efficiency compared with conventional U-OFDM since more streams are transmitted at the same time. When D depths are used, the spectral efficiency of eU-OFDM can be calculated by summing the spectral efficiencies of all the individual streams as

$$\xi_{\mathrm{eU}}(D) = \sum_{d=1}^{D} \frac{\xi_{\mathrm{U}}}{2^{d-1}} = 2\xi_{\mathrm{U}} \left(1 - \left(\frac{1}{2}\right)^{D}\right), \tag{4.71}$$

where ξ_{U} denotes the spectral efficiency of *Depth 1* U-OFDM stream. It can be seen that the spectral efficiency of eU-OFDM increases with the number of depths used. If D is very high, $\xi_{\mathrm{eU}}(D)$ converges to $2\xi_{\mathrm{U}}$, which is also the same as the spectral efficiency of DCO-OFDM. Since DC bias is not required in eU-OFDM, it is more power efficient than DCO-OFDM.

For practical implementation, eU-OFDM has its disadvantages. The frame length of eU-OFDM with D depths is 2^{D-1} times of conventional U-OFDM, and 2^{D} times of DCO-OFDM. The demodulation of all the D depths of eU-OFDM has to be carried out in depth-by-depth manner and, therefore, the detection latency is very large compared with conventional optical OFDM schemes. In addition, since different U-OFDM streams are required at the transmitter, the complexity is also increased.

4.3.3 Layered ACO-OFDM

Layered ACO-OFDM (LACO-OFDM) proposed in [32] achieves the same spectral efficiency as eU-OFDM by simultaneously transmitting multiple ACO-OFDM streams, but it reduces the latency since it has a smaller frame length.

In LACO-OFDM, *Layer 1* ACO-OFDM is the same as conventional ACO-OFDM, whose time-domain samples are rewritten as $x_{\mathrm{ACO},n}^{(1)}$ for $n = 0, 1, \ldots, N-1$, where the superscript represents the layer index. For an OFDM block in which only even subcarriers are modulated, its output is defined as

$$\begin{aligned} x_n &= \frac{1}{\sqrt{N}} \sum\nolimits_{k=0}^{N/2-1} X_{2k} \exp\left(j\frac{2\pi n}{N}2k\right) \\ &= \frac{\sqrt{2}}{2}\frac{1}{\sqrt{N/2}} \sum\nolimits_{k=0}^{N/2-1} X_k^{(2)} \exp\left(j\frac{2\pi n}{N/2}k\right) \\ &= \frac{\sqrt{2}}{2} x^{(2)}{}_{\bmod\ (n,N/2)},\ n = 0, 1,\ ..., N-1, \end{aligned} \tag{4.72}$$

where $X_k^{(2)} = X_{2k}$, $\left\{x_n^{(2)}\right\}$ denotes the $N/2$-point IFFT of $\left\{X_k^{(2)}\right\}$, and $\bmod\ (\cdot, N)$ represents the modulo N operator. It can be seen that x_n is periodic and can be obtained by repeating the signal $x_n^{(2)}$ twice. If only the subcarriers with odd indices of $\left\{X_k^{(2)}\right\}$, i.e., $X_{2(2k+1)}$ for $k = 0, 1,\ ..., N/4-1$, are used for modulation, the time-domain signal $x_n^{(2)}$, $n = 0, 1,\ ..., N/2-1$, follows a half-length symmetry as

$$x_n^{(2)} = -x_{n+N/4}^{(2)},\ n = 0, 1,\ ..., N/4-1, \tag{4.73}$$

which can be clipped at zero without information loss. After asymmetric clipping, another ACO-OFDM stream, *Layer 2* ACO-OFDM, can be obtained, which is denoted as $x_{\mathrm{ACO},n}^{(2)}$, where $x_{\mathrm{ACO},n}^{(2)} = x_{\mathrm{ACO},\ \bmod\ (n,N/2)}^{(2)}$, $n = 0, 1,\ ..., N-1$. Similar to the conventional ACO-OFDM, the negative clipping distortion of $x_{\mathrm{ACO},n}^{(2)}$ falls on the subcarriers with even indices of $X_k^{(2)}$, $k = 0, 1,\ ..., N/2-1$, corresponding to the $4k$th ($k = 0, 1,\ ..., N/4-1$) subcarriers in the conventional ACO-OFDM. Therefore, both the signal and clipping distortion of *Layer 2* ACO-OFDM only fall on the even subcarriers and do not contaminate the *Layer 1* ACO-OFDM signal. By transmitting *Layer 1* and *Layer 2* ACO-OFDM signals simultaneously, $3/4$ of all the subcarriers are utilized, which improves the spectral efficiency of conventional ACO-OFDM by 50%. In *Layer 2* ACO-OFDM, half of the even subcarriers, i.e., $N/4$ subcarriers, are modulated. The $4k$th ($k = 0, 1,\ ..., N/4-1$) subcarriers remain unoccupied, which can be used to further improve the spectral efficiency.

In general, *Layer l* ACO-OFDM for $1 \le l < \log_2 N$ is defined as follows. For an OFDM block in which only the $2^{l-1}k$th ($k = 0, 1,\ ..., N/2^{l-1}-1$) subcarriers are modulated,

$$\begin{aligned} x_n &= \frac{1}{\sqrt{N}} \sum\nolimits_{k=0}^{N/2^{l-1}-1} X_{2^{l-1}k} \exp\left(j\frac{2\pi n}{N}2^{l-1}k\right) \\ &= \frac{1}{\sqrt{2^{l-1}}}\frac{1}{\sqrt{N/2^{l-1}}} \sum\nolimits_{k=0}^{N/2^{l-1}-1} X_k^{(l)} \exp\left(j\frac{2\pi n}{N/2^{l-1}}k\right) \\ &= \frac{1}{\sqrt{2^{l-1}}} x^{(l)}{}_{\bmod\ (n,N/2^{l-1})},\ n = 0, 1,\ ..., N-1, \end{aligned} \tag{4.74}$$

where $X_k^{(l)} = X_{2^{l-1}k}$ and $\left\{x_n^{(l)}\right\}$ denotes the $N/2^{l-1}$-point IFFT of $\left\{X_k^{(l)}\right\}$. It can be seen that x_n is also periodic and can be obtained by repeating the $N/2^{l-1}$-

length signal $x_n^{(l)}$. The *Layer l* ACO-OFDM signal $x_{\mathrm{ACO},n}^{(l)}$, where $x_{\mathrm{ACO},n}^{(l)} = x_{\mathrm{ACO},\ \mathrm{mod}\ (n,N/2^{l-1})}^{(l)}$, $n = 0, 1,\ \ldots, N-1$, can be generated by modulating the subcarriers with odd-indexed $X_k^{(l)}$ and $N/2^l$ subcarriers are utilized. Both the signal and clipping distortion of $x_{\mathrm{ACO},n}^{(l)}$ only fall on the subcarriers of $X_k^{(l)}$, i.e., the $2^{l-1}k$th ($k = 0, 1,\ \ldots, N/2^{l-1} - 1$) subcarriers of the original ACO-OFDM signal. Therefore, different layers of ACO-OFDM signals can be generated, and the signal and clipping distortion of the *Layer l* ACO-OFDM will not interfere with the useful symbols in *Layer 1* to *Layer* $(l-1)$ ACO-OFDM. The examples of *Layer 1*, *Layer 2*, and *Layer 3* ACO-OFDM signals are illustrated in Fig. 4.14, where 16 subcarriers are utilized for modulation.

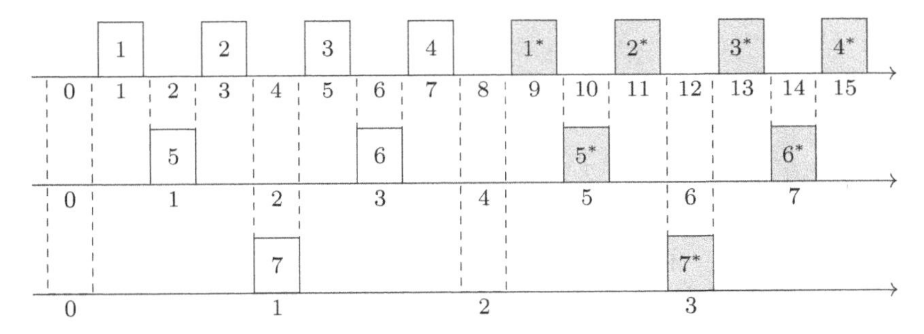

Figure 4.14 Examples of *Layer 1*, *Layer 2*, and *Layer 3* ACO-OFDM signals with 16 subcarriers.

Thus, in the LACO-OFDM scheme, different layers of ACO-OFDM can be combined in the time domain for simultaneous transmission. The time-domain LACO-OFDM signal with L layers is written as

$$x_{\mathrm{LACO},n} = \sum\nolimits_{l=1}^{L} x_{\mathrm{ACO},n}^{(l)} = \sum\nolimits_{l=1}^{L} \left(x_n^{(l)} + i_n^{(l)}\right), n = 0, 1,\ \ldots, N-1, \quad (4.75)$$

where $x_n^{(l)}$ and $i_n^{(l)}$ denote the unclipped signal and the negative clipping distortion of *Layer l* ACO-OFDM, respectively. The number of occupied subcarriers in the LACO-OFDM with L layers can be calculated by

$$N_{\mathrm{LACO}} = \sum\nolimits_{l=1}^{L} N/2^l = (1 - 1/2^L)N, \quad (4.76)$$

which is $(2 - 1/2^{L-1})$ times of conventional ACO-OFDM. When the maximum number of layers $L = \log_2 N - 1$ is used, the spectral efficiency of LACO-OFDM is $(2 - 4/N)$ times of conventional ACO-OFDM, which converges to 2 when N is large.

The block diagram of LACO-OFDM transmitter is summarized in Fig. 4.15, where for simplicity the modulator, Hermitian symmetry, CP insertion and P/S operations are omitted. After bit-to-symbol modulation, the modulated symbol stream is first

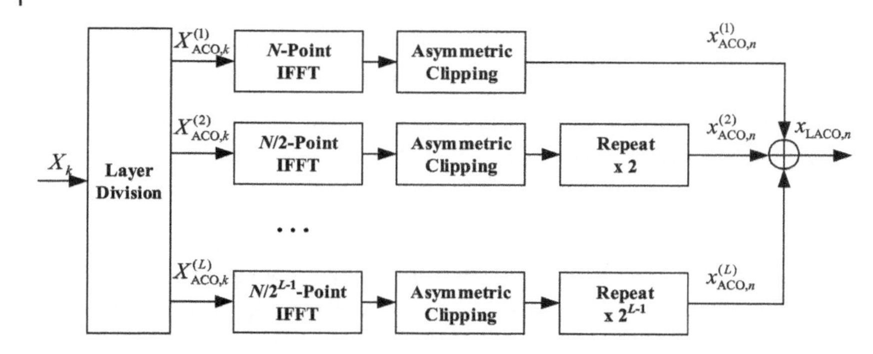

Figure 4.15 Block diagram of LACO-OFDM transmitter with L layers.

divided into L streams for layered transmission. In the *Layer l* ACO-OFDM, only the $2^{l-1}(2k+1)$th ($k = 0, 1, ..., N/2^l - 1$) subcarriers are modulated, which are orthogonal to all the other layers. Due to the periodicity of the time-domain signals as shown in (4.74), the *Layer l* ACO-OFDM can be simply implemented by $N/2^{l-1}$-point IFFT and the $N/2^{l-1}$-length signal is then repeated 2^{l-1} times to obtain the N-length OFDM signal block. Afterwards, the time-domain signals from L layers of ACO-OFDM are combined for simultaneous transmission. Since asymmetric clipping is performed in all the L layers, the combined LACO-OFDM signals are also non-negative and no DC bias is required for obtaining unipolar time-domain signals.

At the receiver, the shot noise and thermal noise are usually modeled by an AWGN, and the received samples are given by $y_n = x_{\text{LACO},n} + w_n$, $n = 0, 1, ..., N-1$, where w_n denote the samples of the AWGN. The received signal samples are fed to the FFT block to generate the frequency-domain symbols as $Y_k = X_{\text{LACO},k} + W_k$ for $k = 0, 1, ..., N-1$, where W_k denotes the frequency-domain representation of the AWGN.

In LACO-OFDM, different layers utilize different subcarriers, which are orthogonal in the frequency domain. However, due to the asymmetric clipping operation at the transmitter, the negative clipping distortion falls on the even subcarriers in each layer, which distorts the higher layers. For *Layer l* ACO-OFDM, the negative clipping distortion falls on the $2^{l-1}(2k)$th ($k = 0, 1, ..., N/2^l - 1$) subcarriers, where the *Layer* $(l+1)$ ACO-OFDM stream is modulated. Therefore, the symbols in *Layer* $(l+1)$ can be recovered only after the negative clipping distortions from the *Layer 1* to *Layer l* have been removed.

For *Layer l* ACO-OFDM, denote the received frequency-domain symbol and negative clipping distortion by $\widehat{Y}_{\text{ACO},k}^{(l)}$ and $\widehat{I}_{\text{ACO},k}^{(l)}$, where $k = 0, 1, ..., N/2^{l-1} - 1$. For *Layer 1* ACO-OFDM, the transmitted symbols can be directly detected by using the odd subcarriers of Y_k according to

$$\widehat{X}_{\text{ACO},k}^{(1)} = \arg\min_{X \in \mathcal{X}} \left| X - 2\widehat{Y}_{\text{ACO},k}^{(1)} \right|, \ k = 1, 3, ..., N/2 - 1, \tag{4.77}$$

where $\widehat{Y}^{(1)}_{\text{ACO},k} = Y_k$ and the factor 2 is due to the fact that the clipping operation reduces the power of ACO-OFDM symbols in the odd subcarriers by half. The estimated time-domain *Layer 1* ACO-OFDM samples $\widehat{x}^{(1)}_{\text{ACO},n}$ can be regenerated from $\widehat{X}^{(1)}_{\text{ACO},k}$, and the frequency-domain negative clipping distortion $\widehat{I}^{(1)}_{\text{ACO},k}$ of *Layer 1* ACO-OFDM can also be estimated by performing the FFT on $\widehat{x}^{(1)}_{\text{ACO},n}$. After subtracting $\widehat{I}^{(1)}_{\text{ACO},k}$ from the received frequency-domain symbols on the even subcarriers, the *Layer 2* ACO-OFDM symbols can be detected.

For *Layer l* ($l > 1$) ACO-OFDM, after removing the negative clipping distortions from *Layer 1* to *Layer* $(l-1)$ ACO-OFDM, the received frequency-domain symbols can be expressed as

$$\begin{aligned}\widehat{Y}^{(l)}_{\text{ACO},k} &= Y_{2^{l-1}k} - \sum\nolimits_{m=1}^{l-1} \widehat{I}^{(m)}_{\text{ACO},2^{l-m}k} \\ &= \widehat{Y}^{(l-1)}_{\text{ACO},2k} - \widehat{I}^{(l-1)}_{\text{ACO},2k},\ k = 0, 1,\ ...,\ N/2^{l-1} - 1,\end{aligned} \tag{4.78}$$

where only one subtraction operation per symbol is required. Similar to (4.77), the transmitted symbols in *Layer l* ACO-OFDM are detected according to

$$\widehat{X}^{(l)}_{\text{ACO},k} = \arg \min_{X \in \mathcal{X}} \left| X - 2\widehat{Y}^{(l)}_{\text{ACO},k} \right|,\ k = 1, 3,\ ...,\ N/2^{l-1} - 1. \tag{4.79}$$

If $l = L$, the detection is completed. Otherwise, $\widehat{I}^{(l)}_{\text{ACO},k}$ and hence $\widehat{Y}^{(l+1)}_{\text{ACO},k}$ are generated to detect the transmitted symbols in *Layer* $l+1$ ACO-OFDM.

The spectral efficiencies of the LACO-OFDM and the conventional ACO-OFDM are compared in Fig. 4.16, where the length of CP is omitted from the calculation of spectral efficiency. It can be seen that the LACO-OFDM scheme improves the spectral efficiency significantly, compared to the ACO-OFDM of the same modulation order. The spectral efficiency of LACO-OFDM improves when the layer number increases and it converges to twice of conventional ACO-OFDM when the layer number is sufficiently large. Even with a small number of layers such as $L = 4$, the 87.5% spectral efficiency improvement is still considerable. When LACO-OFDM with two and three layers modulated by 16QAM are used, they achieve the same spectral efficiencies as ACO-OFDM with 64QAM and 128QAM, respectively. Therefore, lower modulation orders can be adopted in LACO-OFDM to achieve the same spectral efficiency as ACO-OFDM, which has the advantage of lower SNR requirement at the receiver.

To guarantee that the information bits in different layers have similar performance, the modulation scheme and average power of modulated subcarriers should be the same in each layer. In *Layer 1* ACO-OFDM, the symbols are directly detected with the received signals and they are only distorted by the noise at the receiver. In other layers, however, the symbols are also distorted by the estimation error of the negative clipping distortion in previous layers, which could degrade their performance. Fortunately, when the SNR increases, the estimation of negative clipping distortion becomes more accurate and the performance of the other layers converges to that of *Layer 1*, which has been verified by simulations.

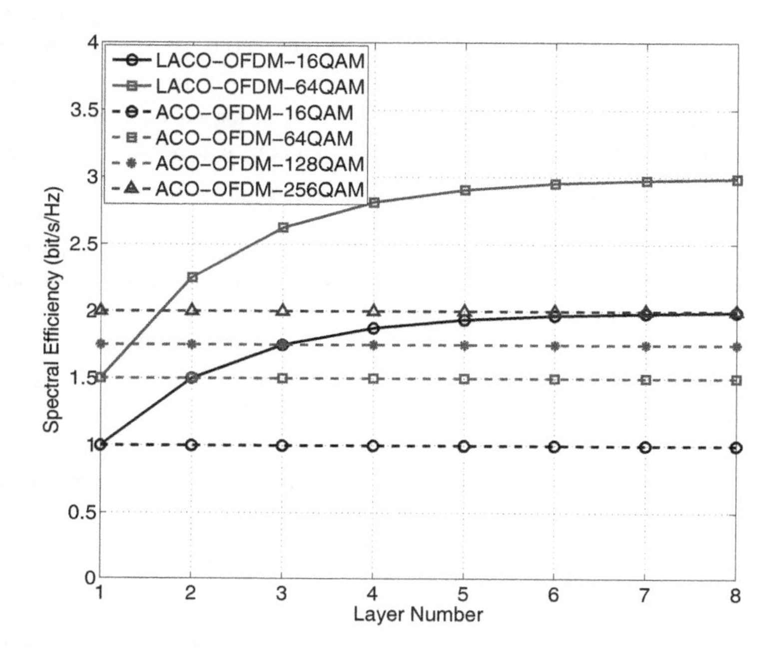

Figure 4.16 Spectral efficiency comparison between LACO-OFDM and ACO-OFDM.

The complexity of an LACO-OFDM transceiver is analyzed in two parts. At the transmitter, only one N-point IFFT is required in conventional ACO-OFDM and its complexity can be written as $\mathcal{O}(N \log_2(N))$. In LACO-OFDM with L layers, however, L different sizes of IFFT blocks are employed, and the total computational complexity increases to $\sum_{l=1}^{L} \mathcal{O}(N/2^{l-1} \log_2(N/2^{l-1})) \approx (2 - 1/2^{L-1})\, \mathcal{O}(N \log_2(N))$. Therefore, the computational complexity of LACO-OFDM transmitter is less than twice of the conventional ACO-OFDM transmitter and this complexity increase is the same as the spectral efficiency improvement. Since the L layers are calculated at the same time, there is no more latency compared with conventional ACO-OFDM. At the receiver, the computational complexity of conventional ACO-OFDM is also $\mathcal{O}(N \log_2(N))$ since only one N-point FFT is utilized. In LACO-OFDM with L layers, the computational complexity is $\mathcal{O}(N \log_2(N)) + 2 \sum_{l=1}^{L-1} \mathcal{O}(N/2^{l-1} \log_2(N/2^{l-1})) \approx (5 - 1/2^{L-3})\, \mathcal{O}(N \log_2(N))$, which is less than five times the conventional ACO-OFDM receiver. In hardware implementation, the complexity of LACO-OFDM receiver is only twice the conventional one since the N-point FFT/IFFT block can be reused. With the assumption that N is the power of 2, for the sequence

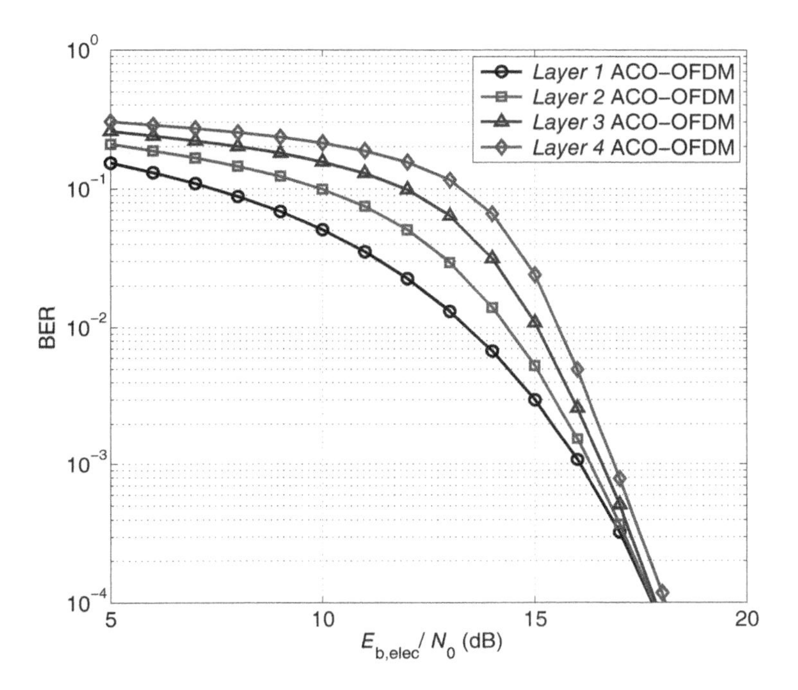

Figure 4.17 BER performance of LACO-OFDM with 16QAM and four layers.

with length $N/2^l$, an N-length sequence can be obtained by zero-padding and the same N-point FFT/IFFT block can be shared. The serial structure also leads to extra latency in the LACO-OFDM receiver. Fortunately, when the number of layers increases, the sizes of FFT/IFFT used in the iterative receiver decrease exponentially and the latency of the iterative receiver is also $(5 - 1/2^{L-3})$ times the conventional ACO-OFDM receiver, which is acceptable considering the improvement of spectral efficiency. Compared with eU-OFDM, LACO-OFDM has much smaller frame size, which significantly reduces the latency.

The BER performance of the LACO-OFDM is evaluated by simulations. The number of IFFT points used at the transmitter is 512, and the subcarriers are modulated by16QAM in all layers. Figure 4.17 shows the BER performance of LACO-OFDM with four layers, where the BER in each layer of LACO-OFDM is calculated separately. For the same $E_{\text{b,elec}}/N_0$ level, the BER increases with the layer number since the symbols in higher layers are distorted by the estimation error of the lower layers. However, as shown in Fig. 4.17, when $E_{\text{b,elec}}/N_0$ increases, the estimation accuracy improves with the reduction of BER in higher layers, and the four BER curves converge to one, which is also the working point for the practical system. This result also matches the performance analysis.

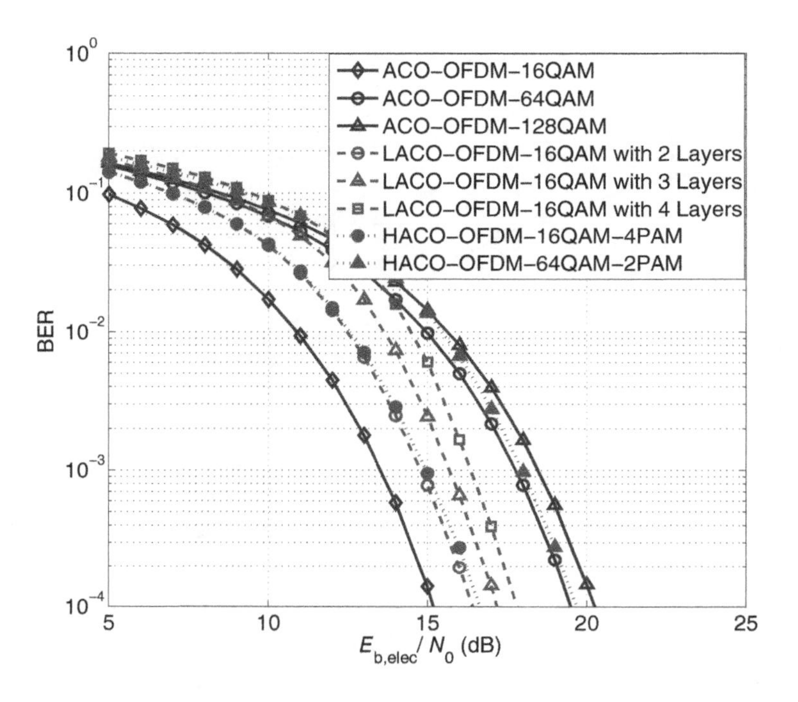

Figure 4.18 BER performance comparison of LACO-OFDM, ACO-OFDM, and HACO-OFDM.

The average BER of LACO-OFDM is compared with that of conventional ACO-OFDM and HACO-OFDM in Fig. 4.18, where BERs of all layers in LACO-OFDM are averaged. In LACO-OFDM, the subcarriers are modulated by 16QAM in all layers. The numbers of layers are set to 2, 3, and 4, and the spectral efficiencies are 1.5 bit/s/Hz, 1.75 bit/s/Hz, and 1.875 bit/s/Hz, respectively. Three different kinds of ACO-OFDM are used for comparison, where the odd subcarriers are modulated by 16QAM, 64QAM, and 128QAM, corresponding to the spectral efficiencies of 1 bit/s/Hz, 1.5 bit/s/Hz, and 1.75 bit/s/Hz, respectively. In HACO-OFDM, the odd subcarriers are modulated by ACO-OFDM with 16QAM and 64QAM and the even subcarriers are modulated by PAM-DMT with 4PAM and 2PAM, corresponding to the spectral efficiencies of 1.5 bit/s/Hz and 1.75 bit/s/Hz. The optimal proportion of optical power for HACO-OFDM is obtained by the method in [27], and the proportions of optical power allocated to ACO-OFDM block are 0.6 and 0.85 for two different HACO-OFDM schemes.

When the same modulation orders are used, the performance of LACO-OFDM is worse than the conventional ACO-OFDM, which is due to the fact that the power of ACO-OFDM in each layer is smaller than that of the conventional ACO-OFDM since it is distributed to different layers. For example, at the BER of 10^{-4}, the performance degradations of LACO-OFDM-16QAM with 2, 3, and 4 layers are 1.2 dB, 2.0 dB, and 2.6 dB compared with the conventional ACO-OFDM-16QAM, but the spectral efficiency is improved by 50%, 75%, and 87.5% as shown in Fig. 4.16.

When the three schemes with the same spectral efficiency are compared, it can be seen that LACO-OFDM significantly outperforms the conventional ACO-OFDM, and it is also better than HACO-OFDM. The LACO-OFDM-16QAM with two layers achieves 3.1 dB gain compared with the ACO-OFDM-64QAM with the same spectral efficiency of 1.5 bit/s/Hz, and it is also slightly better than HACO-OFDM with the same spectral efficiency. For the LACO-OFDM-16QAM with three layers, it also outperforms ACO-OFDM-128QAM by 3.1 dB with the same spectral efficiency of 1.75 bit/s/Hz. Even compared with HACO-OFDM with the same spectral efficiency, the performance gain is still 1.77 dB. When LACO-OFDM-16QAM with four layers is employed, which has 0.125 bit/s/Hz more spectral efficiency compared with ACO-OFDM-128QAM, the performance gain is 2.4 dB.

In order to fully exploit the structure of LACO-OFDM signals, an improved receiver was proposed in [33] to further enhance the performance of LACO-OFDM, where different layers of ACO-OFDM signals are distinguished in the time domain and pairwise clipping is utilized in each layer.

The signals in different layers of LACO-OFDM are first separated in the time domain. The symbols on *Layer* 1 ACO-OFDM are detected after FFT as in (4.77). The time-domain signals in *Layer* 1 ACO-OFDM are reconstructed by

$$\hat{y}_n^{(1)} = \sum_{k=0}^{N-1} \hat{X}_{\mathrm{ACO},k}^{(1)} \exp\left(j\frac{2\pi}{N}nk\right),\ n = 0, 1, \ldots, N-1, \tag{4.80}$$

where $\hat{X}_{\mathrm{ACO},k}^{(1)} = \left(\hat{X}_{\mathrm{ACO},N-k}^{(1)}\right)^*$.

Afterwards, the reconstructed signals $\hat{y}_n^{(1)}$ are discarded from the original received signals y_n, where the remaining signals $\tilde{y}_n^{(2)} = y_n - \hat{y}_n^{(1)}$ can be considered as the combination of the signals in *Layer* $2 \sim L$ ACO-OFDM. Since the signals $\hat{y}_n^{(1)}$ containing both the transmitted symbols and clipping distortion in *Layer* 1 ACO-OFDM are already removed, the symbols in *Layer* 2 ACO-OFDM can be directly detected after FFT as well.

When the symbols in *Layer* $1 \sim l-1$ $(l > 1)$ ACO-OFDM are detected, all of their reconstructed signals are subtracted from the original received signals y_n, and one has

$$\tilde{y}_n^{(l)} = y_n - \sum_{m=1}^{l-1} \hat{y}_n^{(m)} = \tilde{y}_n^{(l-1)} - \hat{y}_n^{(l-1)}, \tag{4.81}$$

where $\tilde{y}_n^{(1)} = y_n$ and the reconstructed time-domain signals in *Layer* l ACO-OFDM are given by

$$\hat{y}_n^{(l)} = \sum_{k=0}^{N/2^{l-1}-1} \hat{X}_{\mathrm{ACO},k}^{(l)} \exp\left(j\frac{2\pi}{N}n \cdot 2^{l-1}k\right), \tag{4.82}$$

for $n = 0, 1, ..., N-1$.

The symbols in *Layer* l ACO-OFDM can be directly detected after the FFT of $\tilde{y}_n^{(l)}$. It can be seen that in LACO-OFDM with L layers, $\tilde{y}_n^{(l)}$ contains more than one layer of ACO-OFDM when $l < L$. However, $\tilde{y}_n^{(L)}$ only includes the signals in *Layer* L ACO-OFDM since all the signals from *Layer* $1 \sim L-1$ have been removed. Therefore, the structure of *Layer* L ACO-OFDM can be utilized to further improve the performance.

Denote the received signals with only *Layer* l ACO-OFDM as $\bar{y}_n^{(l)} = x_n^{(l)} + w_n + e_n^{(l)}$, where $e_n^{(l)}$ is the inter-layer interference and it follows a Gaussian distribution according to the central limit theorem, and we have $\bar{y}_n^{(L)} = \tilde{y}_n^{(L)}$. In *Layer* l ACO-OFDM, the transmitted time-domain signals are periodic. Besides, an asymmetric clipping is imposed on the time-domain signals so that either $x_n^{(l)}$ or $x_{n+N/2^l}^{(l)}$ is zero when they are transmitted, and the remaining signals are non-negative. Therefore, one can estimate which signal should be set to zero according to $\bar{y}_n^{(l)}$, and half of the noise and inter-layer interference are eliminated. In *Layer* l ACO-OFDM, considering the periodicity of the signals, the pairwise clipping is modified as

$$\bar{y}_{n,c}^{(l)} = \begin{cases} \bar{y}_n^{(l)} I_{H(n')}, & n' \leq N/2^l, \\ \bar{y}_n^{(l)} \left(1 - I_{H(n'-N/2^l)}\right), & n' > N/2^l, \end{cases} \tag{4.83}$$

where $n' = \mod(n, N/2^{l-1})$ and $I_{\{A\}}$ is an indicator function with $I_{\{A\}} = 1$ if the event A is true and $I_{\{A\}} = 0$ otherwise. $H(n')$ is defined as

$$H(n') : \sum_{m=0}^{2^{l-1}-1} \bar{y}_{n'+mN/2^{l-1}}^{(l)} \geqslant \sum_{m=0}^{2^{l-1}-1} \bar{y}_{n'+N/2^l+mN/2^{l-1}}^{(l)}. \tag{4.84}$$

Pairwise clipping is first utilized in *Layer* L ACO-OFDM signals since $\bar{y}_n^{(L)}$ can be obtained after signals of all the other layers are removed. Afterwards, the pairwise-clipped signal is used to demodulate the symbols in *Layer* L ACO-OFDM, which can achieve better performance compared with the conventional method since half of the noise and inter-layer interference have been eliminated, and the received time-domain signals in *Layer* L ACO-OFDM are reconstructed.

Since the time-domain signals in each layer have been reconstructed, the received signals with only *Layer* l ACO-OFDM can be obtained by subtracting the signals in other layers from the received signals as

$$\bar{y}_n^{(l)} = y_n - \sum_{m \neq l} \hat{y}_n^{(m)}, \tag{4.85}$$

so that different layers of ACO-OFDM signals are distinguished in the time domain and pairwise clipping can be applied to eliminate half of the noise and inter-layer interference.

The symbols on *Layer* l ACO-OFDM are detected again after pairwise clipping and FFT, where more accurate results are obtained, which could be used to update

Algorithm 4.2 Improved Receiver for LACO-OFDM.

Input: Received signals, y_n; Constellation set, $\mathcal{S}$;
Output: Detected symbols, $\hat{X}^{(l)}_{\text{ACO},k}$;

1: $\tilde{y}^{(1)}_n = y_n$;
2: **for** $l = 1 : L-1$ **do**
3: $\quad \tilde{Y}^{(l)}_{\text{ACO},k} = \text{FFT}\left(\tilde{y}^{(l)}_n\right)$;
4: $\quad \hat{X}^{(l)}_{\text{ACO},k} = \arg\min_{X\in\mathcal{S}} |X - 2\tilde{Y}^{(l)}_{\text{ACO},k}|$;
5: $\quad \hat{y}^{(l)}_n = \sum_{k=0}^{N/2^{l-1}-1} \hat{X}^{(l)}_{\text{ACO},k} \exp\left(j\frac{2\pi}{N}n \cdot 2^{l-1}k\right)$;
6: $\quad \tilde{y}^{(l+1)}_n = \tilde{y}^{(l)}_n - \hat{y}^{(l)}_n$;
7: **end for**
8: $\bar{y}^{(L)}_n = \tilde{y}^{(L)}_n$;
9: **for** $i = 1 : N_{\text{iter}}$ **do**
10: $\quad$ **for** $l = L : -1 : 1$ **do**
11: $\quad\quad$ Calculate $\bar{y}^{(l)}_{n,c}$ according to Eqs. (4.83)–(4.84);
12: $\quad\quad \bar{Y}^{(l)}_{\text{ACO},k,c} = \text{FFT}\left(\bar{y}^{(l)}_{n,c}\right)$;
13: $\quad\quad \hat{X}^{(l)}_{\text{ACO},k} = \arg\min_{X\in\mathcal{S}} |X - 2\bar{Y}^{(l)}_{\text{ACO},k,c}|$;
14: $\quad\quad \hat{y}^{(l)}_n = \sum_{k=0}^{N/2^{l-1}-1} \hat{X}^{(l)}_{\text{ACO},k} \exp\left(j\frac{2\pi}{N}n \cdot 2^{l-1}k\right)$;
15: $\quad\quad \bar{y}^{(l')}_n = y_n - \sum_{m\neq l'} \hat{y}^{(m)}_n,\ l' \neq l$;
16: $\quad$ **end for**
17: **end for**
18: **return** $\hat{X}^{(l)}_{\text{ACO},k}$;

the signals $\hat{y}^{(l)}_n$ and back substitute to (4.85) to update the signals in other layers. Therefore, the receiver operates in an iterative way. In each iteration, the signals in each layer are sequentially detected, and the reconstructed time-domain signals are used to update the signals in other layers. The pseudocode of the proposed receiver is summarized in Algorithm 4.2, where N_{iter} denotes the number of iterations.

4.4 Optical OFDM under lighting constraints

In indoor VLC systems, communication and illumination should be maintained simultaneously, where the light may be dimmed to satisfy different illumination and power requirements [34]. Furthermore, dimming technology is energy efficient. However, the dimming operation could interfere with the communication function of VLC systems since it alters the received optical power and SNR. Conventional optical OFDM schemes concentrate mainly on data transmission and could not support various dimming levels efficiently. For IEEE 802.15.7 Standard, single carrier pulsed modulations such as on-off keying (OOK), pulse-position modulation (PPM) and color shift keying (CSK) are utilized, whilst dimming control is usually realized by combining the existing modulation schemes with PAM or pulse width modulation (PWM) [34].

The inherent nonlinearity of LEDs is a challenge for OFDM implementation since OFDM has high PAPR. The input of LED has a minimum threshold value that can generate current, which is referred to as turn-on voltage (TOV). When the input voltage is above the TOV, the voltage–current and current–power characteristics are nonlinear. Several algorithms have been proposed to mitigate the effect of LED nonlinearity, where the transfer characteristic of LED is regarded as quasi-linear in a limited range after predistortion. That is, if v_{min} and v_{max} are the minimum and maximum allowed signals according to the voltage levels permitted by LED, the transfer characteristic of LED between $[v_{\text{min}}, v_{\text{max}}]$ is assumed to be linear. Specifically, the relationship between the emitted optical power and the input voltage is given by

$$P_{\text{opt}}(t) = \begin{cases} 0, & v(t) < v_{\text{min}}, \\ \eta\left(v(t) - v_{\text{min}}\right), & v_{\text{min}} \leq v(t) \leq v_{\text{max}}, \\ \eta\left(v_{\text{max}} - v_{\text{min}}\right), & v(t) > v_{\text{max}}, \end{cases} \tag{4.86}$$

where η and $v(t)$ denote the voltage-power transfer coefficient and instantaneous input voltage, respectively.

Since the illumination level is proportional to the average optical power, dimming control can be achieved by adjusting the average optical power of LEDs. Thus, the dimming level d is defined as

$$d = E\left(P_{\text{opt}}(t)\right) / \left(\eta\left(v_{\text{max}} - v_{\text{min}}\right)\right), \tag{4.87}$$

which falls in the interval $[0, 1]$. When the required dimming level is adjusted, the received optical power and effective SNR are changed, which will vary the achievable data rate for a given BER requirement. Therefore, any modulation scheme should support high data rate under different dimming targets.

The basic idea of dimming control for DCO-OFDM is to adjust the DC bias to fulfill different illumination requirements. Since the time-domain OFDM signals follow Gaussian distribution with zero mean, they are symmetrically distributed on both sides of the DC bias, and the average optical power is in proportion to the DC bias. If the DC bias is not in the middle of the linear range, the signals on one side will suffer from severer clipping. In order to maintain acceptable clipping distortion, the actual dynamic range of the signals is smaller than the linear range of the LEDs. For example, when the DC bias is set to B_{DC}, the actual dynamic range would be $[\max\left(v_{\text{min}}, 2B_{\text{DC}} - v_{\text{max}}\right), \min\left(v_{\text{max}}, 2B_{\text{DC}} - v_{\text{min}}\right)]$. The transmitted signals should be scaled to satisfy this dynamic range and they would become very small when high or low illumination is required, which degrades the performance significantly. Therefore, several optical OFDM schemes have been proposed under lighting constraints, which aim to fully exploit the dynamic range of LEDs and provide stable data transmission under different illumination requirements.

4.4.1 Pulse width modulation

The PWM signal is periodic, whose pulse width can be adjusted to change the average optical power. Denoting the period of the PWM signal as T_{PWM}, and the pulse width T_{W}, the PWM signal $p_{\text{PWM}}(t)$ in one period is given by

$$p_{\text{PWM}}(t) = \begin{cases} 1, & 0 \leq t \leq T_{\text{W}}, \\ 0, & T_{\text{W}} < t \leq T_{\text{PWM}}. \end{cases} \tag{4.88}$$

The average optical power of $p_{\text{PWM}}(t)$ is equal to $T_{\text{W}}/T_{\text{PWM}}$, which can be adjusted from 0 to 1 by changing the pulse width T_{W}. In Fig. 4.19, three PWM signals are depicted for different dimming levels $d = 20\%$, $d = 50\%$, and $d = 80\%$, where the signal rate $f_{\text{PWM}} = 1/T_{\text{PWM}}$ is 1 MHz.

In order to adjust the average optical power of the OFDM signal, a PWM signal can be superimposed, and the transmitted signal is the product of the DMT and PWM waveforms, i.e.,

$$x_{\text{PWM-DMT}}(t) = x_{\text{OFDM}}(t) \cdot p_{\text{PWM}}(t), \tag{4.89}$$

where $x_{\text{OFDM}}(t)$ denotes the time-domain optical OFDM signal, which can be modulated by any optical OFDM schemes, such as DCO-OFDM, ACO-OFDM, and so on. This combination was referred to as PWM-DMT modulation in [35]. The ratio of the OFDM time-domain signal duration over the PWM period is defined as $R = T_{\text{OFDM}}/T_{\text{PWM}}$, which should not be smaller than 1. If $R < 1$ and $T_{\text{W}} < T_{\text{PWM}} - T_{\text{OFDM}}$ for a certain dimming requirement, $p_{\text{PWM}}(t)$ can be zero within an entire OFDM signal duration, which will cause information loss and inter-carrier interference [35]. Therefore, the PWM signal rate should be at least twice the bandwidth of OFDM signals, which is difficult to be implemented. Moreover, the PWM signal would spread the bandwidth of the combined signals, leading to inefficient current-to-light conversion efficiency.

PWM-DMT is an efficient means for dimming control since the optical power of the transmitted signal is adjusted proportionally to the pulse width. After direct detection and synchronization, the amplitude of received signal is the product of dimming level and original OFDM signals. The dimming level can be regarded as part of the channel state information and can be handled by equalization. Therefore, the demodulation of PWM-DMT is exactly the same as conventional optical OFDM.

Since PWM-DMT requires a very high-frequency PWM signal compared with the OFDM signal, which is infeasible for high-data-rate implementation, variable optical OFDM (VO-OFDM) was proposed in [36]. In VO-OFDM, the OFDM signals are only modulated onto the PWM signals during the on-state, which requires a relatively low-frequency PWM signal. The waveform of VO-OFDM is shown in Fig. 4.20, where the OFDM is transmitted only when the PWM signal is on.

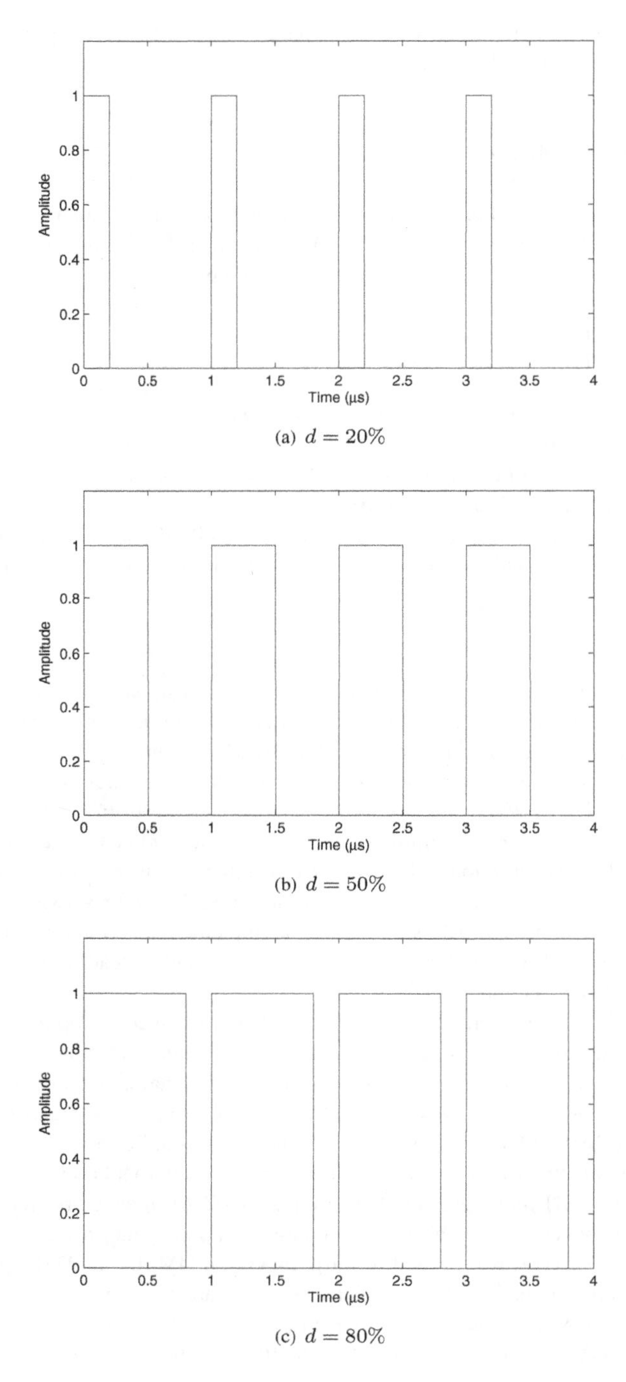

Figure 4.19 PWM signal with different pulse widths for dimming control.

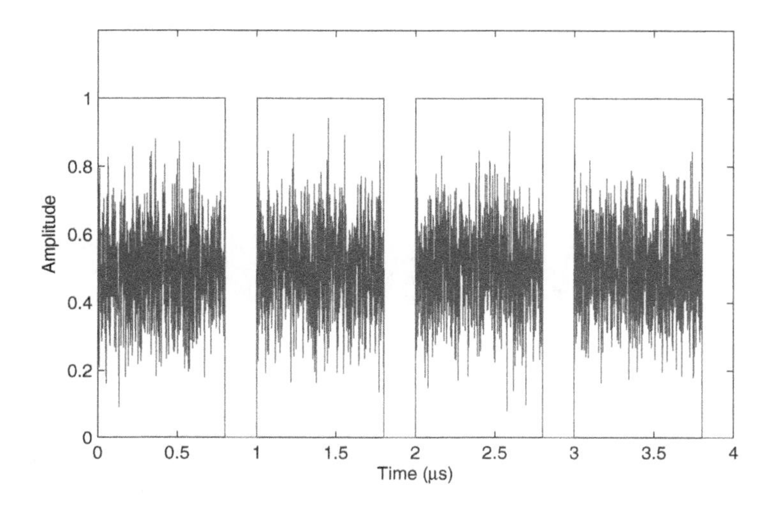

Figure 4.20 Variable optical OFDM for visible light communications with dimming control.

The VO-OFDM signal is not a simple multiplication of OFDM and PWM signals since there should be zero-padding in OFDM during the off-state of PWM. In this way, the period of a PWM signal can be much larger than the OFDM signal duration, which is easier for implementation. However, the period of a PWM signal cannot be too large since it can induce light flicker, which may cause noticeable, negative physiological changes in the human eye. To avoid flicker, the period of PWM signal in VO-OFDM should be smaller than the maximum flickering time period (MFTP). The MFTP is defined as the maximum time period over which the light intensity can change without the human eye perceiving it, which should be less than 5 ms.

For both PWM-DMT and VO-OFDM, the resultant dimming level can be calculated by the product of the dimming levels of PWM and OFDM, which is given by

$$d = d_{\text{OFDM}} \cdot d_{\text{PWM}}, \tag{4.90}$$

where d_{OFDM} and d_{PWM} denote the dimming levels of PWM and OFDM, respectively. When the required dimming level is low, it can be adjusted by d_{PWM} directly. For example, if the DC bias is set to the middle of the linear range, i.e., $(v_{\min} + v_{\max})/2$, then $d_{\text{OFDM}} = 50\%$, and d can be adjusted continuously from 0 to 50%, where d_{PWM} is within the range $[0, 100\%]$. However, the combination of PWM signal is only able to reduce the optical power since part of the signal is set to zero during the PWM signal duration. When the required dimming level is higher than 50%, the dimming level of OFDM needs to be adjusted as the conventional optical OFDM by increasing the DC bias, and d_{PWM} should be set to 100%. Therefore, these PWM-based schemes are more suitable for low illumination requirements.

4.4.2 Reverse polarity optical OFDM

Reverse polarity optical OFDM (RPO-OFDM) also uses a low-frequency PWM signal for dimming control, but it utilizes the entire PWM signal duration for data transmission instead of only the on-state duration [37]. The combined RPO-OFDM signal is given by

$$x_{\text{RPO}}(t) = \begin{cases} p_{\text{PWM}}(t) - m \cdot x_{\text{OFDM}}(t), & 0 \leq t \leq T_{\text{M}}, \\ p_{\text{PWM}}(t) + m \cdot x_{\text{OFDM}}(t), & T_{\text{M}} < t \leq T_{\text{PWM}}, \end{cases} \tag{4.91}$$

where m denotes the scaling factor of optical OFDM signal. When the PWM signal is on, the polarity of optical OFDM signal is reversed for combination, while the optical OFDM signal is directly transmitted when the PWM signal is off. In this way, the combined signal can utilize the full linear range of LEDs for transmission, which can increase the effective electrical power at the receiver. In fact, if the DC component in the PRO-OFDM signal is ignored, the power of received OFDM signal remains the same when the PWM signal changes. Therefore, as long as the electronic power of the OFDM signal remains unchanged, the same performance can be achieved when the PWM signal changes. The dimming level of RPO-OFDM can be given by

$$d = d_{\text{OFDM}} \cdot (1 - d_{\text{PWM}}) + (1 - d_{\text{OFDM}}) \cdot d_{\text{PWM}}. \tag{4.92}$$

Hence, the supportable dimming range is between d_{OFDM} and $1 - d_{\text{OFDM}}$ by changing the pulse width of PWM. If a larger dimming range is required, the dimming level of optical OFDM should be changed by scaling.

For conventional DCO-OFDM with the DC bias in the middle of the linear range, $d_{\text{OFDM}} = 50\%$ and the PWM signal cannot change the dimming level in RPO-OFDM according to (4.92). However, ACO-OFDM and U-OFDM can be used in RPO-OFDM to achieve a larger dimming range, since DC bias is not required in these modulation schemes, and the dimming level is much lower. Figure 4.21 shows the RPO-OFDM signal waveforms with $d_{\text{PWM}} = 20\%$ and $d_{\text{PWM}} = 80\%$, respectively, where ACO-OFDM is used as the optical OFDM scheme for modulation. The dimming level for ACO-OFDM is set to 10%, and the actual dimming levels of RPO-OFDM are 26% and 74%, respectively.

RPO-OFDM can support stable BER performance when the dimming level changes within the range $[d_{\text{OFDM}}, 1 - d_{\text{OFDM}}]$. When the required dimming level is outside this range, the dimming level for conventional OFDM should be reduced, which will decrease the received electrical power and lower modulation order should be used to maintain the BER performance. However, since ACO-OFDM or U-OFDM is used in RPO-OFDM, the spectral efficiency is limited. Some spectrum-efficient optical OFDM schemes introduced in Section 4.3 may be preferred to further improve its spectral efficiency.

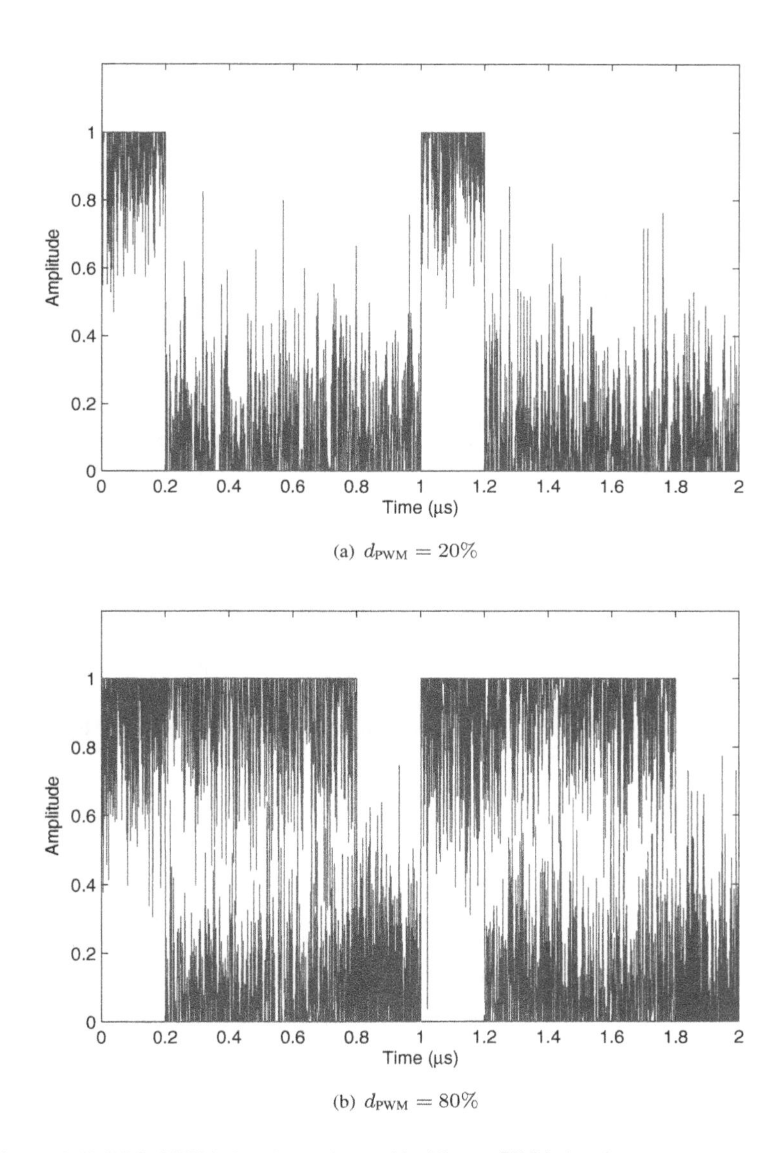

Figure 4.21 RPO-OFDM signal waveform with different PWM signals.

4.4.3 Asymmetrical hybrid optical OFDM

The aforementioned schemes all utilize PWM signal to adjust the average optical power. In PWM-DMT, high-frequency PWM signal is required, which is not suitable for high-speed transmission. VO-OFDM and RPO-OFDM both use a much lower-frequency PWM signal for dimming control. In VO-OFDM, the off-state of PWM is empty, leading to the reduction of data rate. In RPO-OFDM, unipolar optical OFDM schemes are used, which is not spectrum-efficient. Asymmetrical hybrid optical OFDM (AHO-OFDM) was proposed in [38], which does not require extra PWM signal for dimming control.

In AHO-OFDM, ACO-OFDM and PAM-DMT signals are combined to transmit simultaneously. In PAM-DMT, only the imaginary part of even subcarriers is modulated to make sure that the ACO-OFDM symbols are not interfered. Unlike HACO-OFDM where ACO-OFDM and PAM-DMT signals are directly added, either ACO-OFDM or PAM-DMT signal is inverted in AHO-OFDM so that the combined signals are bipolar. Dimming control is achieved by directly adjusting the amplitude of the combined signals so that PWM signal is not required. Given different powers of ACO-OFDM and PAM-DMT signals corresponding to the positive and negative values, the combined AHO-OFDM signal is asymmetrical and therefore the dynamic range of LEDs could be fully utilized. The combined signals are biased by B_{DC} and the transmitted AHO-OFDM signals generated by ACO-OFDM and inverse PAM-DMT are given by

$$x_{\text{AHO},n} = x_{\text{ACO},n} - x_{\text{PAM},n} + B_{\text{DC}},\ n = 0, 1,\ ...,\ N-1, \tag{4.93}$$

where $x_{\text{ACO},n}$ and $x_{\text{PAM},n}$ denote the time-domain signals of ACO-OFDM and PAM-DMT, respectively.

According to Section 4.1.2, the average amplitude of the combined AHO-OFDM signal can be calculated as

$$E_{\text{AHO}} = E\left(x_{\text{AHO},n}\right) = \frac{\sigma_{\text{ACO}}}{\sqrt{2\pi}} - \frac{\sigma_{\text{PAM}}}{\sqrt{2\pi}} + B_{\text{DC}}, \tag{4.94}$$

which is proportional to the average optical power of LEDs since the amplitude of AHO-OFDM signal is used to modulate the instantaneous power of the optical emitter.

The dimming level for AHO-OFDM can be defined as

$$d = \left(E_{\text{AHO}} - v_{\min}\right) / \left(v_{\max} - v_{\min}\right) \tag{4.95}$$

according to (4.87). For a given dimming level d, the average amplitude E_{AHO} can be obtained according to (4.95). It is seen from (4.94) that the average amplitude E_{AHO} is determined by the DC bias B_{DC} as well as σ_{ACO} and σ_{PAM}. Since the amplitude of combined AHO-OFDM signal is constrained by the dynamic range of LEDs, it has to be clipped when it is beyond the dynamic range of LEDs, which results in undesirable clipping distortion. To estimate the clipping distortion of the proposed scheme, the scaling factors of ACO-OFDM and PAM-DMT as β_{ACO} and β_{PAM} are defined, where $\beta_{\text{ACO}} = \left(v_{\max} - B_{\text{DC}}\right)/\sigma_{\text{ACO}}$ and $\beta_{\text{PAM}} = \left(B_{\text{DC}} - v_{\min}\right)/\sigma_{\text{PAM}}$.

The probability of the clipped signal is then given by

$$P\left(x_{\text{AHO},n} > v_{\max}\right) = P\left(x_{\text{ACO},n} - x_{\text{PAM},n} > v_{\max} - B_{\text{DC}}\right)$$
$$= \int_0^{\infty} p_{\text{PAM}}\left(y\right) \int_{\beta_{\text{ACO}}\sigma_{\text{ACO}}+y}^{\infty} p_{\text{ACO}}\left(x\right) dx dy, \tag{4.96}$$

and

$$P\left(x_{\text{AHO},n} < v_{\min}\right) = P\left(x_{\text{PAM},n} - x_{\text{ACO},n} > B_{\text{DC}} - v_{\min}\right)$$
$$= \int_0^{\infty} p_{\text{ACO}}\left(x\right) \int_{\beta_{\text{PAM}}\sigma_{\text{PAM}}+x}^{\infty} p_{\text{PAM}}\left(y\right) dy dx. \tag{4.97}$$

When the scaling factors β_{ACO} and β_{PAM} are large enough, the probability of clipped signal would be very small and clipping distortion can be suppressed. For example, in the case of $\beta_{\text{ACO}} = \beta_{\text{PAM}} = 3$, the probability of the clipped signal would be less than 1‰. However, a large scaling factor will result in low effective power at the receiver, which degrades the system performance. Therefore, a tradeoff has to be made between the effective power and the clipping distortion for a required BER performance. When the scaling factors are chosen, the required DC bias for the desired dimming level d can be derived as

$$B_{\text{DC}} = \frac{\beta_{\text{ACO}}\beta_{\text{PAM}}\sqrt{2\pi}\left(\left(v_{\max} - v_{\min}\right) d + v_{\min}\right) - \beta_{\text{ACO}} v_{\min} - \beta_{\text{PAM}} v_{\max}}{\beta_{\text{ACO}}\beta_{\text{PAM}}\sqrt{2\pi} - \beta_{\text{ACO}} - \beta_{\text{PAM}}} \tag{4.98}$$

according to the definition of scaling factors and (4.94)–(4.95).

Figure 4.22 illustrates an example of AHO-OFDM signal generated by ACO-OFDM and inverse PAM-DMT, where the required dimming level is 70% and the number of subcarriers is set to 128. 16QAM and 16PAM are employed in ACO-OFDM and PAM-DMT, respectively. The scaling factors of both ACO-OFDM and PAM-DMT are set to three. In Fig. 4.22(a), the ACO-OFDM and inverse PAM-DMT signals are depicted separately, where the former is above the DC bias B_{DC} and the latter is below it. Figure 4.22(b) shows the combined AHO-OFDM signals, which are asymmetrical to the DC bias B_{DC}. One can see that B_{DC} is unequal to the desired average amplitude E_{AHO} and the dynamic range of LEDs could be fully utilized.

Simulations are conducted to evaluate the performance of the AHO-OFDM scheme. The maximum and minimum allowed signals of LEDs are $v_{\max} = 1$ and $v_{\min} = 0$, respectively. Figure 4.23 shows the obtained dimming level d as a function of β_{ACO}, β_{PAM}, and B_{DC} according to (4.98). It can be seen that a wide dimming range is achieved. For example, when the scaling factors $\beta_{\text{ACO}} = \beta_{\text{PAM}} = 4$, the achieved dimming range is from 10% to 90%. Larger scaling factors can be used to support a wider dimming range. Therefore, the AHO-OFDM scheme can achieve arbitrary dimming range in theory.

The performance of AHO-OFDM is compared with DCO-OFDM and HACO-OFDM. In DCO-OFDM, dimming control is achieved by simply adjusting the DC

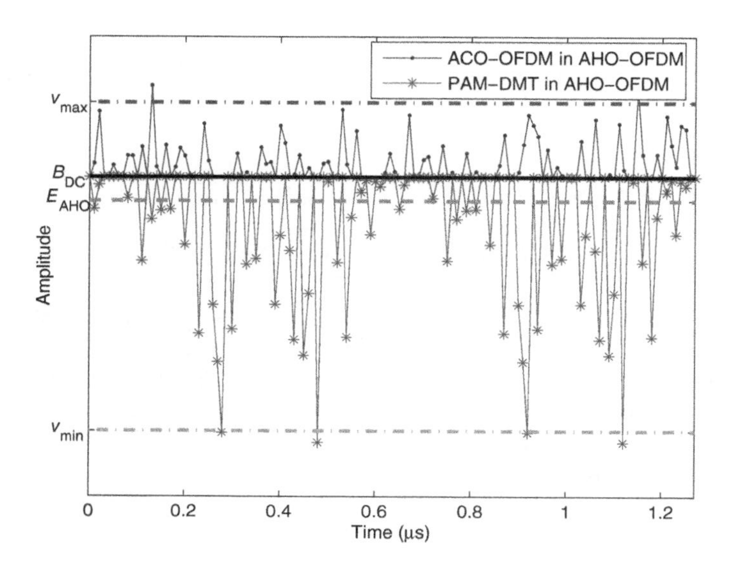

(a) ACO-OFDM and PAM-DMT signals in AHO-OFDM

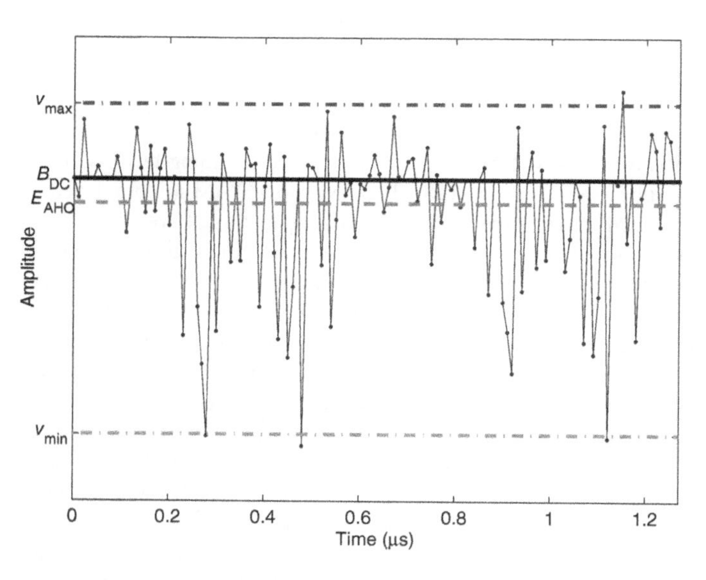

(b) The combined AHO-OFDM signal

Figure 4.22 An example of AHO-OFDM signal with dimming level $d = 70\%$.

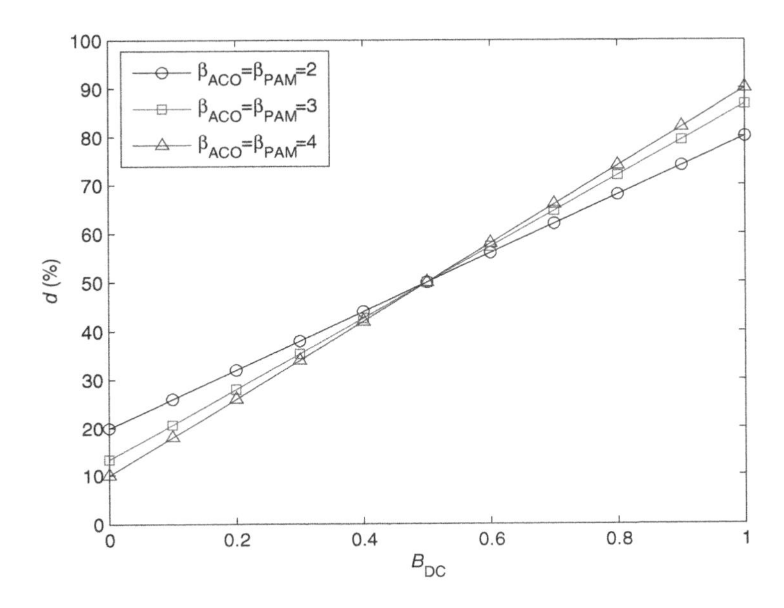

Figure 4.23 Optical dimming level d as a function of β_{ACO}, β_{PAM}, and B_{DC}.

bias, which does not require PWM signals. Adaptive scaling factors can be used to realize different dimming levels in DCO-OFDM to mitigate clipping distortion. In the HACO-OFDM scheme, the optical power is equally distributed to ACO-OFDM and PAM-DMT, and a DC bias still has to be added when the required dimming level is high. The achievable spectral efficiency comparisons of HACO-OFDM, DCO-OFDM, and AHO-OFDM with different dimming levels are shown in Fig. 4.24 for a target BER of 2×10^{-3} with the noise power of -10 dBm. It can be seen that AHO-OFDM can support a much wider dimming range compared with its two counterparts, and its achievable spectral efficiency is relatively stable when the dimming level varies since its asymmetry could fully utilize the dynamic range of LEDs. For example, it can support a wide dimming range from 5% to 95% with the spectral efficiency of at least 1.5 bit/s/Hz. For an extremely small dimming level where DCO-OFDM and HACO-OFDM could not work, i.e., 2%, the achievable spectral efficiency is still 0.75 bit/s/Hz for AHO-OFDM.

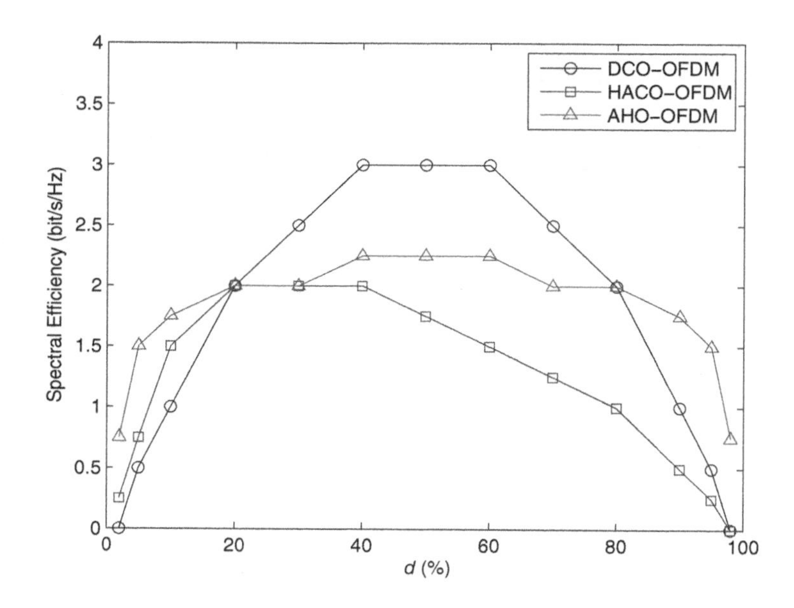

Figure 4.24 Performance comparison of AHO-OFDM, DCO-OFDM, and HACO-OFDM for a target BER of 2×10^{-3} under different dimming levels.

4.5 Conclusion

This chapter has reviewed the optical OFDM techniques for broadband and high-data-rate VLC systems. Since IM/DD scheme is used in VLC, the amplitude of the optical OFDM signals is constrained to be real-valued and non-negative. In this chapter, the comparisons of mainstream optical OFDM schemes, such as DCO-OFDM, ACO-OFDM, PAM-DMT, and U-OFDM, are provided. Optical OFDM suffers from high PAPR as in RF, which is prone to induce severe nonlinear distortion and impair the transmission performance of VLC systems. Therefore, several techniques have been introduced to enhance the performance of optical OFDM by optimization of DC bias and scaling factor, mitigating the nonlinear effect of LED, and PAPR reduction. Besides, recently proposed power- and spectral-efficient optical OFDM schemes, such as hybrid optical OFDM, eU-OFDM, and LACO-OFDM are also discussed. At last, the integration of OFDM and dimming control is discussed for VLC system, which has shown that dimmable OFDM can support a wide dimming range with a small throughput fluctuation.

References

1 R. W. Chang, *Orthogonal Frequency Multiplex Data Transmission System*. U.S. Patent 3,488,445, 1966.
2 J. Armstrong, "OFDM: From copper and wireless to optical," in *Proc. Optical Fiber Communication Conference and Exposition and the National Fiber Optic Engineers Conference (OFC/NFOEC) 2008* (San Diego, CA), Feb. 24–28, 2008, Tutorial, OMM1.
3 3GPP TS 36.211, *Evolved Universal Terrestrial Radio Access (E-UTRA): Physical Channels and Modulation*. v.8.9.0, Dec. 2009.
4 J. Armstrong, "OFDM for optical communications," *J. Lightw. Technol.*, vol. 27, no. 3, pp. 189–204, Feb. 2009.
5 R. Gitlin and E. Ho, "The performance of staggered quadrature amplitude modulation in the presence of phase jitter," *IEEE Trans. Commun.*, vol. 23, no. 3, pp. 348–352, Mar. 1975.
6 J. M. Kahn and J. R. Barry, "Wireless infrared communications," *Proc. IEEE*, vol. 85, no. 2, pp. 265–298, Feb. 1997.
7 J. Armstrong, and A. J. Lowery, "Power efficient optical OFDM," *Electron. Lett.*, vol. 42, no. 6, pp. 370–372, Mar. 2006.
8 S. C. J. Lee, S. Randel, F. Breyer, and A. M. J. Koonen, "PAM-DMT for intensity-modulated and direct-detection optical communication systems," *IEEE Photon. Technol. Lett.*, vol. 21, no. 23, pp. 1749–1751, Dec. 2009.
9 S. D. Dissanayake and J. Armstrong, "Comparison of ACO-OFDM, DCO-OFDM and ADO-OFDM in IM/DD systems," *J. Lightw. Technol.*, vol. 31, no. 7, pp. 1063–1072, Apr. 2013.
10 K. Asadzadeh, A. Dabbo, and S. Hranilovic, "Receiver design for asymmetrically clipped optical OFDM," in *Proc. IEEE Global Communications Conference (GLOBECOM) Workshops 2011* (Houston, TX), Dec. 5–9, 2011, pp. 777–781.
11 D. Tsonev, S. Sinanovic, and H. Haas, "Novel unipolar orthogonal frequency division multiplexing (U-OFDM) for optical wireless," in *Proc. IEEE Vehicular Technology Conference (VTC Spring) 2012* (Yokohama, Japan), May 6–9, 2012, pp. 1–5.
12 N. Fernando, Y. Hong, and E. Viterbo, "Flip-OFDM for optical wireless communications," in *Proc. IEEE Information Theory Workshop (ITW) 2011* (Paraty, Brazil), Oct. 16–20, 2011, pp. 5–9.
13 N. Fernando, Y. Hong, and E. Viterbo, "Flip-OFDM for unipolar communication systems," *IEEE Trans. Commun.*, vol. 60, no. 12, pp. 3726–3733, Dec. 2012.
14 S. Dimitrov and H. Haas, "Information rate of OFDM-based optical wireless communication systems with nonlinear distortion," *J. Lightw. Technol.*, vol. 31, no. 6, pp. 918–929, Mar. 2013.
15 J. Bussgang, "Cross correlation function of amplitude-distorted Gaussian signals," Mass. Inst. Technol., Cambridge, MA, USA, Tech. Rep. 216, Mar. 1952.
16 Z. Wang, Q. Wang, S. Chen, and L. Hanzo, "An adaptive scaling and biasing scheme for OFDM-based visible light communication systems," *Opt. Exp.*, vol. 22, no. 10, pp. 12707–12715, May 2014.
17 M. Zhang and Z. Zhang, "An optimum DC-biasing for DCO-OFDM system," *IEEE Commun. Lett.*, vol. 18, no. 8,

pp. 1351–1354, Jun. 2014.

18 P. Zillmann and G. R. Fettweis, "On the capacity of multicarrier transmission over nonlinear channels," in *Proc. IEEE Vehicular Technology Conference (VTC Spring) 2005* (Stockholm, Sweden), May 30–Jun. 1, 2005, pp. 1148–1152.

19 J. Tan, Z. Wang, Q. Wang, and L. Dai, "Near-optimal low-complexity sequence detection for clipped DCO-OFDM," *IEEE Photon. Technol. Lett*, vol. 28, no. 3, pp. 233–236, Feb. 2016.

20 J. Tan, Z. Wang, Q. Wang, and L. Dai, "BICM-ID scheme for clipped DCO-OFDM in visible light communications," *Opt. Exp.*, vol. 24, no. 5, pp. 4573–4581, 2016.

21 S. H. Han and J H .Lee, "An overview of peak-to-average power ratio reduction techniques for multicarrier transmission," *IEEE Wirel. Commun.*, vol. 12, no. 2, pp. 56–65, Apr. 2012.

22 B. Ranjha and M. Kavehrad, "Precoding techniques for PAPR reduction in asymmetrically clipped OFDM based optical wireless system," in *Proc. SPIE*, vol. 8165, Jan. 2013.

23 W. O. Popoola, Z. Ghassemlooy, and B. G. Stewart, "Pilot-assisted PAPR reduction technique for optical OFDM communication systems, " *J. Lightw. Technol.*, vol. 32, no. 7, pp. 1374–1382, Apr. 2014.

24 J. Tan, Q. Wang, and Z. Wang, "Modified PTS-based PAPR reduction for ACO-OFDM in visible light communications," *Sci. China Inf. Sci.*, vol.58, no.12, Oct. 2015.

25 W. Xu, M. Wu, H. Zhang, X. You, and C. Zhao, "ACO-OFDM-specified recoverable upper clipping with efficient detection for optical wireless communications," *IEEE Photon. J.*, vol. 6, no. 5, p. 7902617, Oct. 2014.

26 S. D. Dissanayake, K. Panta, and J. Armstrong, "A novel technique to simultaneously transmit ACO-OFDM and DCO-OFDM in IM/DD systems," in *Proc. IEEE Global Communications Conference (GLOBECOM) Workshops 2011* (Houston, TX), Dec. 5–9, 2011, pp. 782–786.

27 B. Ranjha and M. Kavehrad, "Hybrid asymmetrically clipped OFDM-based IM/DD optical wireless system," *J. Opt. Commun. Netw*, vol. 6, no. 4, pp. 387–396, Apr. 2014.

28 Q. Wang, Z. Wang, and L. Dai, "Iterative receiver for hybrid asymmetrically clipped optical OFDM," *J. Lightw. Technol.*, vol. 32, no. 22, pp. 4471–4477, Nov. 2014.

29 J. Li, X. Zhang, Q. Gao, Y. Luo, and D. Gu, "Exact BEP analysis for coherent M-ary PAM and QAM over AWGN and Rayleigh fading channels," in *Proc. IEEE Vehicular Technology Conference (VTC Spring) 2008* (Singapore), May 11–14, 2008, pp. 390–394.

30 D. Tsonev and H. Haas, "Avoiding spectral efficiency loss in unipolar OFDM for optical wireless communication," in *Proc. IEEE International Conference on Communications (ICC) 2014* (Sydney, Australia), Jun. 10–14, 2014, pp. 3336–3341.

31 D. Tsonev, S. Videv, and H. Haas, "Unlocking spectral efficiency in intensity modulation and direct detection systems," *IEEE J. Sel. Areas Commun.*, vol. 33, no. 9, pp. 1758–1770, Sep. 2015.

32 Q. Wang, C. Qian, X. Guo, Z. Wang, D. G. Cunningham, and I. H. White, "Layered ACO-OFDM for intensity-modulated direct-detection optical wireless transmission," *Opt. Exp.*, vol. 23, no. 9, pp. 12382–12393, May 2015.

33 Q. Wang, Z. Wang, X. Guo, and L. Dai, "Improved receiver design for layered ACO-OFDM in optical wireless communications," *IEEE Photon. Technol. Lett.*, vol. 28, no. 3, pp. 319–322, Feb. 2016.

34 *IEEE 802.15.7, Part 15.7: Short-Range Wireless Optical Communication Using Visible Light*, IEEE Std. 802.15.7-2011, Sep. 2011.

35 G. Ntogari, T. Kamalakis, J. Walewski, and T. Sphicopoulos, "Combining illumination dimming based on pulsewidth modulation with visible-light communications based on discrete multitone," *J. Opt. Commun. Netw.*, vol. 3, no. 1, pp. 56–65, Jan. 2011.

36 Z. Wang, W. D. Zhong, C. Yu, J. Chen, C. P. S. Francois, and W. Chen, "Performance of dimming control scheme in visible light communication system," *Opt. Exp.*, vol. 20, no. 17, pp. 18861–18868, Aug. 2012.

37 H. Elgala and T. D. C. Little, "Reverse polarity optical-OFDM (RPO-OFDM): Dimming compatible OFDM for gigabit VLC links," *Opt. Exp.*, vol. 21, no. 20,

pp. 24288–24299, Oct. 2013.

38 Q. Wang, Z. Wang, and L. Dai, "Asymmetrical hybrid optical OFDM for visible light communications with dimming control," *IEEE Photon. Technol. Lett.*, vol. 27, no. 9, pp. 974–977, May 2015.

5
Multicolor Modulation

Generally, white light emitting diodes (LEDs) are classified into two types, which are single-chip LEDs and RGB-type LEDs. The single-chip LED uses a single blue LED that excites a yellow phosphor to create an overall white emission. However, the slow response of yellow phosphor limits the transmission bandwidth in visible light communications (VLCs). The RGB-type LEDs combine light from LEDs of the three primary colors, which are red, green, and blue, and three LEDs emit their corresponding colors simultaneously. They are preferable to single-chip LEDs to improve the transmission rate due to their faster response time. Moreover, these three wavelengths can be used to carry multiple data streams independently and thus offer the possibility of the wavelength division multiplexing (WDM).

This chapter discusses multicolor modulation schemes to satisfy both communication and illumination requirements. In Section 5.1, we introduce color shift keying (CSK), which has been adopted in the IEEE 802.15.7 standard. Besides the CSK constellation in the standard, the optimal design rules of CSK constellation as well as Quad-LED CSK are provided. In practical communications, channel coding is widely used before modulation to achieve better performance. In Section 5.2, CSK with coded modulation is discussed with both hard and soft detections. In Section 5.3, WDM is introduced, which modulates signals on different optical sources for both communication and illumination requirements. Particularly, a receiver-side predistortion is used before channel decoding to improve the performance of WDM systems.

5.1
Color shift keying

Figure 5.1 illustrates the diagram of CSK transmitter with three light sources of bands i, j, and k. A scrambler is utilized before channel coding in order to ensure pseudo-random data and flicker-free illumination, whose polynomial generator is given by

$$g(D) = 1 + D^{14} + D^{15}, \tag{5.1}$$

where D is a single bit delay element [1]. After scrambling and channel coding, the coded bits are mapped into the color coordinates (x, y) by the color coding block.

Visible Light Communications: Modulation and Signal Processing. First edition. Zhaocheng Wang, Qi Wang, Wei Huang, and Zhengyuan Xu.
Published 2017 by John Wiley & Sons, Inc.

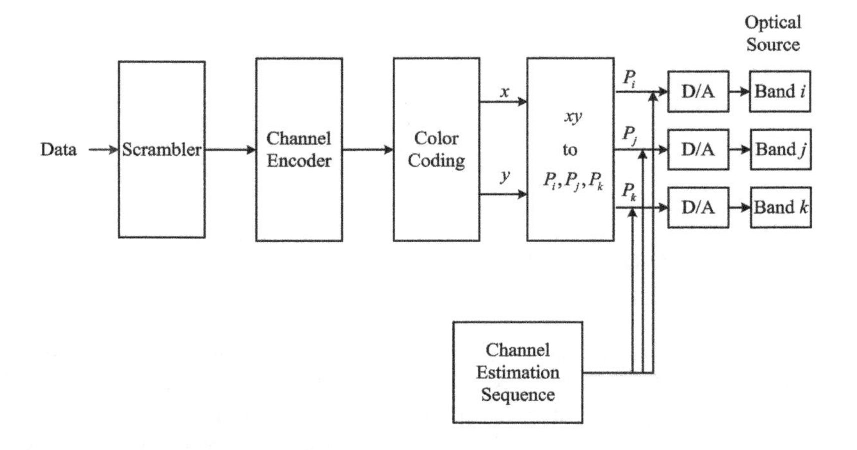

Figure 5.1 Diagram of CSK transmitter.

The color coordinates are then transformed to a three-dimensional intensity vector $[P_i\, P_j\, P_k]^{\mathrm{T}}$ to modulate the three light sources of bands i, j, and k. We denote the color coordinates of the three light sources as (x_i, y_i), (x_j, y_j), (x_k, y_k), respectively. For a given color coordinate (x_p, y_p), the intensities for the three light sources can be calculated by

$$x_p = P_i x_i + P_j x_j + P_k x_k, \tag{5.2}$$

$$y_p = P_i y_i + P_j y_j + P_k y_k, \tag{5.3}$$

$$P_i + P_j + P_k = 1. \tag{5.4}$$

At the receiver, the color coordinates are reconstructed by the received optical powers of the three colors, which are then used for decoding.

5.1.1 Constellation

The visible light spectrum is defined from 380 nm to 780 nm in wavelength, which is divided into seven frequency bands in the IEEE 802.15.7 standard [1]. In CSK, three color light sources are used to generate the transmitted signal, where the three vertices of the CSK constellation triangle are determined by the center wavelength of the corresponding color bands on (x, y) color coordinates. Table 5.1 gives the color coordinate values for the seven color bands whose spectral peaks are at the center of each color band, and Fig. 5.2 illustrates the centers of the seven color bands on color coordinates.

Table 5.1 Color coordinates for the seven color bands [1].

Band (nm)	Code	Center (nm)	(x, y)
380-478	000	429	$(0.169, 0.007)$
478-540	001	509	$(0.011, 0.733)$
540-588	010	564	$(0.402, 0.597)$
588-633	011	611	$(0.669, 0.331)$
633-679	100	656	$(0.729, 0.271)$
679-726	101	703	$(0.734, 0.265)$
726-780	110	753	$(0.734, 0.265)$

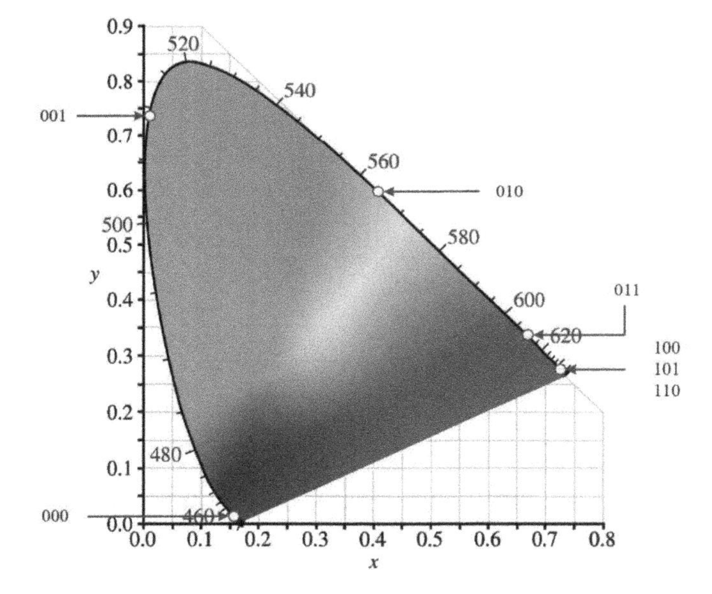

Figure 5.2 Centers of the seven color bands on x-y color coordinates.

The constellations of 4-CSK, 8-CSK, and 16-CSK recommended by the IEEE 802.15.7 standard are shown in Fig. 5.3, where the points I, J, and K denote the center of the three color bands on the color coordinates defined in Table 5.1. In order to make sure the triangle is large enough to maximize the distances between the adjacent symbols, nine valid combinations of the color bands are provided in Table 5.2 [2]. In this way, the output color for illumination can be guaranteed [3].

In 4-CSK, the symbols are denoted as S_0 to S_3, where S_1, S_2, and S_3 are the vertices of the triangle $\triangle IJK$ and S_0 is the centroid of the triangle $\triangle IJK$.

In 8-CSK, the symbols are denoted as S_0 to S_7, where S_0, S_4, and S_7 are the vertices of the triangle $\triangle IJK$. S_1 and S_2 are the trisection points of the lines JK

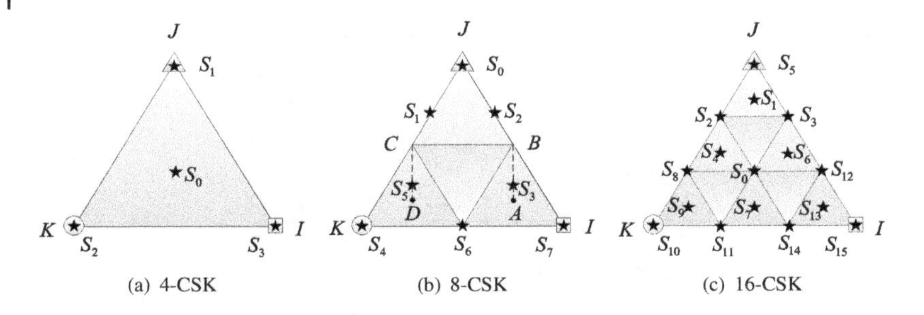

Figure 5.3 Constellations of 4-CSK, 8-CSK, and 16-CSK provided by IEEE 802.15.7 standard.

Table 5.2 Valid color band combinations for CSK [2].

Combination	Band i	Band j	Band k
1	110	010	000
2	110	001	000
3	101	010	000
4	101	001	000
5	100	010	000
6	100	001	000
7	011	010	000
8	011	001	000
9	010	001	000

and JI, where we have $JS_1 = 1/3JK$ and $JS_2 = 1/3JI$. The midpoints of the lines JI, JK, and KI are denoted as B, C, and S_6, respectively. Point A is the centroid of the triangle $\triangle BS_6I$, while D is the centroid of the triangle $\triangle CKS_6$. S_3 and S_5 are the trisection points of the lines AB and CD that satisfy $AS_3 = 1/3AB$ and $DS_5 = 1/3CD$.

In 16-CSK, the symbols are denoted as S_0 to S_{15}, where S_5, S_{10}, and S_{15} are the vertices of the triangle $\triangle IJK$. The symbols S_2, S_8, S_3, S_{12}, S_{11}, and S_{14} divide the sides JK, JI, and KI into three parts, respectively. S_0 is the centroid of the triangle $\triangle IJK$, while S_1, S_4, S_6, S_7, S_9, and S_{13} are the centroids of the smaller triangles $\triangle JS_2S_3$, $\triangle S_2S_8S_0$, $\triangle S_3S_0S_{12}$, $\triangle S_8KS_{11}$, $\triangle S_0S_{11}S_{14}$, and $\triangle S_{14}S_{12}I$, respectively.

Although the intensities of the light sources are different in different symbols, the total power of each symbol remains constant. Therefore, flicker is avoided completely. The dimming control of CSK can be implemented by adjusting the current drivers of each light sources isometrically, which maintains the constellation since the proportions of the three colors remain unchanged.

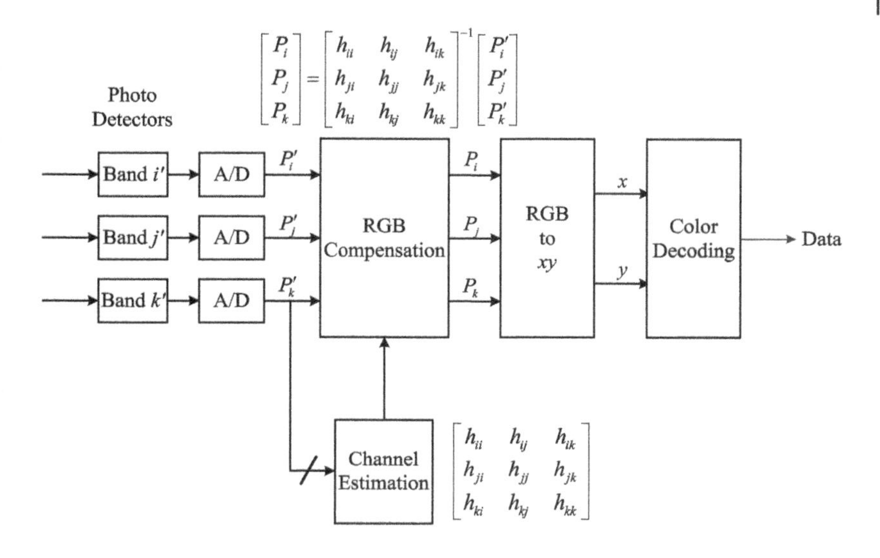

Figure 5.4 Diagram of CSK receiver with color calibration.

5.1.2
Color calibration

It should be noted that, the optical source may have a different spectral peak or even multiple spectral peaks in practice. Therefore, the implementation of CSK can use the color band based on the center wavelength of the actual optical source. At the receiver, color calibration should be conducted to compensate the color coordinate errors and cancel the interference among different colors. Besides, other light devices and ambient light may cause multicolor imbalance and multicolor interference as well, which can be compensated by a color calibration at the same time. The diagram of CSK receiver with color calibration is shown in Fig. 5.4.

Before data transmission, Walsh codes modulated by OOK are used to estimate the channel propagation matrix. Three Walsh code sequences with length-4 are used for the three color bands i, j, and k, namely, $W(1,4) = 1, -1, 1, -1$, $W(2,4) = 1, 1, -1, -1$, and $W(3,4) = 1, -1, -1, 1$. In order to maintain the accuracy of the estimation, each bit of the Walsh code is transmitted twice by repetition coding. The estimated channel propagation matrix is a 3×3 square matrix defined as

$$\mathbf{H} = \begin{bmatrix} h_{ii} & h_{ij} & h_{ik} \\ h_{ji} & h_{jj} & h_{jk} \\ h_{ki} & h_{kj} & h_{kk} \end{bmatrix}. \tag{5.5}$$

At the receiver, the received signals can be compensated by the color calibration module, which multiplies the signal vector with inverted estimated channel matrix

as

$$\begin{bmatrix} P_i \\ P_j \\ P_k \end{bmatrix} = \begin{bmatrix} h_{ii} & h_{ij} & h_{ik} \\ h_{ji} & h_{jj} & h_{jk} \\ h_{ki} & h_{kj} & h_{kk} \end{bmatrix}^{-1} \begin{bmatrix} P'_i \\ P'_j \\ P'_k \end{bmatrix}. \tag{5.6}$$

5.1.3 Constellation optimization

The IEEE 802.15.7 standard provides a simple design rule for CSK constellation. However, the constellation is not optimized for the best performance. Several methods have been proposed to redesign the CSK constellation [4–7].

Assuming that the shaping pulse is rectangular and additive white Gaussian noise (AWGN) is induced at the receiver, the equivalent discrete-time channel model can be written as [7]

$$\mathbf{r} = \mathbf{H}\mathbf{p}_m + \mathbf{n}, \tag{5.7}$$

where $\mathbf{p}_m = [P_{i,m}\, P_{j,m}\, P_{k,m}]^{\mathrm{T}}$ denotes the mth symbol in the M-CSK constellation, $\mathbf{r}$, $\mathbf{H}$, and $\mathbf{n}$ are the received symbol, the channel response, and the AWGN noise vector, respectively.

In the IEEE 802.15.7 standard, the optical power of each symbol is constant to minimize light intensity fluctuations since the fluctuations may do harm to human health [1]. Denoting the optical gain of each optical source as η_i , η_j , and η_k in units of lumens per ampere (lm/A), the optical gain vector is given by $\boldsymbol{\eta} = [\eta_i\, \eta_j\, \eta_k]^{\mathrm{T}}$. The optical gains η_i , η_j , and η_k may change with temperature and usage time, and this can be reflected by variation in $\mathbf{H}$. The instantaneous optical power of the mth symbol is defined as

$$\mathcal{L}(\mathbf{p}_m) = \langle \boldsymbol{\eta}, \mathbf{p}_m \rangle, \tag{5.8}$$

where $\langle \cdot, \cdot \rangle$ denotes the inner product operator. The IEEE 802.15.7 standard states that

$$\mathcal{L}(\mathbf{p}_m) = L,\ m = 1, 2, \ldots, M, \tag{5.9}$$

where L is the fixed optical power.

In practical data transmission, the symbol rate is usually very high to support high data rate. Therefore, the fluctuations of light intensity negligible for human eyes and it is acceptable for small fluctuations in the design of CSK constellation to provide larger degree of freedom [8]. The dynamic range of the instantaneous optical power is given by

$$L_{\min} \leq \mathcal{L}(\mathbf{p}_m) \leq L_{\max}, \tag{5.10}$$

where $L_{\min}$ and $L_{\max}$ are the minimum and maximum allowable total optical power, respectively. When $L_{\min} = L_{\max}$, the total optical power is constant for each symbol as in the IEEE 802.15.7 standard. Moreover, since intensity modulation is used and the optical sources have limited dynamic ranges, the symbols should also satisfy the following constraint as

$$\mathbf{0} \preceq \mathbf{p}_m \preceq [I_i \, I_j \, I_k]^{\mathrm{T}}, \tag{5.11}$$

where I_i, I_j , and I_k are the maximum allowable currents for the optical sources of bands i, j, and k, respectively.

It is also possible to design the CSK constellation with constant current instead of optical power, where we define the total current of symbol $\mathbf{p}_m$ as

$$\mathcal{I}(\mathbf{p}_m) = P_{i,m} + P_{j,m} + P_{k,m}. \tag{5.12}$$

If $I_{\min}$ and $I_{\max}$ are the minimum and maximum allowable total currents, then (5.10) can be rewritten as

$$I_{\min} \leq \mathcal{I}(\mathbf{p}_m) \leq I_{\max}, \tag{5.13}$$

which can be used to limit radio frequency (RF) radiation caused by high current switching through stray inductance [7].

For a given color coordinate (x_p, y_p), denote the relative luminous flux as $\mathbf{p}$, then

$$\begin{bmatrix} \frac{x_i}{y_i} & \frac{x_j}{y_j} & \frac{x_k}{y_k} \\ 1 & 1 & 1 \\ \frac{1-x_i-y_i}{y_i} & \frac{1-x_j-y_j}{y_j} & \frac{1-x_k-y_k}{y_k} \end{bmatrix} \mathbf{p} = \begin{bmatrix} \frac{x_p}{y_p} \\ 1 \\ \frac{1-x_p-y_p}{y_p} \end{bmatrix}, \tag{5.14}$$

according to (5.2)–(5.4), and the sum of the elements of $\mathbf{p}$ is equal to one. If the perceived chromaticity is required to be (x_p, y_p), the constellation should satisfy the following constraint

$$\frac{1}{M}\sum_{m=1}^{M} \mathcal{L}(\mathbf{p}_m) = L. \tag{5.15}$$

If $L_{\min} < L_{\max}$, the value of L should be specified in the constellation design. Alternatively, it can be used as a design variable.

At high signal-to-noise ratios (SNRs), the symbol error rate (SER) of CSK is dominated by the minimum pairwise Euclidean distance between all received symbols when AWGN channel is assumed with uniform constellation. Therefore, the optimization of the CSK constellation is achieved by maximizing the minimum pairwise Euclidean distance for a given SNR, which is written as

$$\begin{aligned} \mathcal{A} = \arg\max_{\{\mathbf{p}_m\}} \min_{s \neq t} & \left|\mathbf{H}(\mathbf{p}_s - \mathbf{p}_t)\right|^2 \\ \text{subject to} \quad & L_{\min} \leq \mathcal{L}(\mathbf{p}_m) \leq L_{\max}, \\ & \mathbf{0} \preceq \mathbf{p}_m \preceq [I_i \, I_j \, I_k]^{\mathrm{T}}, \\ & \frac{1}{M}\sum_{m=1}^{M} \mathcal{L}(\mathbf{p}_m) = L. \end{aligned} \tag{5.16}$$

If the total current is constrained, the first constraint in (5.16) should be replaced by (5.13).

The objective in (5.16) is both nonconvex and non-differentiable, which cannot be solved by conventional gradient-based numerical optimization tools. For nonconvex problems, the conventional optimization algorithms can only find local minima, which depends on the initial estimate of the constellation used to start the optimization process. Therefore, the optimization problem can be solved multiple times with random starting points, so that it is more likely to find the global optimal solution. Besides, the following approximation may be used to eliminate the non-differentiability of (5.16) [6, 7]

$$\min_m d_m \approx -\ln\left(\sum_{m=1}^{N} \exp\left(-\beta d_m\right)\right)/\beta, \tag{5.17}$$

where $\{d_m\}$ is the set of pairwise distances and β is a large positive number to maintain accuracy of the approximation. If one of the distance is very small compared to the others, it is obvious that (5.17) is a good approximation. With this approximation, the objective in (5.16) is rewritten as

$$\arg\max_{\{\mathbf{p}_m\}} -\ln\left(\sum_{s\neq t} \exp\left(-\beta|\mathbf{H}\left(\mathbf{p}_s-\mathbf{p}_t\right)|^2\right)\right)/\beta, \tag{5.18}$$

which can be readily solved by any optimization toolkit under the constraints given in (5.16), and the smallest distances between all symbols approach the same value when they are near a local optimum. However, for a given β, the approximation of the minimum in (5.17) becomes less reliable near a local minimum. If β is set to a fixed large value, there will be large gradients, which may result in numeric instability and convergence problems. Therefore, the value of β is increased progressively in a series of sequential optimizations [7] to obtain a reliable approximation without the convergence issue. In each iteration, the value of β is doubled until the mean squared error between the constellations in the current and past iterations is less than a specified threshold. According to [7], the optimized constellations have much larger minimum pairwise Euclidean distance compared with the IEEE 802.15.7 standard, which is summarized in Table 5.3.

Table 5.3 The minimum pairwise Euclidean distance of CSK constellation with different design rules.

Design Rule	4-CSK	8-CSK	16-CSK
IEEE 802.15.7 standard	0.4245	0.1758	0.1400
Optimization in [7]	0.5141	0.2725	0.1755

5.1.4 CSK with Quad-LED

In Quad-LED (QLED), four optical sources of bands i, j, k, and v are used for illumination, which provides better illumination performance. Furthermore, the additional dimension also increases the degree of freedom in the constellation design, leading to improved transmission performance. Similar to the conventional CSK, the design of QLED CSK is based on the CIE 1931 color space. The four optical sources are in the bands of blue, cyan, yellow, and red (BCYR), respectively. When QLED is used, a quadrilateral instead of a triangular constellation can be obtained, which allows a simple symbol mapping and constellation design as in squared QAM.

The relationship of the color coordinate and the intensities for the four light sources can be written as

$$x_p = P_i x_i + P_j x_j + P_k x_k + P_l x_v, \tag{5.19}$$

$$y_p = P_i y_i + P_j y_j + P_k y_k + P_k y_v, \tag{5.20}$$

$$P_i + P_j + P_k + P_v = 1, \tag{5.21}$$

which are extensions of (5.2)–(5.4). However, (5.19)–(5.21) are under-determined since there are only three equations with four unknown variables, and do not have a unique solution when the color coordinates are given. Therefore, only up to three optical sources are used in each QLED CSK symbol, which can be solved similar to the conventional CSK as in (5.2)–(5.4). The constellation of QLED CSK can be regarded as a subset of four conventional CSK constellations with different colors. In order to generate any color within the quadrilateral bounded by the four bands, the QLED system requires at least three LEDs to illuminate at specific intensities, two LEDs for any color on the border lines and one at the central wavelength position or the vertices [9]. Hence, the quadrilateral can be divided into four smaller triangles, each illuminated by the optical sources corresponding to its three vertices, and only up to three out of four optical sources will be "on" at any time instance in the QLED CSK system, which maintains the same total optical power as conventional CSK. However, the additional optical source increases electrical power since a certain level of biasing is required to satisfy the switching requirements.

The constellations of 4-CSK, 8-CSK, and 16-CSK for QLED are illustrated in Fig. 5.5. Similar to the squared QAM constellation, Gray mapping can be imposed on the CSK constellation to further reduce the bit error rate. In QLED 4-CSK constellation, the symbols are located at vertices of the quadrilateral. In QLED 8-CSK constellation, the symbols S_0, S_3, S_5, and S_6 are located at the vertices of the quadrilateral as well, while the symbols S_1, S_2, S_4, and S_7 are the midpoints of the four sides. In QLED 16-CSK constellation, the symbols are located as in conventional 16-QAM, which divide the quadrilateral into 16 smaller quadrilaterals as their vertices. Similarly, 64-CSK constellation can be obtained with the locations of 64-QAM symbols.

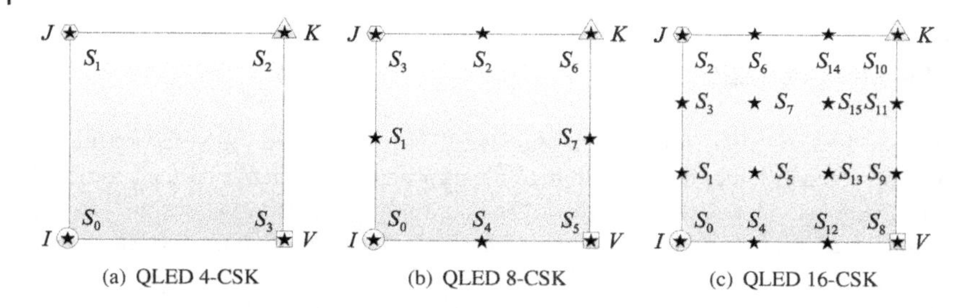

Figure 5.5 Constellations of 4-CSK, 8-CSK, and 16-CSK for QLED.

The color coordinates of the symbols for QLED 4-CSK, 8-CSK, and 16-CSK are shown in Table 5.4, which also provides the corresponding intensity value in each band calculated by (5.2)–(5.4). Depending on the symbol positions, the color coordinates utilized in (5.2)–(5.4) vary for different symbols. As in the conventional CSK, the total optical power of each symbol is set to one to eliminate flicker due to intensity variations.

The constellation proposed in [9] gives a simple design rule for QLED CSK constellation. Furthermore, the constellation can be optimized with the method introduced in Section 5.1.3.

5.2 CSK with coded modulation

When perfect channel state information (CSI) is assumed at the receiver, the estimated intensity vector $\hat{\mathbf{p}}$ can be obtained by the maximum likelihood (ML)-based detection given by

$$\hat{\mathbf{p}} = \arg\min_{\tilde{\mathbf{p}} \in \mathcal{S}} \|\mathbf{r} - \mathbf{H}\tilde{\mathbf{p}}\|^2, \tag{5.22}$$

which is then mapped to the bit vector by a channel decoder. Since the detection in (5.22) estimates the optimal intensity vector directly, it is referred to as hard detection.

The performance of the ML-based hard detection can be approximated by the union bound, where the average bit error probability can be expressed as [10]

$$P_{\text{e,union}} \leqslant \frac{1}{MN_b} \sum_{\xi=1}^{M} \sum_{\zeta=1,\zeta\neq\xi}^{M} d\left(\tilde{\mathbf{p}}^{(\xi)}, \tilde{\mathbf{p}}^{(\zeta)}\right) P\left(\tilde{\mathbf{p}}^{(\xi)} \mapsto \tilde{\mathbf{p}}^{(\zeta)}\right), \tag{5.23}$$

where $d\left(\tilde{\mathbf{p}}^{(\xi)}, \tilde{\mathbf{p}}^{(\zeta)}\right)$ denotes the Hamming distance between the intensity vectors $\tilde{\mathbf{p}}^{(\xi)}$ and $\tilde{\mathbf{p}}^{(\zeta)}$. $P\left(\tilde{\mathbf{p}}^{(\xi)} \mapsto \tilde{\mathbf{p}}^{(\zeta)}\right)$ is the pairwise error probability (PEP) of the two

Table 5.4 Constellations of QLED 4-CSK, 8-CSK, and 16-CSK and their intensity values in each band [9].

	Symbol	x	y	I	J	K	V
4-CSK	S_0	0.169	0.007	1	0	0	0
	S_1	0.011	0.460	0	1	0	0
	S_2	0.734	0.265	0	0	0	1
	S_3	0.402	0.597	0	0	1	0
8-CSK	S_0	0.169	0.007	1	0	0	0
	S_1	0.09	0.2335	0.5	0.5	0	0
	S_2	0.2065	0.5285	0	0.5	0.5	0
	S_3	0.011	0.460	0	1	0	0
	S_4	0.4515	0.1360	0.5	0	0	0.5
	S_5	0.734	0.265	0	0	0	1
	S_6	0.402	0.597	0	0	1	0
	S_7	0.568	0.431	0	0	0.5	0.5
16-CSK	S_0	0.1690	0.0070	1	0	0	0
	S_1	0.1163	0.1580	0.6667	0.3333	0	0
	S_2	0.0110	0.4600	0	1	0	0
	S_3	0.0637	0.3090	0.3333	0.6667	0	0
	S_4	0.3573	0.0930	0.6667	0	0	0.3333
	S_5	0.2853	0.2306	0.3787	0.3247	0	0.2966
	S_6	0.1413	0.5057	0	0.6667	0.3333	0
	S_7	0.2134	0.3681	0.3202	0.2915	0.3882	0
	S_8	0.7340	0.2650	0	0	0	1
	S_9	0.6223	0.3757	0	0	0.3333	0.6667
	S_{10}	0.4020	0.5970	0	0	1	0
	S_{11}	0.5127	0.4853	0	0	0.6667	0.3333
	S_{12}	0.5457	0.1790	0.3333	0	0	0.6667
	S_{13}	0.4544	0.3031	0.2934	0	0.3428	0.3638
	S_{14}	0.2717	0.5513	0	0.3333	0.6667	0
	S_{15}	0.3630	0.4272	0	0.3955	0.2563	0.3483

intensity vectors given by

$$\begin{aligned}
&P\left(\tilde{\mathbf{p}}^{(\xi)} \mapsto \tilde{\mathbf{p}}^{(\zeta)}\right) \\
=&P\left(\left\|\mathbf{r}-\mathbf{H}\tilde{\mathbf{p}}^{(\xi)}\right\|^2 > \left\|\mathbf{r}-\mathbf{H}\tilde{\mathbf{p}}^{(\zeta)}\right\|^2\right) \\
=&P\left(\left\|\mathbf{H}\tilde{\mathbf{p}}^{(\xi)}\right\|^2/2-\mathbf{r}^{\mathrm{T}}\mathbf{H}\tilde{\mathbf{p}}^{(\xi)} > \left\|\mathbf{H}\tilde{\mathbf{p}}^{(\zeta)}\right\|^2/2-\mathbf{r}^{\mathrm{T}}\mathbf{H}\tilde{\mathbf{p}}^{(\zeta)}\right) \\
=&P\left(\left(\mathbf{H}\tilde{\mathbf{p}}^{(\xi)}+\mathbf{n}\right)^{\mathrm{T}}\mathbf{H}\left(\tilde{\mathbf{p}}^{(\zeta)}-\tilde{\mathbf{p}}^{(\xi)}\right) > \frac{\left\|\mathbf{H}\tilde{\mathbf{p}}^{(\zeta)}\right\|^2-\left\|\mathbf{H}\tilde{\mathbf{p}}^{(\xi)}\right\|^2}{2}\right) \\
=&P\left(\mathbf{n}^{\mathrm{T}}\mathbf{H}\left(\tilde{\mathbf{p}}^{(\zeta)}-\tilde{\mathbf{p}}^{(\xi)}\right) > \Psi\right),
\end{aligned} \tag{5.24}$$

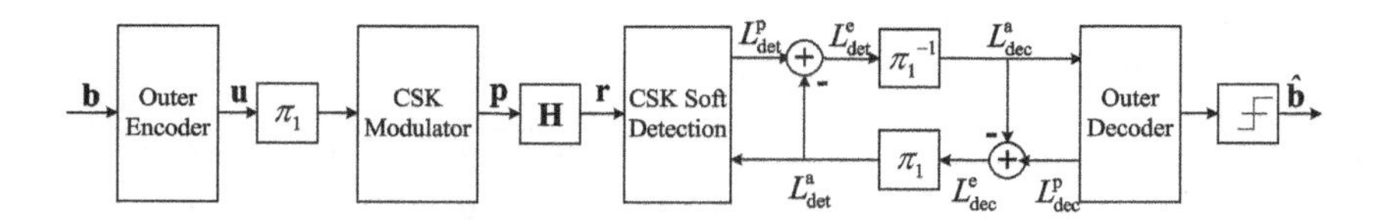

Figure 5.6 Diagram of MAP-based soft detection for coded CSK system.

where Ψ is defined as

$$\Psi = \frac{\left\|\mathbf{H}\tilde{\mathbf{p}}^{(\zeta)}\right\|^2 - \left\|\mathbf{H}\tilde{\mathbf{p}}^{(\xi)}\right\|^2}{2} - \left(\mathbf{H}\tilde{\mathbf{p}}^{(\xi)}\right)^{\mathrm{T}}\mathbf{H}\left(\tilde{\mathbf{p}}^{(\zeta)} - \tilde{\mathbf{p}}^{(\xi)}\right). \tag{5.25}$$

Since $\mathbf{H}\left(\tilde{\mathbf{p}}^{(\zeta)} - \tilde{\mathbf{p}}^{(\xi)}\right)$ is a constant matrix and the noise term $\mathbf{n}$ is Gaussian-distributed, the term $\mathbf{n}^{\mathrm{T}}\mathbf{H}\left(\tilde{\mathbf{p}}^{(\zeta)} - \tilde{\mathbf{p}}^{(\xi)}\right)$ also follows the Gaussian distribution with zero mean and variance of $\left\|\mathbf{H}\left(\tilde{\mathbf{p}}^{(\zeta)} - \tilde{\mathbf{p}}^{(\xi)}\right)\right\|^2 \sigma_0^2$, where σ_0^2 is the variance of the noise. Therefore, the PEP between $\tilde{\mathbf{p}}^{(\xi)}$ and $\tilde{\mathbf{p}}^{(\zeta)}$ is expressed as

$$P\left(\tilde{\mathbf{p}}^{(\xi)} \mapsto \tilde{\mathbf{p}}^{(\zeta)}\right) = Q\left(\frac{\Psi}{\left\|\mathbf{H}\left(\tilde{\mathbf{p}}^{(\xi)} - \tilde{\mathbf{p}}^{(\zeta)}\right)\right\|\sigma_0}\right), \tag{5.26}$$

where $Q(x) = \frac{1}{\sqrt{2\pi}}\int_x^\infty \exp\left(-\frac{u^2}{2}\right)du$ represents the standard tail probability function of the Gaussian distribution with zero mean and unity variance.

A joint maximum *a posteriori* (MAP)-based soft detection was proposed in [10], whose diagram is shown in Fig. 5.6. The information bit vector $\mathbf{b}$ is firstly encoded by the channel encoder, and the coded vector $\mathbf{u}$ is sent to the interleaver π_1. The interleaved bits are used for CSK mapping to modulate the optical sources. At the receiver, a joint MAP-based soft detection is used to generate the soft information as the input for channel decoder, which exchanges *extrinsic* information with the channel decoder, and hard decision is only performed when the channel decoder reaches its maximum number of iterations. In Fig. 5.6, $L^{\mathrm{p}}_{\mathrm{det}}$, $L^{\mathrm{a}}_{\mathrm{det}}$, and $L^{\mathrm{e}}_{\mathrm{det}}$ denote the *a posteriori*, *a priori*, and *extrinsic* log-likelihood ratio (LLR) of the detection module, while $L^{\mathrm{p}}_{\mathrm{dec}}$, $L^{\mathrm{a}}_{\mathrm{dec}}$, and $L^{\mathrm{e}}_{\mathrm{dec}}$ represent the *a posteriori*, *a priori*, and *extrinsic* LLR of the channel decoder. For the bit vector $\mathbf{u}$ with $\log_2 M$ bits, the bit-wise *a posteriori* information of its vth bit is given by the Max-Log approximation as [10]

$$\begin{aligned} L^{\mathrm{p}}_{\mathrm{det}}(u_v) = & L^{\mathrm{a}}_{\mathrm{det}}(u_v) + \max_{\tilde{\mathbf{p}} \in \mathcal{S}^1_{u_v}} \left(-\|\mathbf{r} - \mathbf{H}\tilde{\mathbf{p}}\|^2/2\sigma^2 + A\right) \\ & - \max_{\tilde{\mathbf{p}} \in \mathcal{S}^0_{u_v}} \left(-\|\mathbf{r} - \mathbf{H}\tilde{\mathbf{p}}\|^2/2\sigma^2 + A\right), \end{aligned} \tag{5.27}$$

where A is defined as $A = \sum_{\tau=1,\tau\neq v}^{N_b} u_v L^{\mathrm{a}}_{\mathrm{det}}(u_\tau)$, $\mathcal{S}^0_{u_v}$ and $\mathcal{S}^1_{u_v}$ are the subsets of $\mathcal{S}$ given by $\mathcal{S}^0_{u_v} = \{\tilde{\mathbf{p}} \in \mathcal{S} | u_v = 0\}$ and $\mathcal{S}^1_{u_v} = \{\tilde{\mathbf{p}} \in \mathcal{S} | u_v = 1\}$. The *extrinsic*

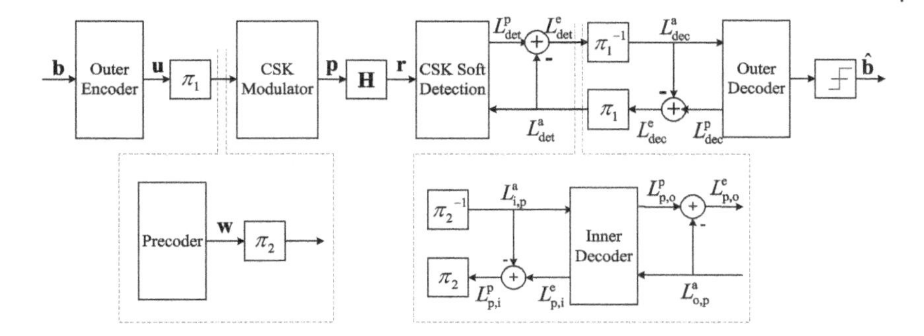

Figure 5.7 Diagram of three-stage CSK system with MAP-based soft detection and URC precoder.

LLRs $L^{\mathrm{e}}_{\mathrm{det}}$ are firstly deinterleaved before fed into the outer decoder as the *a priori* LLRs $L^{\mathrm{a}}_{\mathrm{dec}}$. Afterwards, the updated *extrinsic* LLRs $L^{\mathrm{e}}_{\mathrm{dec}}$ are fed back and reinterleaved, and then they are used as the *a priori* LLRs $L^{\mathrm{a}}_{\mathrm{det}}$ for the detection block.

In order to reduce the error floor of a coded system, a unity-rate code (URC) is usually utilized after the channel encoder since it can spread the extrinsic information without increasing the system's interleaver delay with its infinite impulse response. Besides, it keeps the throughput unchanged with negligible increase of complexity. Therefore, a three-stage CSK system was also proposed in [10], which inserts a URC encoder and interleaver at the transmitter. At the receiver, the iterations are firstly performed between the joint MAP-based soft detection and the URC decoder referred to as inner iterations. Afterwards, its output $L^{\mathrm{p}}_{\mathrm{p,o}}$ is fed to the outer decoder, which calculates *extrinsic* information $L^{\mathrm{a}}_{\mathrm{o,p}}$ for the inner decoder in return. The diagram of the three-stage CSK system with MAP-based soft detection and URC precoder is illustrated in Fig. 5.7.

5.3
Wavelength division multiplexing with predistorion

Another multicolor modulation scheme is WDM, which utilizes colors with different wavelengths for data transmission. Unlike the CSK modulation where one symbol is might be mapped to different colors, the streams modulated on different optical sources in WDM are independent to provide parallel transmission. Recently, WDM has been employed in VLC links to boost the data rate, and several gigabits/s transmissions have been reported [11–13].

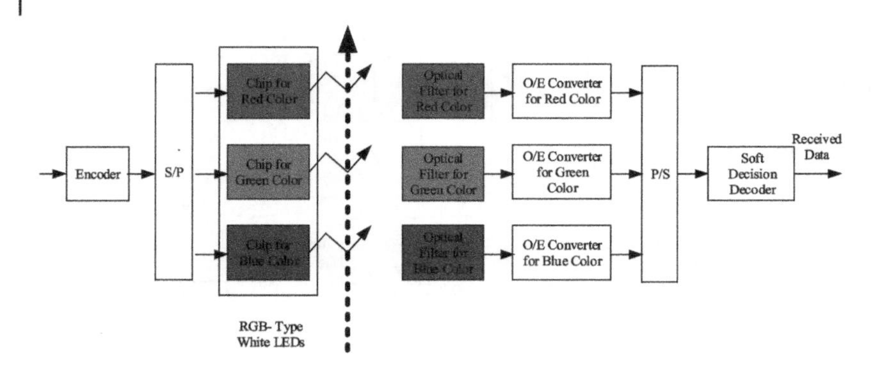

Figure 5.8 VLC system with RGB-type white LEDs.

5.3.1
System model

A VLC system using RGB-type white LEDs is illustrated in Fig. 5.8, where the information bits are encoded by the channel encoder and then serial-to-parallel (S/P) converted into three streams. Each bit-stream is sent to the chip of a primary color LED for simple OOK modulation. In order to generate white light, the modulated optical signals of the three primary color LEDs are mixed using specific optical power mixing ratios. At the receiver, the optical filters for the three primary colors are employed to attenuate the ambient light and separate the three primary color signals, which are converted to their electronic counterparts by the optical to electronic (O/E) converters. The three received electronic signals are then parallel to serial (P/S) converted into the single signal stream for the soft-decision decoder.

Since illuminance takes priority over communication for indoor LEDs, one should make sure that the appropriate optical power mixing ratio of the three primary colors are used for white light. Let M_r, M_g, and M_b denote the optical power emitted from the red, green, and blue LEDs. In [14], four combinations of the three primary colors and their optical power mixing ratios $M_r : M_g : M_b$ were provided to generate the white light, whose wavelengths and the mixture ratios are listed in Table 5.5. Obviously, unequal optical power ratios are required for three primary colors to produce the white light. The O/E conversion efficiency η is calculated as $\eta = \gamma\frac{e\lambda}{hc}$, where γ is the quantum efficiency of the photo detector, e is the electron charge, λ is the signal wavelength, h is Plank's constant, and c is the speed of light [15]. For different colors, both the wavelengths and the quantum efficiencies of the photo detector are different. Therefore, the conversion efficiencies between the optical and electronic signals are also different for the three primary colors, which are denoted as η_r, η_g, and η_b for the red, green, and blue colors, respectively. The amplitude of the electronic signal at the receiver is proportional to the intensity of the received light. If the electronic energies received per symbol from the red, green, and blue LEDs are defined as E_r, E_g, and E_b, we have

$$\sqrt{E_r} : \sqrt{E_g} : \sqrt{E_b} = M_r\eta_r : M_g\eta_g : M_b\eta_b. \tag{5.28}$$

Table 5.5 Combination of three primary colors and the corresponding optical power mixing ratio $M_r : M_g : M_b$ for white light [14].

Type	Red	Green	Blue
1. Wavelength (nm)	600	555	480
Mixture ratio	1	0.89	2.51
2. Wavelength (nm)	610	555	475
Mixture ratio	1	1.43	2.29
3. Wavelength (nm)	610	555	450
Mixture ratio	1	2.62	1.96
4. Wavelength (nm)	610	565	450
Mixture ratio	1	11.17	7.19

5.3.2 Receiver-side predistortion

Since the energies received from different LEDs not equal, the reliability of received signal from each O/E converter corresponding to each primary color is also different, which causes the performance degradation when soft-decision decoder is used at the receiver. In order to improve the system performance, a predistorion module can be used to predistort the received signal before it is passed to the soft-decision decoder [16]. We denote the serial received signal from the P/S converter as $r(t)$, which combines the signals from the red, green, and blue color converters. The weight parameters of the predistortion block for different LED signals are defined as F_r, F_g, and F_b. The predistortion block gives the signal $r(t)$ different weights according to which O/E converter the signal comes from. Therefore, the output of the predistortion block $r'(t)$ is given by

$$r'(t) = \begin{cases} r(t) \cdot F_r, & r(t) \in \text{red light;} \\ r(t) \cdot F_g, & r(t) \in \text{green light;} \\ r(t) \cdot F_b, & r(t) \in \text{blue light.} \end{cases} \tag{5.29}$$

With the predistortion module, a higher weight is assigned to the more reliable signal, while a less reliable signal has a lower weight. However, the weighting factors of the predistortion block must be carefully selected to optimize the achievable system performance.

Consider a classic convolutional code as an example. When the soft-decision Viterbi decoder [17] is used, a common technique of estimating the attainable performance of the decoder is to use the union upper bound of the first event error probability, which is given by [18]

$$P_e \leq \sum_{d=d_{free}}^{\infty} n_d P_d, \tag{5.30}$$

where d_{free} is the free distance of the convolutional code, n_d denotes the number of trellis-paths having a distance d from the all-zero path that merge with the all-zero path for the first time, and P_d represents the pairwise error probability. Furthermore, n_d is the coefficient of the polynomial $T(B, D)|_{B=1}$ derived from the transfer function $T(B, D)$ [17]. The VLC channel may be modeled by an AWGN channel under the line of sight user scenario of high-rate optical wireless communication systems using white LEDs [19]. For the AWGN channel having the noise power spectral density of $N_0/2$, the pairwise error probability P_d for OOK modulation can be expressed as

$$P_d = Q\left(\sqrt{\frac{d\,E_s}{N_0}}\right), \tag{5.31}$$

where E_s denotes the energy per symbol.

Since the energy received from each O/E converter corresponding to each primary color is different, the overall energy of each path is also different. Thus, the union upper bound of the first event error probability in the absence of the predistortion block can be rewritten as

$$P_e \leq \sum_{d=d_{free}}^{\infty} \sum_{k=1}^{n_d} Q\left(\sqrt{\frac{\left(r_{d,k}\sqrt{E_r} + g_{d,k}\sqrt{E_g} + b_{d,k}\sqrt{E_b}\right)^2}{d\,N_0}}\right), \tag{5.32}$$

where k represents the kth path at a distance of d from the all-zero path that merges with the all-zero path for the first time, while the numbers of ones transmitted from the red, green, and blue LEDs in the kth path are defined as $r_{d,k}$, $g_{d,k}$, and $b_{d,k}$, respectively. Apparently, $r_{d,k} + g_{d,k} + b_{d,k} = d$, and the upper bound in (5.32) is larger than that given in (5.30).

In order to improve the performance of the soft-decision Viterbi decoder, the predistortion block allows more reliable signals to contribute more to both the branch- and path-metric calculation, while reducing the contribution of the less reliable signals, by weighting the three signals gleaned from the three O/E converters corresponding to the primary colors of red, green, and blue with the weighting factors of F_r, F_g, and F_b. The weighting factor changes both the desired signal energy and the noise power gleaned from each O/E converter simultaneously, which is proportional to the squared value of the weight applied, and therefore the union upper bound of the first event error probability can be expressed as

$$P_e \leq \sum_{d=d_{free}}^{\infty} \sum_{k=1}^{n_d} Q\left(\sqrt{\frac{\left(r_{d,k}F_r\sqrt{E_r} + g_{d,k}F_g\sqrt{E_g} + b_{d,k}F_b\sqrt{E_b}\right)^2}{\left(r_{d,k}F_r^2 + g_{d,k}F_g^2 + b_{d,k}F_b^2\right)N_0}}\right).$$

(5.33)

Based on Cauchy–Schwarz inequality, the union upper bound in (5.33) is minimized as

$$\sum_{d=d_{free}}^{\infty}\sum_{k=1}^{n_d} Q\left(\sqrt{\frac{\left(r_{d,k}F_r\sqrt{E_r}+g_{d,k}F_g\sqrt{E_g}+b_{d,k}F_b\sqrt{E_b}\right)^2}{\left(r_{d,k}F_r^2+g_{d,k}F_g^2+b_{d,k}F_b^2\right)N_0}}\right)$$
$$\geq \sum_{d=d_{free}}^{\infty}\sum_{k=1}^{n_d} Q\left(\sqrt{\frac{\left(r_{d,k}F_r^2+g_{d,k}F_g^2+b_{d,k}F_b^2\right)\cdot\left(r_{d,k}E_r+g_{d,k}E_r+b_{d,k}E_r\right)}{\left(r_{d,k}F_r^2+g_{d,k}F_g^2+b_{d,k}F_b^2\right)N_0}}\right)$$
$$= \sum_{d=d_{free}}^{\infty}\sum_{k=1}^{n_d} Q\left(\sqrt{\frac{\left(r_{d,k}E_r+g_{d,k}E_r+b_{d,k}E_r\right)}{N_0}}\right), \tag{5.34}$$

where the last equality holds only when

$$F_r : F_g : F_b = \sqrt{E_r} : \sqrt{E_g} : \sqrt{E_b}. \tag{5.35}$$

To keep the signal power unchanged after predistortion, the optimal weighting parameters are given by

$$F_r = \frac{\sqrt{3}M_r\eta_r}{\sqrt{M_r^2\eta_r^2+M_g^2\eta_g^2+M_b^2\eta_b^2}}, \tag{5.36}$$

$$F_g = \frac{\sqrt{3}M_g\eta_g}{\sqrt{M_r^2\eta_r^2+M_g^2\eta_g^2+M_b^2\eta_b^2}}, \tag{5.37}$$

$$F_b = \frac{\sqrt{3}M_b\eta_b}{\sqrt{M_r^2\eta_r^2+M_g^2\eta_g^2+M_b^2\eta_b^2}}. \tag{5.38}$$

According to (5.36)–(5.38), it is interesting to note that the weighting factors depend only on the emitted optical powers and on the conversion efficiencies of the O/E converters regardless of the structure of the convolutional codes. Therefore, they are optimal for any convolutional codes. The computational complexity of the predistortion module is extremely low, which requires only a single accumulation and a multiplication operation. For other soft-decision decoders, such as the belief propagation (BP) decoder of low density parity check (LDPC) codes [20] and the BCJR decoder of turbo codes [21], the error performance bounds are less straightforward to obtain, since their code structures are complex and the decoding algorithms are iterative. Therefore, finding the optimal weighting parameters of the predistortion module for these soft-decision decoders is challenging. Fortunately, the optimal weighting parameters of the predistortion block derived for the soft-decision Viterbi decoder may nonetheless be beneficial for employment in soft-decision aided LDPC and turbo decoders.

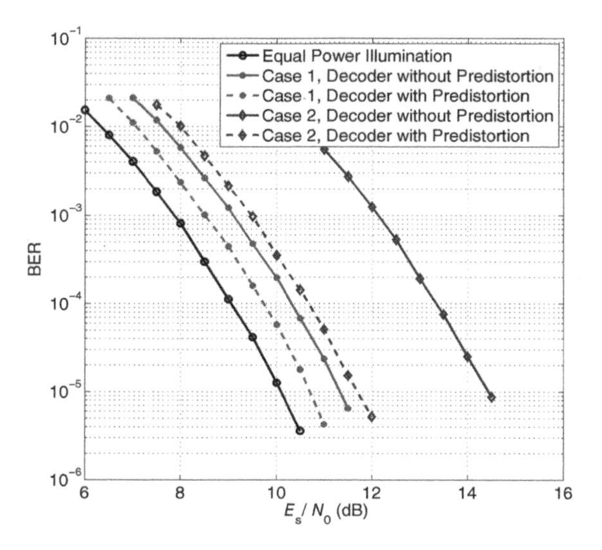

Figure 5.9 BER performance comparison of the RGB-type LED-based VLC system with and without the predistortion module, where the convolutional code with the soft-decision Viterbi decoder is used.

5.3.3 Performance evaluation

Two cases of different mixture ratios of the optical powers emitted from the red, green, and blue LEDs are considered in order to create white light according to Table 5.5. For Case 1, the mixing ratio of the optical emitted powers is $M_r : M_g : M_b = 1 : 2.62 : 1.96$ and the corresponding wavelengths of the red, green, and blue colors are 610 nm, 555 nm, and 450 nm, respectively. By contrast, for Case 2, the mixing ratio is $M_r : M_g : M_b = 1 : 11.17 : 7.19$ and the corresponding wavelengths of the red, green, and blue colors are 610 nm, 565 nm, and 450 nm, respectively. For simplicity, the O/E conversion efficiencies for the three primary colors, namely η_r, η_g, and η_b, are assumed to be equal. Naturally, this does not alter the fundamental nature of the unequal powers of the three received electronic signals corresponding to the red, green, and blue LEDs. Case 2 represents a much more uneven optical power mixture of the three primary colors for emitting white light.

A convolutional code, an LDPC code, and a turbo code are used in the simulations. The convolutional code employed has a code rate of 1/2, while its constraint length is seven and the generator polynomials are the same as $[171, 133]_8$. Soft-decision Viterbi decoder is used with the trace back length of five times of the constraint length. The LDPC code used in the simulations is that of the IEEE 802.11 standard

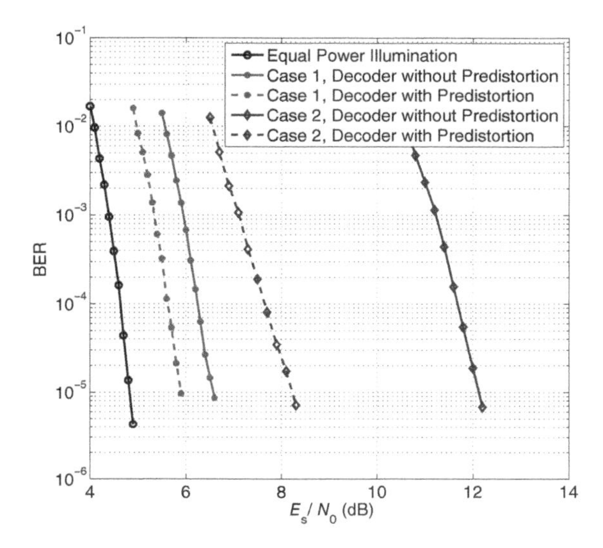

Figure 5.10 BER performance comparison of the RGB-type LED-based VLC systems with and without the predistortion module, where the LDPC code with the BP decoder is used.

with a codeword-length of 1944 bits and a code rate of 1/2 [22], while the BP decoder is employed at the receiver and the maximum number of iterations is 30. At last, the turbo code for the 3rd Generation Partnership Project (3GPP) is used with a code rate of 1/3 and an interleaver length of 1440 bits [23]. The BCJR decoder is used at the receiver and the number of iterations is set to six.

The BER performance obtained both with and without the predistortion block are illustrated in Figs. 5.9–5.11 for the convolutional code, LDPC code, and turbo code, respectively. The BER curve obtained from the case of the equal optical power radiations from the red, green, and blue LEDs is also included as the benchmark, which represents the lower bound of BER obtained with the aid of an equal optical power mixture of $M_r : M_g : M_b = 1 : 1 : 1$.

From Figs. 5.9–5.11, it can be seen that a significant performance gain can be attained with the predistortion block. Quantitatively, at the BER level of 10^{-5} and for the unequal optical power mixture of the three primary colors of Case 1, the performance gains of 0.6 dB, 0.7 dB, and 0.5 dB are achieved, respectively, for the convolutional code, LDPC code, and turbo code over the corresponding systems operating without predistortion. As expected, the attainable BER performance of Case 2 is considerably poorer than that of Case 1 since the optical power distribution of the red, green, and blue LEDs in Case 2 is much more uneven. However, the performance gain attained by the predistortion block is significantly higher in Case 2 due to larger imbalance. Specifically, at the BER level of 10^{-5}, the predistortion scheme

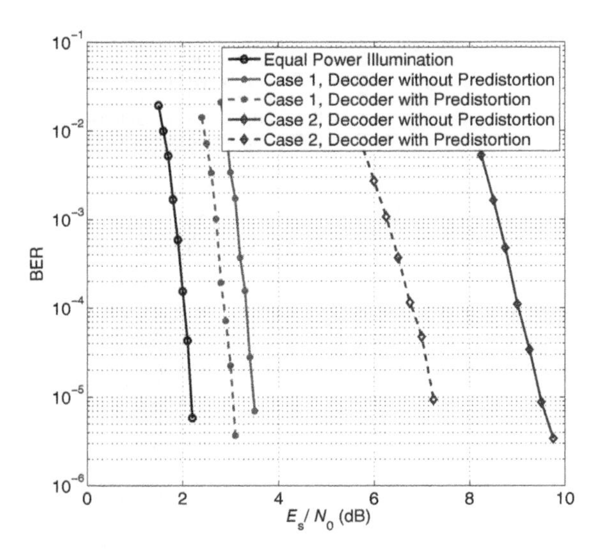

Figure 5.11 BER performance comparison of the RGB-type LED-based VLC systems with and without the predistortion module, where the turbo code with the BCJR decoder is used.

attains SNR gains of 2.8 dB, 3.9 dB, and 2.2 dB, respectively, for the convolutional code, LDPC code, and turbo code. The results of Figs. 5.9–5.11 also confirm that the best BER performance is obtained with the equal optical illumination powers of the red, green, and blue LEDs. However, the equal optical power mixture of the three primary colors cannot produce white light and, therefore, it cannot carry out the primary lighting function.

5.4 Conclusion

This chapter discusses multicolor modulation schemes for VLC systems with RGB-type LEDs, which combine light from red, green, and blue LEDs for white light. Due to its good property, CSK was developed and adopted in the IEEE 802.15.7 standard. Besides the CSK constellation in the standard, the optimal design rules of CSK constellation as well as Quad-LED CSK are provided to achieve superior capacity, while CSK with coded modulation is introduced for practical transmission. Finally, WDM VLC system is discussed with channel coding, and a receiver-side predistortion is proposed before channel decoding, which has shown significant performance gain for WDM systems.

References

1 IEEE Std. 802.15.7-2011, *Part 15.7: Short-Range Wireless Optical Communication Using Visible Light*, Sep. 2011.
2 A. Yokoi, J. Son, and T. Bae, "More description about CSK constellation," March 2011, IEEE 802.15 contribution 15-11-0247-00-0007.
3 S. Rajagopal, R. D. Roberts, and S. K. Lim, "IEEE 802.15.7 visible light communication: Modulation schemes and dimming support," *IEEE Commun. Mag.*, vol. 50, no. 3, pp. 72–82, Mar. 2012.
4 R. J. Drost and B. M. Sadler, "Constellation design for color-shift keying using billiards algorithms," in *Proc. IEEE Global Communications Conference (GLOBECOM) Workshops 2010* (Miami, FL), Dec. 6–10, 2010, pp. 980–984.
5 E. Monteiro and S. Hranilovic, "Constellation design for color-shift keying using interior point methods," in *Proc. IEEE Global Communications Conference (GLOBECOM) Workshops 2012* (Anaheim, CA), Dec. 3–7, 2012, pp. 1224–1228.
6 R. J. Drost and B. M. Sadler, "Constellation design for channel precompensation in multi-wavelength visible light communications," *IEEE Trans. Commun.*, vol. 62, no. 6, pp. 1995–2005, Jun. 2014.
7 E. Monteiro and S. Hranilovic, "Design and implementation of color-shift keying for visible light communications," *J. Lightw. Technol.*, vol. 32, no. 10, pp. 2053–2060, May 2014.
8 A. Wilkins, J. Veitch, and B. Lehman, "LED lighting flicker and potential health concerns: IEEE standard PAR1789 update," in *Proc. IEEE Energy Conversion Congress and Exposition 2010* (Atlanta, GA), Sep. 12–16, 2010, pp. 171–178.
9 R. Singh, T. O'Farrell, and J. P. R. David, "An enhanced color shift keying modulation scheme for high-speed wireless visible light communications," *J. Lightw. Technol.*, vol. 32, no. 14, pp. 2852–2592, Jul. 2014.
10 J. Jiang, R. Zhang, and L. Hanzo, "Analysis and design of three-stage concatenated color-shift keying," *IEEE Trans. Veh. Technol.*, vol. 64, no. 11, pp. 5126–5136, Nov. 2015.
11 W. Y. Lin, C. Y. Chen, H. H. Lu, C. H. Chang, Y. P. Lin, H. C. Lin, and H. W. Wu, "10m/500Mbps WDM visible light communication systems," *Opt. Exp.*, vol. 20, no. 9, pp. 9919–9924, Apr. 2012.
12 F. M. Wu, C. T. Lin, C. C. Wei, C. W. Chen, Z. Y. Chen, and K. Huang, "3.22-Gb/s WDM visible light communication of a single RGB LED employing carrier-less amplitude and phase modulation," in *Proc. Optical Fiber Communication Conference and Exposition and the National Fiber Optic Engineers Conference (OFC/NFOEC) 2013* (Anaheim, CA), Mar. 17–21, 2013, OTh1G.4.
13 Y. Wang, Y. Wang, N. Chi, J. Yu, and H. Shang, "Demonstration of 575-Mb/s downlink and 225-Mb/s uplink bi-directional SCM-WDM visible light communication using RGB LED and phosphor-based LED," *Opt. Exp.*, vol. 21, no. 1, pp. 1203–1208, Jan. 2013.
14 Y. Tanaka, T. Komine, S. Haruyama, and M. Nakagawa, "Indoor visible light data transmission system utilizing white LED lights," *IEICE Trans. Commun.*, vol. E86-B, no. 8, pp. 2440–2454, Aug. 2003.
15 X. Zhu and J. M. Kahn, "Free-space optical

communication through atmospheric turbulence channels," *IEEE Trans. Commun.*, vol. 50, no. 8, pp. 1293–1300, Aug. 2002.

16 Q. Wang, Z. Wang, S. Chen, and L. Hanzo, "Enhancing the decoding performance of optical wireless communication systems using receiver-side predistortion," *Opt. Exp.*, vol. 21, no. 25, pp. 30295–30305, Dec. 2013.

17 A. J. Viterbi, "Convolutional codes and their performance in communication systems," *IEEE Trans. Commun. Technol.*, vol. 19, no. 5, pp. 751–772, Oct. 1971.

18 J. G. Proakis and M. Salehi, *Digital Communications*, 5th Edition. McGraw-Hill: New York, 2008.

19 T. Komine, J. H. Lee, S. Haruyama, and M. Nakagawa, "Adaptive equalization system for visible light wireless communication utilizing multiple white LED lighting equipment," *IEEE Trans. Wirel. Commun.*, vol. 8, no. 6, pp. 2892–2900, Jun. 2009.

20 D. J. C. MacKay, "Good error-correcting codes based on very sparse matrices," *IEEE Trans. Inf. Theory*, vol. 45, no. 2, pp 399–431, Mar. 1999.

21 L. Bahl, J. Cocke, F. Jelinek, and J. Raviv, "Optimal decoding of linear codes for minimizing symbol error rate," *IEEE Trans. Inf. Theory*, vol. 20, no. 2, pp. 284–287, Mar. 1974.

22 IEEE Std. 802.11-2012, *Part 11: Wireless LAN Medium Access Control (MAC) and Physical Layer (PHY) Specifications*, Mar. 2012.

23 3GPP2 C.S0002-F v1.0, *Physical Layer Standard for CDMA2000 Spread Spectrum Systems*, Dec. 2012.

6
Optical MIMO

In this chapter, we discuss optical multiple-input multiple-output (MIMO) techniques for imaging and non-imaging visible light communication (VLC) systems, including modern optical MIMO, optical spatial modulation (OSM), optical space shift keying (OSSK), and optical MIMO combined with orthogonal frequency division multiplexing (MIMO-OFDM). Multiuser precoding techniques for VLC systems are also introduced under lighting constraints.

Despite the fact that visible light spectrum is as wide as several terahertz (THz), the bandwidth of off-the-shelf light emitting diodes (LEDs) is limited, which makes it very challenging to achieve high-data-rate transmission. Meanwhile, in order to provide sufficient illumination, multiple LED units are usually installed in a single room. Therefore, MIMO techniques can be naturally employed in indoor VLC systems to boost the data rate. Typically, there are two optical MIMO approaches for VLC, namely, non-imaging and imaging MIMO. In non-imaging MIMO systems, each receiver collects light with its own optical concentrator, while the imaging MIMO systems employ an imaging diversity receiver structure to distinguish the light from different transmitters. Two optical MIMO systems are introduced in Section 6.1 and Section 6.2, respectively.

Meanwhile, MIMO VLC systems usually support data transmission for multiple users simultaneously. In order to eliminate the inter-user interference, precoding techniques can be utilized at the transmitters. Although precoding techniques have been widely investigated in RF communications, these schemes cannot be applied to VLC systems directly since intensity modulation is used in VLC systems. Therefore, several precoding schemes for multiuser MIMO VLC under lighting constraints are provided in Section 6.3.

Finally, MIMO-OFDM is a promising technique to provide high-speed VLC transmission, which inherits the advantages of both MIMO and OFDM. In Section 6.4, MIMO-OFDM for VLC systems under both single user and multiple user scenarios are introduced.

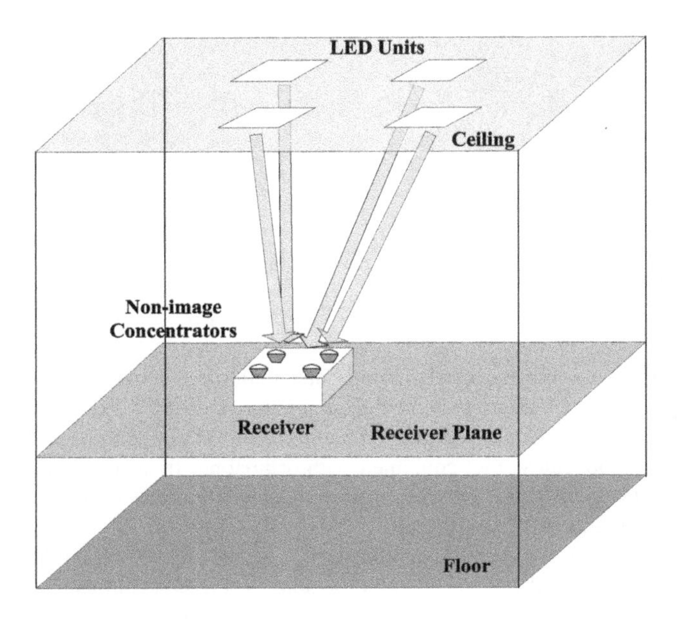

Figure 6.1 Non-imaging optical MIMO VLC system.

6.1 Non-imaging optical MIMO techniques

A non-imaging MIMO VLC system is illustrated in Fig. 6.1, where a single room is equipped with multiple LED units for illumination, which can cooperate to transmit information simultaneously. We assume N_t LED units are at the ceiling, and N_r non-imaging concentrators and photodiodes (PDs) are at the receiver to collect light.

6.1.1 Channel response

The DC gain of the subchannel $h_{p,q}$ between the qth LED unit and the pth PD is expressed as [1]

$$h_{p,q} = \begin{cases} \frac{\rho_p A_p}{d_{p,q}^2} R(\phi_q) \cos(\varphi_{p,q}), & \varphi_{p,q} \leq \Psi_{c,p}, \\ 0, & \varphi_{p,q} > \Psi_{c,p}, \end{cases} \tag{6.1}$$

where ρ_p is the responsivity coefficient of the PD, $d_{p,q}$ is the distance between the qth LED unit and the pth PD, ϕ_q is the emission angle of the qth LED unit, $\varphi_{p,q}$ is the incidence angle of the light, and $\Psi_{c,p}$ is the receiver field of view (FOV) of the pth PD. Unlike conventional RF channels, the channel matrix $\mathbf{H} = \{h_{p,q}\}_{N_r \times N_t}$ is real-valued and commonly highly correlated when the users locate closely [2]. For the pth PD, the receiver collection area A_p can be calculated as

$$A_p = \gamma^2 A_{\mathrm{PD},p} / \sin^2\left(\Psi_{c,p}\right), \tag{6.2}$$

where γ is the concentrator refractive index of the PD, $A_{\mathrm{PD},p}$ is the area of the pth PD. In (6.1), $R\left(\phi_q\right)$ denotes the generalized Lambertian radiant intensity given by [1]

$$R\left(\phi_q\right) = \left((m+1)\cos^m\left(\phi_q\right)\right)/2\pi, \tag{6.3}$$

where m is the order of Lambertian emission.

At the receiver, the optical signal is directly detected by the corresponding PD, which generates an electric signal proportional to the received optical power. Besides that, shot noise and thermal noise are induced at the receiver, which can be modeled as real-valued additive white Gaussian noise (AWGN) with zero mean, and the variance of the noise at the pth PD can be written as [1]

$$\sigma_p^2 = 2eP_pB + 2e\rho_p\chi_{\mathrm{amb}}A_p\left(1-\cos\left(\Psi_{c,p}\right)\right)B + i_{\mathrm{amp}}^2, \tag{6.4}$$

where e is the electronic charge, χ_{amb} is the ambient light photocurrent, B is the bandwidth of receiver, i_{amp} is the preamplifier noise current density, and P_p is the average received optical power at the pth PD collected from all the LED units.

6.1.2
Optical MIMO techniques

In this part, different optical MIMO techniques for non-imaging VLC systems will be introduced and compared, and maximum likelihood (ML) detection is used at the receiver assuming perfect channel estimation and synchronization. Denote the transmitted and received signal vector as $\mathbf{x} = [x_1\, x_2\, \ldots\, x_{N_t}]^{\mathrm{T}}$ and $\mathbf{y} = [y_1\, y_2\, \cdots\, y_{N_r}]^{\mathrm{T}}$, respectively, the channel matrix $\mathbf{H}$ can be calculated according to (6.1). Therefore, the transmitted signal vector can be estimated by

$$\hat{\mathbf{x}} = \arg\max_{\mathbf{x}} p_{\mathbf{y}}\left(\mathbf{y} \mid \mathbf{x}, \mathbf{H}\right) = \arg\min_{\mathbf{x}} \left\|\mathbf{y} - \mathbf{H}\mathbf{x}\right\|^2, \tag{6.5}$$

where $p_{\mathbf{y}}\left(\mathbf{y} \mid \mathbf{x}, \mathbf{H}\right)$ is the conditional probability density function.

The simplest technique of optical MIMO is the repetition coding (RC), where all the transmitters convey the same information: $x_1 = x_2 = \cdots = x_{N_t}$. In this way, multiple transmitters can provide transmit-diversity gain since all the lights can be accumulated at the receiver to enhance the received optical power, leading to better performance. If M-ary pulse amplitude modulation (PAM) is employed at the transmitter, RC can achieve the spectral efficiency of $\log_2(M)$ bit/s/Hz. In VLC system, the transmitted signal should be non-negative real-valued since intensity modulation is used. Therefore, unipolar M-ary PAM can be used whose intensity level is given by [2]

$$I_m^{\mathrm{PAM}} = \frac{2I}{M-1}m,\; m = 0, 1, \ldots, M-1, \tag{6.6}$$

where I is the average optical power. The lower bound of its bit error rate (BER) can be calculated as

$$P_{e,\text{PAM}} \geq \frac{2(M-1)}{M\log_2(M)} Q\left(\frac{1}{M-1}\sqrt{\gamma_{\text{elec,r}}}\right), \tag{6.7}$$

where $Q(x) = \frac{1}{\sqrt{2\pi}}\int_x^{\infty} \exp\left(-\frac{u^2}{2}\right)du$ represents the standard tail probability function of the Gaussian distribution with zero mean and unity variance, and $\gamma_{\text{elec,r}}$ denotes the received electrical signal-to-noise ratio (SNR).

In order to maintain the total optical power constant, the optical power for each transmitter is set to I/N_t and the optical-to-electrical conversion coefficient is normalized. Therefore, the received optical power at the pth PD from all the transmitters can be calculated as

$$I_{r,p} = \sum_{q=1}^{N_t} I h_{p,q}/N_t. \tag{6.8}$$

Denote $E_s = I^2 T_s$ as the mean emitted electrical energy of the intensity-modulated optical signals and T_s as the symbol period, when maximum ratio combining (MRC) is used, the electrical SNR after combining is given by

$$\gamma_{\text{elec,r}} = \frac{T_s}{N_0}\sum_{p=1}^{N_r}\left(\sum_{q=1}^{N_t}\frac{I}{N_t}h_{p,q}\right)^2 = \frac{E_s}{N_0 N_t^2}\sum_{p=1}^{N_r}\left(\sum_{q=1}^{N_t}h_{p,q}\right)^2. \tag{6.9}$$

With (6.7) and (6.9), the BER for RC is bounded by [2]

$$P_{e,\text{RC}} \geq \frac{2(M-1)}{M\log_2(M)} Q\left(\frac{1}{M-1}\sqrt{\frac{E_s}{N_0 N_t^2}\sum_{p=1}^{N_r}\left(\sum_{q=1}^{N_t}h_{p,q}\right)^2}\right). \tag{6.10}$$

RC can provide reliable transmission performance with transmit-diversity. However, it is not spectrally efficient since all the LEDs transmit the same information. If different LEDs convey independent information, enhanced spectral efficiency can be achieved, which is referred to as spatial multiplexing (SMP) [3]. When the unipolar M-PAM is used at each transmitter, and the optical power is also distributed equally to each LED as RC, the spectral efficiency of SMP is $N_t \log_2(M)$ bit/s/Hz. In order to estimate the BER of SMP, pairwise error probability (PEP) is used, which calculates the probability that the detector wrongly estimates the signal vector $\mathbf{x}_{m^{(1)}}$ as $\mathbf{x}_{m^{(2)}}$, and we have

$$\begin{aligned} \text{PEP}_{\text{SMP}} &= \text{PEP}\left(\mathbf{x}_{m^{(1)}} \to \mathbf{x}_{m^{(2)}} | \mathbf{H}\right) \\ &= Q\left(\sqrt{\frac{T_s}{4N_0}\left\|\mathbf{H}\left(\mathbf{x}_{m^{(1)}} \to \mathbf{x}_{m^{(2)}}\right)\right\|^2}\right). \end{aligned} \tag{6.11}$$

When all the possible error situations are considered, the BER of SMP can be upper-bounded by [2]

$$P_{e,\text{SMP}} \leq \frac{1}{M^{N_t}\log_2(M^{N_t})} \sum_{m^{(1)}=1}^{M^{N_t}} \sum_{m^{(2)}=1}^{M^{N_t}} d_H\left(\mathbf{b}_{m^{(1)}}, \mathbf{b}_{m^{(2)}}\right)$$
$$Q\left(\sqrt{\frac{T_s}{4N_0}\left\|\mathbf{H}\left(\mathbf{x}_{m^{(1)}} \rightarrow \mathbf{x}_{m^{(2)}}\right)\right\|^2}\right), \tag{6.12}$$

where $\mathbf{b}_{m^{(1)}}$ and $\mathbf{b}_{m^{(2)}}$ denote the bit vectors corresponding to the signal vectors $\mathbf{x}_{m^{(1)}}$ and $\mathbf{x}_{m^{(2)}}$, respectively, and $d_H\left(\mathbf{b}_{m^{(1)}}, \mathbf{b}_{m^{(2)}}\right)$ is the Hamming distance between $\mathbf{b}_{m^{(1)}}$ and $\mathbf{b}_{m^{(2)}}$.

As an energy-efficient modulation scheme, spatial modulation (SM) was proposed in [4], which transmits information not only by the amplitude and phase of the signals, but also by the indices of antennas. In SM, only one antenna is activated during one symbol duration, where some bits are transmitted by conventional constellations such as quadrature amplitude modulation (QAM) and PAM, and the index of the activated antenna represents the rest of the information. In this way, spatial dimension can be used to improve the spectral efficiency without extra transmitting power, leading to improved energy efficiency. Moreover, inter-channel interference is completely avoided since only one channel is used at a time. Due to its good performance, SM has been applied to the optical domain termed as optical SM (OSM) [5, 6]. Since intensity modulation is employed, unipolar PAM can be used at each transmitter. When N_t transmitters are used and M-ary unipolar PAM is mapped to the activated transmitter, the spectral efficiency of OSM is $\log_2(N_t) + \log_2(M)$ bit/s/Hz. Unlike RC and SMP, where the zero intensity is used in PAM, zero intensity cannot be used in SM since the receiver cannot distinguish which transmitter is activated if all the signals are zero. Therefore, modified PAM constellation is used in OSM, which is given by [2]

$$I_m^{\text{mPAM}} = \frac{2I}{M+1}m,\ m = 1, 2, \ldots, M. \tag{6.13}$$

Similar to SMP, the PEP of SM can be calculated by

$$\begin{aligned}\text{PEP}_{\text{SM}} &= \text{PEP}\left(\mathbf{x}_{m^{(1)}} \rightarrow \mathbf{x}_{m^{(2)}} \middle| \mathbf{H}\right) \\ &= Q\left(\sqrt{\frac{T_s}{4N_0}\sum_{p=1}^{N_r}\left|I_{m^{(1)}}^{\text{SM}} h_{pq^{(1)}} - I_{m^{(2)}}^{\text{SM}} h_{pq^{(2)}}\right|^2}\right).\end{aligned} \tag{6.14}$$

When all the MN_t signal vectors of OSM are considered, the upper bound of its

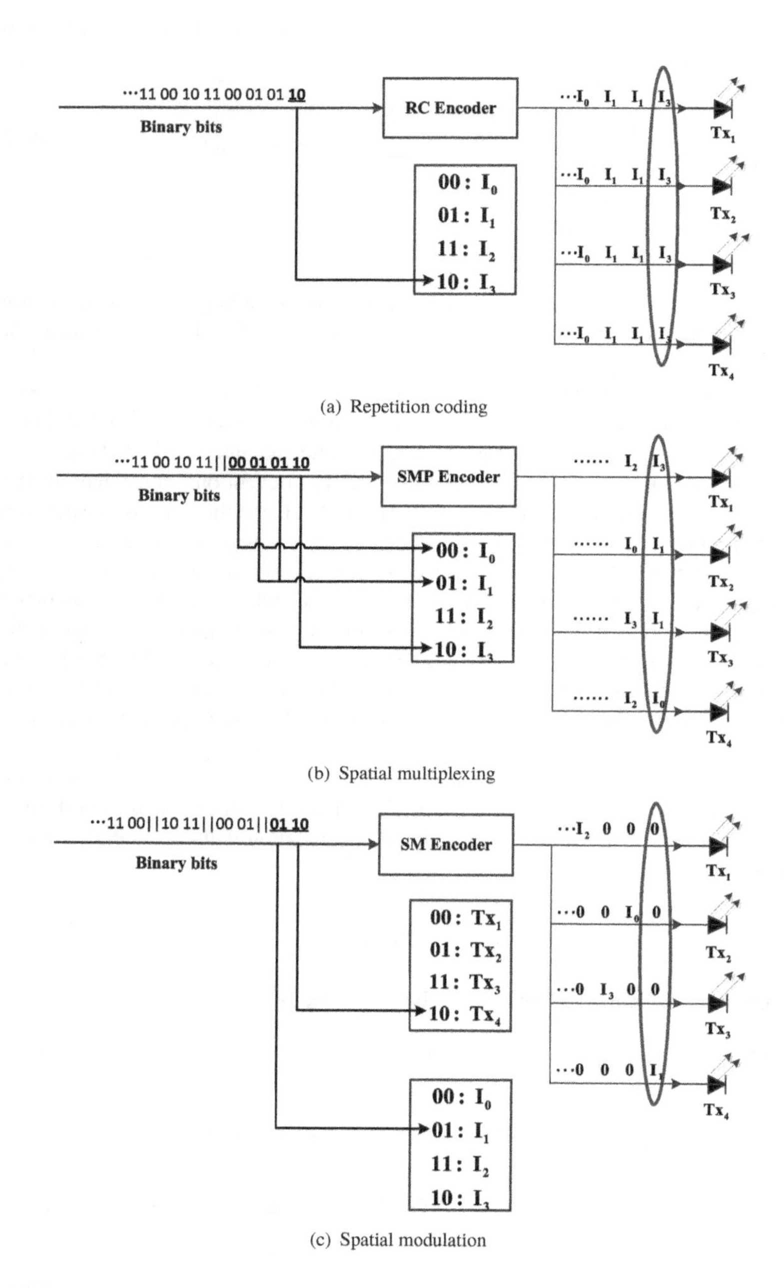

Figure 6.2 Examples of optical MIMO encoding.

BER is given by

$$P_{e,\mathrm{SM}} \leq \frac{1}{MN_t \log_2(MN_t)} \sum_{m^{(1)}=1}^{M} \sum_{q^{(1)}=1}^{N_t} \sum_{m^{(2)}=1}^{M} \sum_{q^{(2)}=1}^{N_t} d_{\mathrm{H}}\left(\mathbf{b}_{m^{(1)}q^{(1)}}, \mathbf{b}_{m^{(2)}q^{(2)}}\right) \tag{6.15}$$

$$Q\left(\sqrt{\frac{T_s}{4N_0} \sum_{p=1}^{N_r} \left| I_{m^{(1)}}^{\mathrm{SM}} h_{pq^{(1)}} - I_{m^{(2)}}^{\mathrm{SM}} h_{pq^{(2)}} \right|^2}\right),$$

where $\mathbf{b}_{m^{(1)}q^{(1)}}$ is the bit vector corresponding to the situation where transmitter $q^{(1)}$ is activated and its intensity is $I_{m^{(1)}}^{\mathrm{SM}}$, $\mathbf{b}_{m^{(2)}q^{(2)}}$ is the demodulated bit vector corresponding to the situation where transmitter $q^{(2)}$ is activated and its intensity is $I_{m^{(2)}}^{\mathrm{SM}}$. In Fig. 6.2, some examples of encoding for RC, SMP, and SM are given.

Due to the high channel correlation in line-of-sight (LOS) scenarios, OSM and SMP cannot perform well in visible light communications [2]. Therefore, power imbalance is proposed to reduce the channel correlation [7], which introduces independent amplification factors to different LEDs. The amplification matrix is denoted as $\mathbf{A} = \mathrm{diag}\{a_1, \ldots, a_{N_t}\}$, which is normalized by $\frac{\sum_{q=1}^{N_t} a_q}{N_t} = 1$. Therefore, $N_t - 1$ variables should be considered in the optimization, leading to high complexity. In [2], a heuristic solution was given with exponential function

$$a_q = \begin{cases} \frac{N_t}{\sum_{i=0}^{N_t-1} \alpha^i}, & q = 1, \\ \alpha a_{q-1}, & 1 < q \leq N_t, \end{cases} \tag{6.16}$$

where $\alpha = 10^{\frac{\beta}{10}}$ is the optical power imbalance factor and β is the optical power imbalance factor in dB. At the receiver, power imbalance factor can be regarded as part of the channel gain, which can be obtained by conventional channel estimation. Therefore, the receiver does not need to know the actual power imbalance factor, and power imbalance will not increase the complexity of the receiver.

6.1.3 Performance comparison

In this section, simulations are conducted to compare the MIMO techniques for non-imaging VLC system. The simulation parameters for the VLC system configuration are listed in Table 6.1. Both transmitters and receivers are aligned in a quadratically 2×2 manner as shown in Fig. 6.1, which are in the middle of the room. The vertical distance between the transmitters and receivers are 1.75 m. The spacing between the adjacent receivers is assumed to be $d_{\mathrm{RX}} = 0.1$ m on both the x- and y-axis. Different spacings between adjacent transmitters are investigated on the x- and y-axis, which are given by $d_{\mathrm{TX}} = 0.2$ m, 0.4 m, and 0.6 m. The corresponding channel matrices are obtained by (6.1).

Figure 6.3 illustrates the BER performance of RC, SMP, and OSM for the three transmitter spacings at the spectral efficiency of $R = 4$ bit/s/Hz, where the SNR is

Table 6.1 Simulation parameters for VLC system configuration.

Room Size (length × width × height)	5 m × 5 m × 3 m
LED emission angle ϕ_q	15°
PD area $A_{\text{PD},p}$	1 cm^2
PD responsivity coefficient ρ_p	1 A/W
PD concentrator refractive index γ	1
Lambertian emission mode number m	1
Receiver FOV angle $\Psi_{c,p}$	15°
Pre-amplifier noise density i_{amp}	5 pA/Hz$^{-1/2}$
Ambient light photocurrent χ_{amb}	10.93 A/m^2/Sr

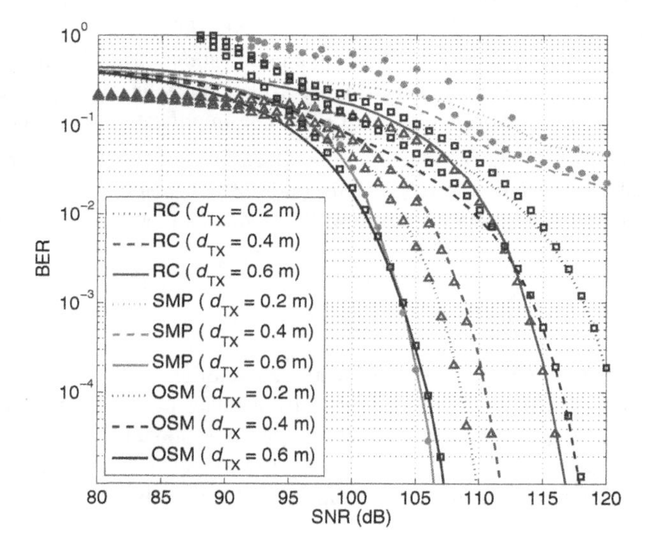

Figure 6.3 Performance comparison of RC, SMP, and OSM for non-imaging VLC system with spectral efficiency of 4 bit/s/Hz, where all the LEDs have equal power.

defined as the energy ratio of transmitted signal to noise. The theoretical lower and upper error bounds according to (6.10) and (6.12) are shown by markers, while the simulation results are shown in different lines. It can be seen that theoretical bounds match well with the simulation results especially at high SNR regions. In order to achieve the required spectral efficiency, 16PAM, 2PAM, and 4PAM are used for RC, SMP, and OSM, respectively. When $d_{\text{TX}} = 0.2$ m and 0.4 m, RC achieves the best performance while SMP performs worst due to the high channel correlation. When $d_{\text{TX}} = 0.6$ m and the channel matrix is well-conditioned, however, RC suffers 7 dB

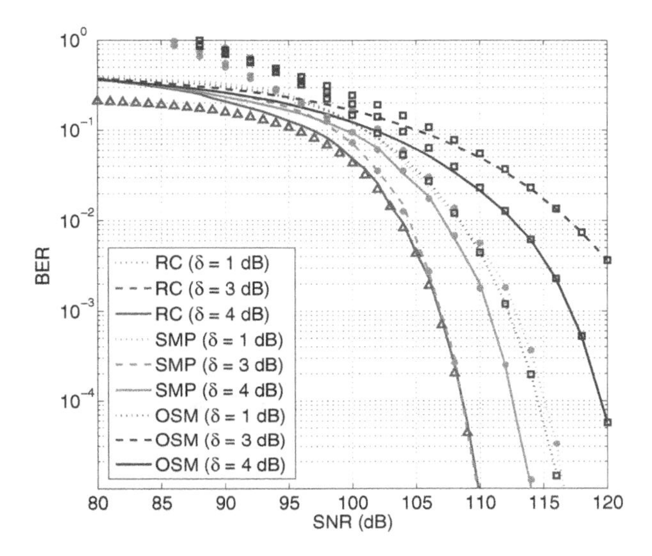

Figure 6.4 Performance comparison of RC, SMP, and OSM for 4×4 non-imaging VLC system with spectral efficiency of 4 bit/s/Hz and different power imbalance factors.

performance degradation compared with the scenario with $d_{\mathrm{TX}} = 0.2$ m since less energy is received. Interestingly, both SMP and OSM outperform RC significantly, and they are even better than RC for the other two scenarios with more received energy. At high SNR regions, SMP outperforms SM since a lower modulation order is used with its multiplexing gain. However, when the SNR is lower than 103 dB, OSM achieves the best performance because OSM can convey information in the spatial domain.

Figure 6.4 compares the BER of RC, SMP, and OSM with $d_{\mathrm{TX}} = 0.2$ m at the spectral efficiency of $R = 4$ bit/s/Hz. Power imbalance is considered with different power imbalance factors δ = 1 dB, 3 dB, and 4 dB. With imbalanced power, the correlation of the channel matrix can be reduced. It can be observed that considerable performance gain can be achieved with power imbalance for both SMP and OSM, while the performance of RC remains the same when power imbalance factor varies. For RC, its performance is only related to the absolute channel gains rather than the channel correlation as shown in (6.10). Therefore, power imbalance has no influence on its performance. Although larger power imbalance factor may lead to lower channel correlation, it also reduces the transmitted power for some of the links, which may also degrade the system performance. Therefore, there should be a tradeoff between the correlation and the transmitted power. For OSM, the best power imbalance factor is about δ = 1 dB, while for SMP, δ = 3 dB provides the best performance.

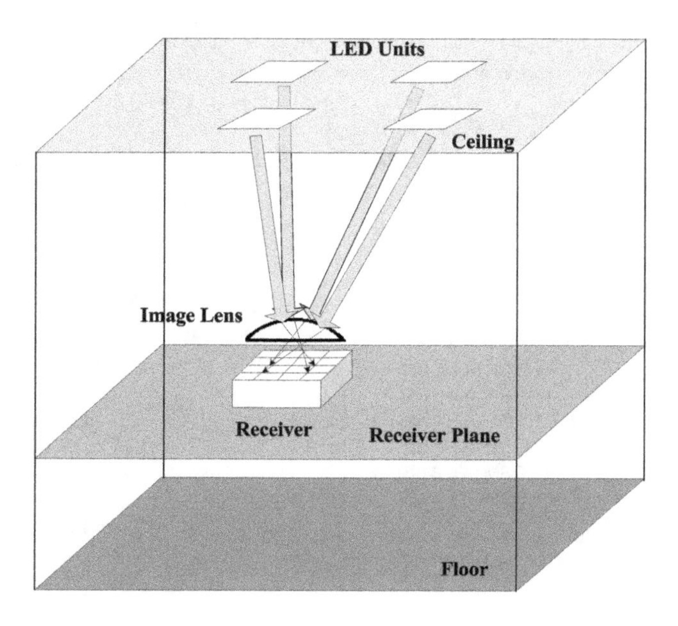

Figure 6.5 Imaging optical MIMO VLC system.

6.2 Imaging optical MIMO techniques

An imaging MIMO VLC system is illustrated in Fig. 6.5, where an imaging diversity receiver is utilized to isolate the signals from different transmitters. Each LED unit is imaged onto a detector array, where the images may be on any pixels. Different from the non-imaging approach, the channel gain of imaging MIMO can be written as [1]

$$h_{\text{image},p,q} = \kappa_{p,q} h'_q, \tag{6.17}$$

where h'_q is the channel gain for the qth LED unit at the aperture of imaging lens when the receiver is at a particular position, which can be written as

$$h'_q = \begin{cases} \frac{\rho' A'}{d'^2_q} R\left(\phi'_q\right) \cos\left(\varphi'_q\right), & \varphi'_q \leq \Psi'_c, \\ 0, & \varphi'_q > \Psi'_c, \end{cases} \tag{6.18}$$

where ρ' is the responsivity coefficient of the receiver, A' is the imaging receiver collection area, d'_q is the distance between the qth LED unit and the center of the receiver collection lens, ϕ'_q is the emission angle of the qth LED unit, φ'_q is the incidence angle of the light, and Ψ'_c is the FOV of the receiver. In (6.17), $\kappa_{p,q}$ denotes

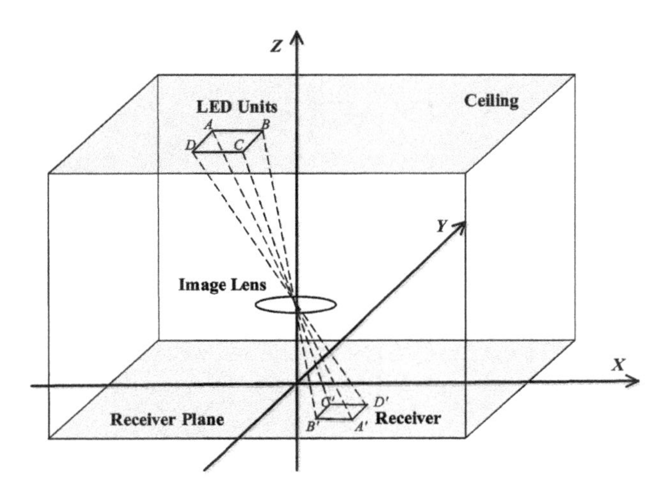

Figure 6.6 An example of the image on the receiver array.

the proportion of the image area that falls on the pth detector pixel in the array

$$\kappa_{p,q} = \frac{A_{q,s(s=p)}}{\sum_{s=1}^{N_r} A_{q,s}}, \tag{6.19}$$

where $A_{q,s}$ represents the area of the image for the qth LED unit on the sth detector pixel in the array.

An example of the image on the receiver array for the qth LED unit is illustrated in Fig. 6.6. A paraxial optics approach is used where the system magnification is independent of the angle of incidence of the rays, and image distortion is not considered as in [1]. Denote the diameter and f-number of the image lens as δ and $f_{\#}$, respectively, the focal length is given by $L = \delta f_{\#}$. Therefore, the magnification of the system can be calculated as $\mathcal{M} = d_z/L$, where d_z is the vertical distance from the ceiling to the receiver. As shown in Fig. 6.6, the four vertices of one LED unit in the ceiling are A, B, C, and D, and their corresponding image coordinates at the receiver are A', B', C', and D'. According to the imaging principle, we have $A'B' = \mathcal{M}AB$, $B'C' = \mathcal{M}BC$, $C'D' = \mathcal{M}CD$, and $D'A' = \mathcal{M}DA$.

Figure 6.7 illustrates three typical scenarios of the images on the detector array. In Fig. 6.7(a), the images for different LED units fall on different detectors. The corresponding channel matrix is diagonal, and the signals from different LED units can be separated perfectly. In Fig. 6.7(b), however, all the received signals fall on the same detector. Only one row of the channel matrix is nonzero, and the signals from different LED units cannot be recovered. Therefore, the number of detectors should be increased to distinguish different signals. Alternatively, the signals fall on the same detector can carry the same information similar to repetition coding. If parts of each signal are received by different detectors, as shown in Fig. 6.7(c), the

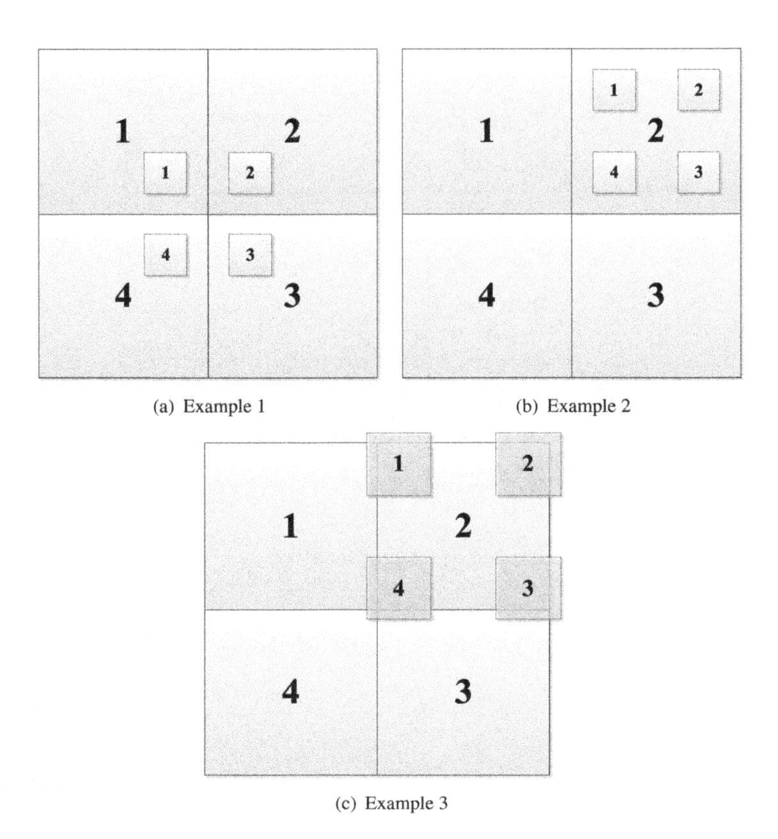

Figure 6.7 Examples of images on the detector array.

signals can still be recovered even though parts of different signals are mixed at the same detector. In order to make sure the channel matrix is full-rank, the number of receivers should be always no smaller than that of transmitters, and a pseudo-inverse operation can be used to estimate the transmitted data.

6.3 Multiuser precoding techniques

In a typical indoor scenario, multiple users usually need to be served simultaneously, where the elimination of multiuser interference is crucial. The precoders and decoders for MU-MIMO in RF communications have been widely studied to cancel the interference [8, 9]. However, they cannot be applied to VLC systems directly. For VLC systems, the channels and the transmitted signals are both real-valued and non-negative. Thus, the constellation is limited within one dimension and DC bias

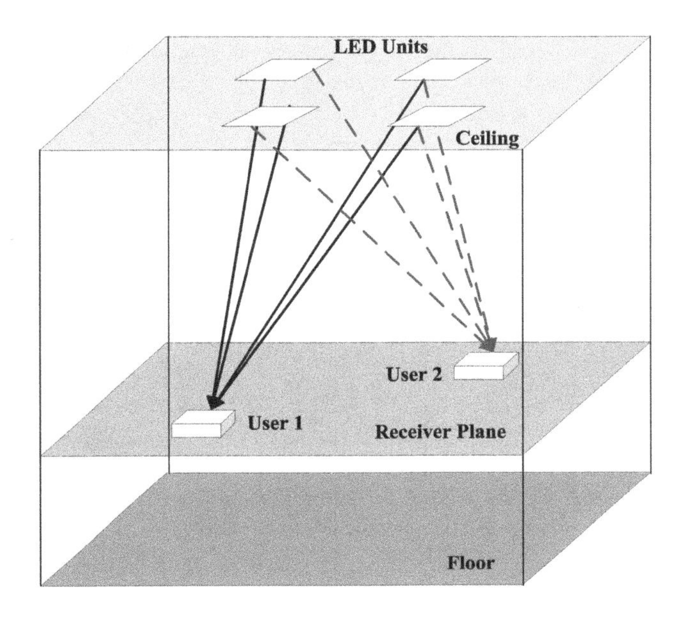

Figure 6.8 Illustration of MU-MIMO VLC system.

is required. Meanwhile, the power constraints for VLC are on optical domain rather than electrical domain.

An MU-MIMO VLC system is illustrated in Fig. 6.8, where a single room is equipped with N_t LED units for illumination, which can cooperate to transmit information for N_r users each with a single PD. In order to eliminate multiuser interference, the transmitted $N_r \times 1$ data vector $\mathbf{d}$ is firstly precoded into an $N_t \times 1$ transmitted vector $\mathbf{x}$. Since VLC systems utilize intensity modulation, the transmitted vector should be real-valued and nonnegative. The channel gain between the LED units and receivers can be calculated according to Section 6.1.1.

Denote d_p as the real-valued symbol for the pth user, which is zero-mean and normalized to the range $[-1, 1]$, at the qth LED unit, d_p is multiplied by a precoding weight $w_{p,q}$, which is also real-valued. By adding up all the weighted symbols from N_r users at the qth LED unit, we have

$$x_q = \sum_{p=1}^{N_r} d_p w_{p,q}, \tag{6.20}$$

which is real-valued but not always non-negative. Therefore, $P_{\text{DC},q}$ is added to x_q for modulating the LED unit and the transmitted signal is given by

$$y_q = x_q + P_{\text{DC},q}. \tag{6.21}$$

Let $\mathbf{w}_p = [w_{p,1}\ w_{p,2}\ \cdots\ w_{p,N_t}]^{\text{T}}$ as the precoding vector and $\mathbf{h}_p = [h_{p,1}\ h_{p,2}\ \cdots\ h_{p,N_t}]^{\text{T}}$

as the channel gain vector for the pth user, after removing the DC component, the received signal at the pth user can be written as

$$r_p = \sum_{q=1}^{N_t} x_q h_{p,q} + z_p = \mathbf{h}_p^{\mathrm{T}} \mathbf{w}_p d_p + \sum_{l \neq p} \mathbf{h}_p^{\mathrm{T}} \mathbf{w}_l d_l + z_p, \tag{6.22}$$

where the first term $\mathbf{h}_p^{\mathrm{T}} \mathbf{w}_p d_p$ denotes the desired signal for the pth user, while the second term $\sum_{l \neq p} \mathbf{h}_p^{\mathrm{T}} \mathbf{w}_l d_l$ denotes the inter-user interference to the pth user, which should be eliminated by precoding.

(6.22) can be rewritten in matrix form as

$$\mathbf{x} = \mathbf{HWd} + \mathbf{z}, \tag{6.23}$$

where $\mathbf{H} = [\mathbf{h}_1\, \mathbf{h}_2\, \cdots\, \mathbf{h}_{N_r}]^{\mathrm{T}}$ and $\mathbf{W} = [\mathbf{w}_1\, \mathbf{w}_2\, \cdots\, \mathbf{w}_{N_r}]$ denote the corresponding channel and precoding matrices.

When linear zero-forcing is used for precoding, the interference $\sum_{l \neq p} \mathbf{h}_p^{\mathrm{T}} \mathbf{w}_l d_l$ from other users is eliminated completely, where we have [10]

$$\mathbf{HW} = \mathrm{diag}(\boldsymbol{\lambda}), \tag{6.24}$$

where $\lambda_p > 0$, and the precoding matrix is calculated by

$$\mathbf{W} = \mathbf{H}^{\mathrm{T}}(\mathbf{HH}^{\mathrm{T}})^{-1}\mathrm{diag}(\boldsymbol{\lambda}). \tag{6.25}$$

Zero-forcing (ZF) is a good precoding approach for high power or low noise scenarios. However, when the channel matrix is ill-conditioned, zero-forcing requires a large normalization factor, which will dramatically reduce the received power [9]. Therefore, when the SNR is low at the receiver, zero-forcing cannot achieve a good performance since noise instead of interference is the dominant impairment of the system.

Linear minimum mean squared error (MMSE) precoding, with which the interference at the receivers is not identically zero, however, can achieve a tradeoff between interference and noise based on which one is the dominant part in the signal-to-interference-plus-noise ratio (SINR) at the receivers. The MMSE precoding matrix is given by [11]

$$\mathbf{W} = \mathbf{H}^{\mathrm{H}}\left(\mathbf{HH}^{\mathrm{H}} + \mathrm{diag}\left(\boldsymbol{\sigma}_{\mathbf{Z}}^2\right)\right)^{-1}\mathrm{diag}\left(\boldsymbol{\lambda}\right), \tag{6.26}$$

where $\boldsymbol{\sigma}_{\mathbf{Z}}^2$ denotes the variance vector of $\mathbf{Z}$.

When more than one PD is located at each user, the ZF and MMSE schemes cannot work well since the channels of PDs at one user are highly correlated and these PDs could cooperate to decode the signal. Meanwhile, the number of LEDs is usually assumed to be no less than the number of PDs to eliminate the inter-user interference effectively [10–13], which may not hold in some specific cases where the number of activated LEDs are determined by the illumination requirements. Therefore, leakage-based precoding is proposed in [14], which utilizes the criterion that maximizes the

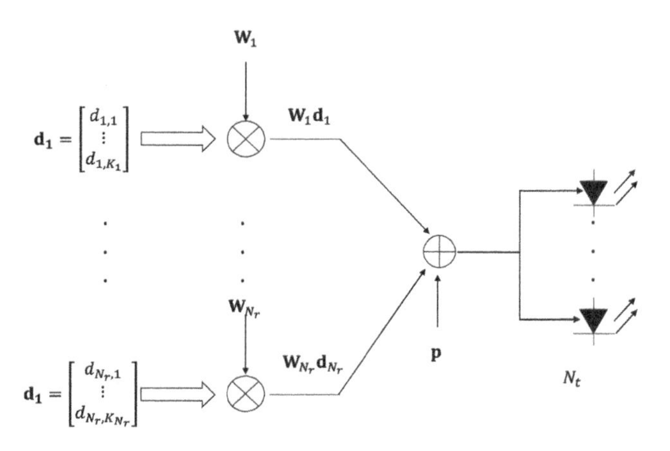

Figure 6.9 Diagram of the MU-MIMO VLC transmitter.

signal-to-leakage-plus-noise ratio (SLNR) and does not have the restriction on the number of LEDs and PDs.

At the ith user, $1 \leq i \leq N_r$, K_i PDs are equipped to receive k_i data streams. The total number of receiving PDs is $\sum_i K_i = K$, while the total number of data streams is $\sum_i k_i = k$. Figure 6.9 illustrates the diagram of the MU-MIMO VLC transmitter, and the corresponding transmitting vector is given by

$$\mathbf{r} = \mathbf{W}\mathbf{d} + \mathbf{p} = \sum_{i=1}^{N_r} \mathbf{W}_i\mathbf{d}_i + \mathbf{p}, \tag{6.27}$$

where $\mathbf{d} = [\mathbf{d}_1{}^{\mathrm{T}}\, \mathbf{d}_2{}^{\mathrm{T}} \cdots \mathbf{d}_{N_r}{}^{\mathrm{T}}]^{\mathrm{T}}$ is the real-valued source symbol vector with $\mathbf{d}_i = [d_{i,1}\, d_{i,2} \cdots d_{i,k_i}]^{\mathrm{T}}$ denoting the k_i data streams for the ith user. $\mathbf{W} = [\mathbf{W}_1\, \mathbf{W}_2 \cdots \mathbf{W}_{N_r}]$ is the precoding matrix, where $\mathbf{W}_i \in \mathbb{R}^{N_t \times k_i}$ is the precoding matrix for the ith user and $\mathbf{p}$ is the $N_t \times 1$ DC vector to ensure the non-negativity of the modulated signals.

Since the source symbol vector $\mathbf{d}$ is zero-mean and independent of the precoding matrix, the precoded signal vector is also zero-mean, and we have

$$\mathrm{E}(\mathbf{W}\mathbf{d}) = \mathrm{E}(\mathbf{W})\mathrm{E}(\mathbf{d}) = \mathbf{0}. \tag{6.28}$$

Therefore, the expectation of the transmitted signal is equal to the DC bias, and the average optical power emitted from the transmitter is proportional to the DC bias determined by the illumination requirements.

Let $\mathbf{H}_i \in \mathbb{R}^{K_i \times N_t}$ denote the channel matrix between the LEDs and the ith user, which is given by

$$\mathbf{H}_i = \begin{bmatrix} h_{i,1,1} & \cdots & h_{i,1,N_t} \\ \vdots & \ddots & \vdots \\ h_{i,K_i,1} & \cdots & h_{i,K_i,N_t} \end{bmatrix}, \tag{6.29}$$

where $h_{i,n,m}$ represents the channel gain between the mth LED unit and the nth PD of the ith user, and can be obtained by (6.1).

When direct detection is used at receivers, the received signal at the ith user can be written as

$$\mathbf{r}_i = \mathbf{H}_i\mathbf{W}\mathbf{d} + \mathbf{H}_i\mathbf{p} + \mathbf{z}_i = \mathbf{H}_i\mathbf{W}_i\mathbf{d}_i + \sum_{j\neq i}\mathbf{H}_i\mathbf{W}_j\mathbf{d}_j + \mathbf{H}_i\mathbf{p} + \mathbf{z}_i, \tag{6.30}$$

where the first term is the desired signal, the second part is the inter-user interference, the third term is the DC bias and the last part $\mathbf{z}_i = [z_{i,1}\ z_{i,2}\ \cdots\ z_{i,K_i}]^{\mathrm{T}}$ is the additive white Gaussian noise with the covariance of $\mathbf{K}_i = \mathrm{diag}([\sigma_{i,1}^2\ \sigma_{i,2}^2\ \cdots\ \sigma_{i,R_i}^2])$. $z_{i,n}$ denotes the noise at the nth PD of the ith user with the variance of $\sigma_{i,n}^2$, which is given by (6.4). Since the PDs within one user are close and have similar parameters, it is assumed that $\sigma_{i,n}^2 = \sigma_{i,j}^2 = \sigma_i^2$, $\forall n, j \in [1, ..., K_i]$.

At the ith user, a linear decoder $\mathbf{G}_i$ is applied after DC removal to recover the transmitted signal, and the decoded signal vector is given by

$$\hat{\mathbf{d}}_i = \mathbf{G}_i\mathbf{H}_i\mathbf{W}_i\mathbf{d}_i + \mathbf{G}_i\sum_{j\neq i}\mathbf{H}_i\mathbf{W}_j\mathbf{d}_j + \mathbf{G}_i\mathbf{z}_i. \tag{6.31}$$

For simplicity, the precoding can be split into two parts

$$\mathbf{W} = \mathbf{Q}\mathrm{diag}(\mathbf{a}), \tag{6.32}$$

where $\mathbf{Q} = [\mathbf{Q}_1\ \mathbf{Q}_2\ \cdots\ \mathbf{Q}_{N_r}] \in \mathbb{R}^{N_t\times k}$ is used to cancel the interference with $\mathbf{Q}_i \in \mathbb{R}^{N_t\times k_i}$ denoting the interference elimination matrix for the ith user, while $\mathbf{a}$ is the $k \times 1$ power scaler vector to satisfy the optical power constraint. Specifically, to ensure the non-negativity of the transmitted signal, we have

$$\mathbf{W}\mathbf{d} \geqslant -\mathbf{p}. \tag{6.33}$$

The source symbol is assumed to be modulated by PAM with optical power normalization that the constellation is bounded within $[-1, 1]$. Therefore, we have $-1 \leqslant d_n \leqslant 1$, where d_n is the nth element of the source symbol vector $\mathbf{d}$. Consequently, the optical power constraint can be written as

$$\mathrm{abs}(\mathbf{W})\mathbf{1} \leqslant \mathbf{p}, \tag{6.34}$$

where abs$(\cdot)$ denotes the element-wise absolute operator and $\mathbf{1}$ is a $k \times 1$ vector whose elements are all unit.

Accordingly, the magnitude of the transmitted signal $\mathbf{x}$ is also bounded. The simple double-sided signal clipping distortion model is considered, where the signals within the linear range are transferred linearly while the signals out of the linear range are clipped [15]. By adjusting the boundary and limiting the signals within the linear range, the clipping distortion can be avoided. Obviously, the power scalers are positive and the optical power constraint can be rewritten as

$$\text{abs}(\mathbf{W})\mathbf{1} = \text{abs}(\mathbf{Q})\mathbf{a} \leq \mathbf{p}. \tag{6.35}$$

When ZF scheme is used in MU-MIMO systems, the matrix $\mathbf{Q}_i$ is calculated as

$$\mathbf{Q}_i = \mathbf{F}_i\mathbf{V}_i, \tag{6.36}$$

where $\mathbf{F}_i$ is used to eliminate the inter-user interference and calculated as $\mathbf{F}_i = \mathbf{I} - \widetilde{\mathbf{H}}_i^{\dagger}\widetilde{\mathbf{H}}_i$. $\mathbf{I}$ is the N_t-dimensional identity matrix, $\widetilde{\mathbf{H}}_i$ is defined by $\widetilde{\mathbf{H}}_i = [\mathbf{H}_1^{\mathrm{T}} \cdots \mathbf{H}_{i-1}^{\mathrm{T}}\, \mathbf{H}_{i+1}^{\mathrm{T}} \cdots \mathbf{H}_{N_r}^{\mathrm{T}}]^{\mathrm{T}}$, and $\widetilde{\mathbf{H}}_i^{\dagger}$ is the pseudo inverse of $\widetilde{\mathbf{H}}_i$. $\mathbf{V}_i$ is used to eliminate the intra-user interference further. $\mathbf{H}_i\mathbf{F}_i$ can be rewritten as the singular value decomposition (SVD) form that $\mathbf{H}_i\mathbf{F}_i = \mathbf{U}\Lambda\mathbf{V}^{\mathrm{T}}$, where $\mathbf{V}_i$ is composed of the k_i columns of $\mathbf{V}$ that correspond to the k_i largest singular values in Λ while the decoder matrix $\mathbf{G}_i$ is composed of the corresponding k_i rows of $\mathbf{U}^{\mathrm{T}}$. It is easy to verify that interference can be eliminated completely when the number of LEDs is no less than the number of PDs.

ZF scheme imposes a constraint on the number of transmitters and receiving PDs in order to cancel all inter-user interference. Moreover, the performance of ZF scheme degrades when the channels of different users are highly correlated, which occurs frequently in VLC systems. Therefore, a leakage-based precoding method can be applied to solve this problem. Instead of considering the interference at a specific user that is caused by the data streams of other users, the leakage, which represents the interference caused by a specific user to other users, is considered and the interference elimination matrix is designed by maximizing the SLNR as in [16].

Maximizing SLNR under optical power constraint directly has high complexity, and does not employ the feature of SLNR that only the ith user's precoding matrix is included in the expression of the ith user's SLNR, since the optical power constraint connects the precoding matrices together again. Therefore, the optical power constraint is placed on the design of power scaler and the constraint on the calculation of matrix $\mathbf{Q}$ is not considered. As a result, the interference elimination matrix $\mathbf{Q}_i$ can be constituted by k_i generalized eigenvectors of $(\mathbf{H}_i^{\mathrm{T}}\mathbf{H}_i, ((K_i\sigma_i^2)\mathbf{I} + \widetilde{\mathbf{H}}_i^{\mathrm{T}}\widetilde{\mathbf{H}}_i))$ and the elements of $\mathbf{Q}_i$ satisfy the real-valued constraints naturally. According to the property of generalized eigenvector, it can be shown that

$$\mathbf{Q}_i^{\mathrm{T}}\mathbf{H}_i^{\mathrm{T}}\mathbf{H}_i\mathbf{Q}_i = \Lambda_i, \tag{6.37}$$

and

$$\mathbf{Q}_i^{\mathrm{T}}((K_i\sigma_i^2)\mathbf{I} + \widetilde{\mathbf{H}}_i^{\mathrm{T}}\widetilde{\mathbf{H}}_i)\mathbf{Q}_i = \mathbf{I}. \tag{6.38}$$

At the ith receiver, decoder matrix is given by

$$\mathbf{G}_i = c_i\mathbf{W}_i^T\mathbf{H}_i^T, \tag{6.39}$$

where c_i is the normalization scaler $c_i = \sqrt{1/Tr(\mathbf{W}_i^{\mathrm{T}}\mathbf{H}_i^{\mathrm{T}}\mathbf{H}_i\mathbf{W}_i)}$. We can see that intra-user interference is also eliminated. Thus, the decoded signal can be split into k_i separate data streams given by

$$\begin{aligned}\hat{\mathbf{d}}_i &= \mathbf{G}_i\mathbf{H}_i\mathbf{W}_i\mathbf{d}_i + \mathbf{G}_i\sum_{j\neq i}\mathbf{H}_i\mathbf{W}_j\mathbf{d}_j + \mathbf{G}_i\mathbf{z}_i \\ &= c_i\Lambda_i(\mathrm{diag}(\mathbf{a}_i))^2\mathbf{d}_i + \mathbf{G}_i\sum_{j\neq i}\mathbf{H}_i\mathbf{W}_j\mathbf{d}_j + \tilde{\mathbf{z}}_i,\end{aligned} \tag{6.40}$$

where $\mathbf{a}_i$ is the $k_i \times 1$ power scaler for the ith user. The second part in (6.40) is the interference after decoding. Unlike ZF scheme, the interference cannot be eliminated completely since the noise is considered in the design of precoding and decoding matrix. However, the interference is reduced greatly and the residual interference can be ignored compared with the noise. $\widetilde{\mathbf{z}}_i$ is the equivalent noise with the covariance matrix of

$$\widetilde{\mathbf{R}}_i = \mathbf{G}_i\mathbf{R}_i\mathbf{G}_i^{\mathrm{T}} = \sigma_i^2\mathbf{G}_i\mathbf{G}_i^{\mathrm{T}}. \tag{6.41}$$

In [16], the power allocated on different data streams is assumed to be equal. Under this assumption, the power scaler $\mathbf{a}$ can be expressed as $\mathbf{a} = a\mathbf{1}$. According to the optical power constraint, a can be calculated as

$$a = \min_n \frac{\mathbf{p}_n}{(\mathrm{abs}(\mathbf{Q})\mathbf{1})_n}, \tag{6.42}$$

where $\mathbf{v}_n$ denotes the nth element of an arbitrary vector $\mathbf{v}$.

The inter-user interference elimination matrix $\mathbf{Q}_i$ can suppress the multiuser interference. Meanwhile, the multiple data streams for one user are also separated. However, the performance of leakage-based precoding is limited if the difference between the diagonal elements of Λ_i in (6.40) is large. Since the PDs of one user are close such that their channels are highly correlated, the diagonal elements of Λ_i can be of great difference. Therefore, equal power allocation on different data streams is not a good option in this scenario and the optimal power allocation has to be considered to optimize the throughput of the system.

Since the source signal symbol $\mathbf{d}_i$ is modulated by PAM constellation, given the BER constraint $\mathrm{BER_T}$, the achievable rate of the ith user can be calculated as [10]

$$\zeta_i = \sum_{n=1}^{r_i}\log_2(1 + \eta\sqrt{\mathrm{SINR}_{i,n}}), \tag{6.43}$$

where $\eta = \sqrt{(-\log(5\mathrm{BER_T}))^{-1}}$. The optimization problem to maximize the throughput of the system can be written as

$$\begin{aligned}\max_{\mathbf{a}} \quad & \sum_{i=1}^{K}\sum_{n=1}^{r_i}\log_2(1 + \eta\sqrt{\mathrm{SINR}_{i,n}}) \\ \text{subject to} \quad & \mathrm{abs}(\mathbf{Q})\mathbf{a} \leqslant \mathbf{p}, \\ & \mathbf{a} \geqslant \mathbf{0}.\end{aligned} \tag{6.44}$$

Table 6.2 Simulation parameters for MU-MIMO VLC systems.

Room Size	5 m × 5 m × 3 m
Lambertian emission mode number m	1
PD responsibility $\rho_{i,n}$	0.4 A/W
PD area $A_{\mathrm{PD},i,n}$	1 cm^2
Receiver FOV $\theta_{c,i,n}$	62°
Refractive index of optical concentrator $q_{i,n}$	1.5
Pre-amplifier noise density i_{amp}	5 pA/Hz$^{-1/2}$
Ambient light photocurrent χ_{amb}	10.93 A/m^2/Sr
Bandwidth of receiver B	100 MHz
BER constraint $\mathrm{BER_T}$	10^{-3}

For the leakage-based precoding, the inter-user interference can be ignored when $k \leq N_t$. Therefore, the received signal is approximated as

$$\hat{\mathbf{d}}_i \approx c_i \Lambda_i (\mathrm{diag}(\mathbf{a}_i))^2 \mathbf{d}_i + \widetilde{\mathbf{z}}_i. \tag{6.45}$$

As a result, the optimization problem can be simply rewritten as

$$\begin{aligned} \max_{\mathbf{a}} \quad & \sum_{i=1}^{K} \sum_{n=1}^{r_i} \log_2 \left(1 + \eta \frac{\lambda_{i,n} a_{i,n}}{\widetilde{\sigma}_{i,n}}\right) \\ \text{subject to} \quad & \mathrm{abs}(\mathbf{Q})\mathbf{a} \leqslant \mathbf{p}, \\ & \mathbf{a} \geqslant \mathbf{0}, \end{aligned} \tag{6.46}$$

where $\widetilde{\sigma}_{i,n}^2 = \widetilde{R_i}(n,n)$ is the nth diagonal element of $\widetilde{R_i}$. This is a convex optimal problem and can be solved by CVX, a package for specifying and solving convex problems [17]. It should be noted that (6.46) is suitable for both leakage-based precoding and ZF precoding.

To verify the performance of leakage-based precoding combined with power allocation, simulations are conducted with the parameters listed in Table 6.2 where the optical device parameters are obtained from [1, 18]. Without loss of generality, all the transmitting LEDs and PDs are assumed to share identical configurations. The system throughput as a function of the optical power constraint is considered. As shown in (6.46), the system throughput is the sum of the achievable rates for all users. The optical power constraints for different LEDs are assumed the same and denoted as p.

The locations of LEDs and PDs are listed in the Table 6.3 and Table 6.4, respectively. Meanwhile, the spacing between different PDs at one user is small and a value of 0.1 m is adopted in the simulations [12]. To evaluate the performance of the precoding schemes, four scenarios are discussed in the simulations. In scenario 1, the locations of LEDs are listed in Table 6.3 and the PDs are located as Case 1 in Table

Table 6.3 Locations of LEDs.

LED 1 coordinate	[1.25 1.25 3]
LED 2 coordinate	[1.25 3.75 3]
LED 3 coordinate	[3.75 1.25 3]
LED 4 coordinate	[3.75 3.75 3]

Table 6.4 Locations of PDs.

	Case 1	Case 2
PD 1 of User 1	[1.5 1.5 0.85]	[1.5 1.5 0.85]
PD 2 of User 1	[1.4 1.5 0.85]	[1.4 1.5 0.85]
PD 1 of User 2	[2.0 3.0 0.85]	[1.0 1.5 0.85]
PD 2 of User 2	[1.9 3.0 0.85]	[1.1 1.5 0.85]

6.4. Scenario 2 considers the situation that users are close to each other whose PDs are located as Case 2 in Table 6.4 without changing the locations of LEDs. Scenarios 3 and 4 discuss the situation that the number of LEDs is less than the number of PDs. Therefore, only LED 1 and LED 4 are activated in these two scenarios, while the locations of PDs are the same as in scenario 1 and scenario 2, respectively.

Figure 6.10 shows that the optimal power allocation by maximizing the system throughput outperforms the equal power allocation. As the channel correlation increases, the performance gain of optimal power allocation also increases. Therefore, only the optimal power allocation is adopted in the simulations afterwards.

When the number of LEDs is no less than that of PDs, Fig. 6.11 shows the comparison of the leakage-based and ZF precoding schemes under scenario 1 and scenario 2. It is shown that when the two users are relatively far from each other (scenario 1), the advantage of leakage-based precoding is unobvious, which indicates that ZF precoding scheme is suitable for this scenario since the channel correlation is not that high. However, when the two users are close to each other (scenario 2), leakage-based precoding outperforms ZF precoding significantly. It is also shown that the throughput of the system in scenario 2 diminishes since the elimination of the inter-user interference is more difficult when the two users are close to each other. However, leakage-based precoding could alleviate the performance loss compared to ZF precoding in this scenario.

When the number of transmitters is less than that of receivers, Fig. 6.12 shows that leakage-based precoding still performs well (scenario 3 and 4). On the contrary, ZF precoding fails to cancel the interference completely when the number of transmitters is less than that of receivers and the system throughput does not increase with the average optical power.

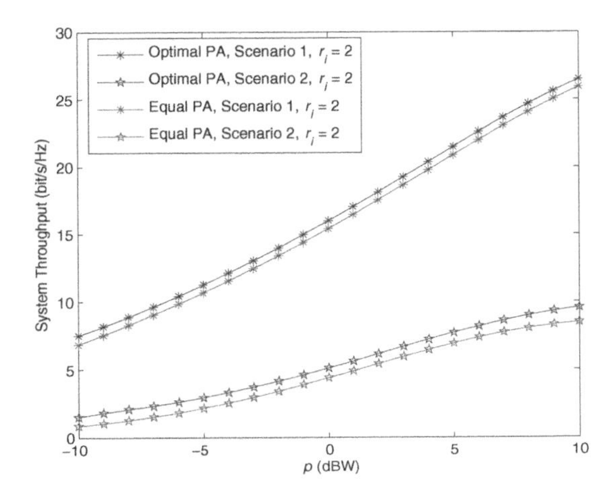

Figure 6.10 System throughput comparison between equal power allocation and optimal power allocation.

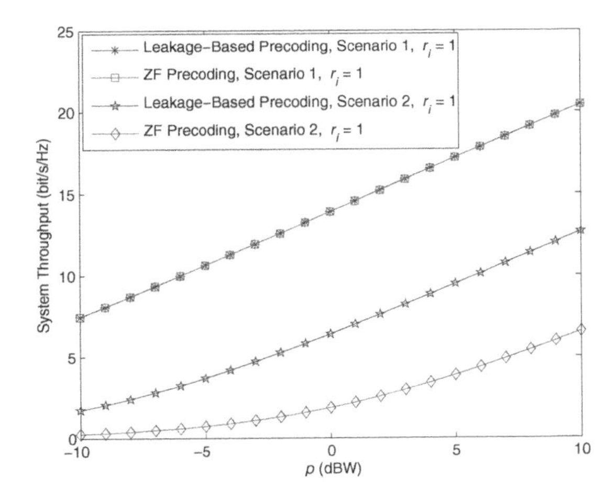

Figure 6.11 System throughput comparison when the number of LEDs is no less than that of PDs.

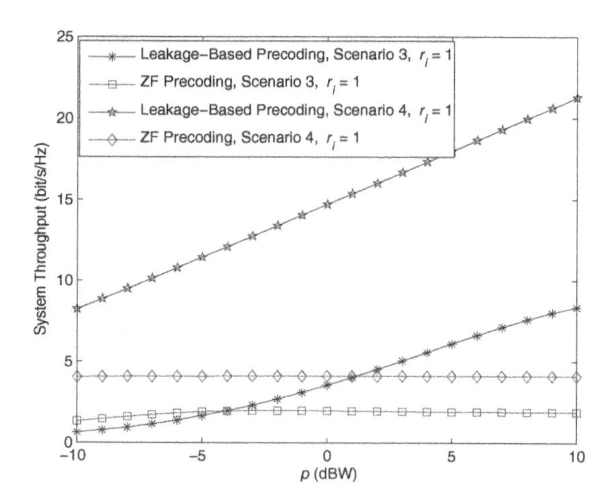

Figure 6.12 System throughput comparison when the constraint about the number of LEDs and PDs is not satisfied.

6.4 Optical MIMO-OFDM

The combination of MIMO and optical orthogonal frequency-division multiplexing (OFDM) yields the MIMO-OFDM, which is a popular technique in RF systems to support multiuser service and provide high-data-rate transmission [19, 20]. It can be extended to the optical domain in VLC systems to improve the performance. In [21], a MIMO-OFDM VLC system is demonstrated, but it requires an imaging diversity receiver to distinguish signals from different LEDs, which is infeasible for multiuser scenarios.

In most studies on MU-MIMO VLC systems, single-carrier modulations are considered with limited bandwidth [10–12], which conduct precoding in the time domain and only the DC channel gain in (6.1) is considered. Since the distances of the multiple transmitter–receiver links are different, their temporal delays are varied, resulting in complex channel gains and phase differences when transformed to the frequency domain. Considering this phenomenon, the time-domain channel response in (6.1) can be rewritten as [22, 23]

$$\tilde{h}_{p,q}(t) = h_{p,q}\delta\left(t - \frac{d_{p,q}}{c}\right), \tag{6.47}$$

where $\delta(\cdot)$ denotes the Dirac delta function and c is the speed of light. Correspondingly, the frequency-domain channel response for the kth subcarrier is given by

$$\tilde{H}_{p,q,k} = h_{p,q}\exp\left(-\frac{j2\pi kBd_{p,q}}{Nc}\right), \tag{6.48}$$

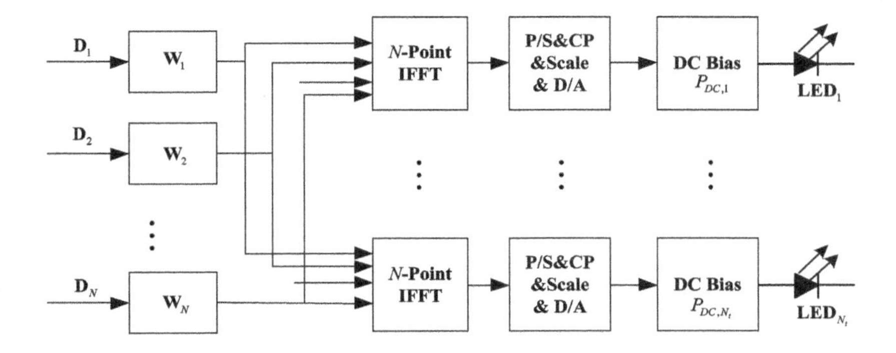

Figure 6.13 MU-MIMO-OFDM VLC transmitter with N_r users, N_t LED units, and N subcarriers.

where B is the system bandwidth and N is the size of fast Fourier transform (FFT). j is the imaginary unit $j = \sqrt{-1}$. According to the expression, the phase of the frequency-domain channel gain is proportional to the bandwidth. Moreover, when the temporal delay is considered, the frequency-domain channel response is complex-valued, which provides another dimension and reduces the channel correlation with the phase differences of multiple links. When up to 100 Gbps high data rate is required in future VLC networks, wide bandwidth optical components are used [24–26], the phase in the complex channel gain cannot be neglected. Therefore, MU-MIMO-OFDM scheme is proposed for VLC system and precoding is conducted on different subcarriers individually [22, 23].

For the pth user, the information bit stream is firstly mapped onto the complex-valued symbols $D_{p,k}$, $k = 0, 1, \cdots, N-1$. Since intensity modulation requires real-valued output, Hermitian symmetry should be imposed on the OFDM subcarriers where $D_{p,k} = D^*_{p,N-k}$, $k = 1, 2, \cdots, N/2-1$, and the subcarriers $D_{p,0}$ and $D_{p,N/2}$ are set to zero.

The diagram of the MU-MIMO-OFDM VLC system is illustrated in Fig. 6.13. At the transmitter of MU-MIMO-OFDM VLC system, precoding is performed on each subcarrier to eliminate multiuser interference. Let $\{W_{p,q,k}, 1 \leq p \leq N_r, 1 \leq q \leq N_t\}$ denote the precoding weights for the kth ($k = 0, 1, \cdots, N-1$) subcarrier. By adding up all the weighted symbols from N_r users at the qth LED unit, the frequency-domain signal can be written as

$$X_{q,k} = \sum_{p=1}^{N_r} W_{p,q,k} D_{p,k},\ k = 0, 1, \cdots, N-1, \tag{6.49}$$

which is also complex-valued.

Afterwards, the frequency-domain signals for the qth LED unit are converted to the time domain by inverse fast Fourier transform (IFFT) as

$$x_{q,n} = \frac{1}{\sqrt{N}} \sum_{k=0}^{N-1} X_{q,n} \exp\left(j\frac{2\pi}{N}nk\right),\ n = 0, 1, \cdots, N-1, \tag{6.50}$$

which are real-valued since the symbols on different subcarriers satisfy the Hermitian symmetry. Since $x_{q,n}$ may be negative, a DC bias $P_{\mathrm{DC},q}$ is added to the qth transmitter in DC-biased optical OFDM (DCO-OFDM) to obtain nonnegative signals for emission.

At the receiver of the pth user, FFT is performed on the received signals to generate frequency-domain symbols given by

$$\begin{aligned} R_{p,k} &= \sum_{q=1}^{N_t} \tilde{H}_{p,q,k} X_{q,k} + Z_{p,k} \\ &= \tilde{\mathbf{H}}_{p,k}^{\mathrm{T}} \mathbf{W}_{p,k} D_{p,k} + \sum_{l \neq p}^{N_t} \tilde{\mathbf{H}}_{l,k}^{\mathrm{T}} \mathbf{W}_{l,k} D_{l,k} + Z_{p,k},\ k = 0, 1, \cdots, N-1, \end{aligned} \tag{6.51}$$

where $\tilde{\mathbf{H}}_{p,k}$, $\tilde{\mathbf{H}}_{l,k}$, $\mathbf{W}_{p,k}$, and $\mathbf{W}_{l,k}$ are $N_t \times 1$ vectors of channel gains and precoding weights for the kth subcarrier, $Z_{p,k}$ denotes the equivalent noise on the kth subcarrier. The first term $\tilde{\mathbf{H}}_{p,k}^{\mathrm{T}} \mathbf{W}_{p,k} D_{p,k}$ in (6.51) denotes the desired signal for the pth user, while the second term $\sum_{l \neq p}^{N_t} \tilde{\mathbf{H}}_{l,k}^{\mathrm{T}} \mathbf{W}_{l,k} D_{l,k}$ is the inter-user interference to the pth user, which should be eliminated by precoding.

(6.51) can be rewritten in the matrix form when all the N_r users are considered, which is given by

$$\mathbf{R}_k = \tilde{\mathbf{H}}_k \mathbf{W}_k \mathbf{D}_k + \mathbf{Z}_k,\ k = 0, 1, \cdots, N-1, \tag{6.52}$$

where $\mathbf{D}_k = [D_{1,k}\, D_{2,k} \cdots D_{N_r,k}]^{\mathrm{T}}$ and $\mathbf{R}_k = [R_{1,k}\, R_{2,k} \cdots R_{N_r,k}]^{\mathrm{T}}$ denote the transmitted and received symbol vectors on the kth subcarrier, $\tilde{\mathbf{H}}_k = \left[\tilde{\mathbf{H}}_{1,k}\, \tilde{\mathbf{H}}_{2,k} \cdots \tilde{\mathbf{H}}_{N_r,k}\right]^{\mathrm{T}}$ and $\mathbf{W}_k = [\mathbf{W}_{1,k}\, \mathbf{W}_{2,k} \cdots \mathbf{W}_{N_r,k}]$ represent the corresponding channel and precoding matrices, and $\mathbf{Z}_k$ is the noise vector on the kth subcarrier. For different subcarriers, the channel matrices are different due to temporal delays, thus their corresponding precoding matrices should be calculated separately. Several precoding schemes have been proposed for MU-MIMO in RF systems [8, 9]. Here, two well-known techniques are employed to eliminate the inter-user interference, namely, ZF and MMSE algorithms.

ZF is a simple method to eliminate the inter-user interference by directly forcing the interference terms to be zeros, i.e., the matrix $\mathbf{H}_k \mathbf{W}_k$ is diagonal as [8]

$$\mathbf{H}_k \mathbf{W}_k = \mathrm{diag}\left(\boldsymbol{\lambda}_k\right), \tag{6.53}$$

where all the entries in $\boldsymbol{\lambda}_k$ are positive and the precoding matrix is given by

$$\mathbf{W}_k = \tilde{\mathbf{H}}_k^{\dagger} \mathrm{diag}\left(\boldsymbol{\lambda}_k\right) = \tilde{\mathbf{H}}_k^{\mathrm{H}} \left(\tilde{\mathbf{H}}_k \tilde{\mathbf{H}}_k^{\mathrm{H}}\right)^{-1} \mathrm{diag}\left(\boldsymbol{\lambda}_k\right). \tag{6.54}$$

In linear MMSE precoding, the interference at the receiver is not completely removed, while a tradeoff between interference and noise is achieved based on which one is the dominant part in the SINR at the receiver. The MMSE-based precoding matrix is calculated as [9]

$$\mathbf{W}_k = \tilde{\mathbf{H}}_k^{\mathrm{H}} \left(\tilde{\mathbf{H}}_k \tilde{\mathbf{H}}_k^{\mathrm{H}} + \mathrm{diag}\left(\boldsymbol{\sigma}_{\mathbf{Z}_k}^2 \right) \right)^{-1} \mathrm{diag}\left(\boldsymbol{\lambda}_k \right), \tag{6.55}$$

where $\boldsymbol{\sigma}_{\mathbf{Z}_k}^2$ denotes the variance vector of $\mathbf{Z}_k$.

6.4.1 DCO-OFDM-based MU-MIMO VLC

According to the central limit theorem, $x_{q,n}$ approximates a Gaussian distribution when $N \geq 64$, which might have a very large absolute value [27]. Therefore, a DC bias cannot necessarily guarantee the non-negativity for all the signals and some signals should be clipped. The DC bias for the qth transmitter can be represented as [28]

$$\bar{P}_{\mathrm{DC},q} = \eta \sqrt{E\left\{ x_{q,n}^2 \right\}}, \tag{6.56}$$

where η denotes the DC bias ratio. When a larger DC bias is used, less signals are clipped, leading to smaller clipping distortion. However, DC bias does not carry useful information, hence it is inefficient in terms of power. A tradeoff should be made between the DC bias and clipping distortion, here we set the minimum DC bias ratio to η_0 [28].

Due to the precoding matrix, the electric power of the N_t transmitters is different, which requires different minimum DC bias. For the qth LED unit, the DC bias is

$$\bar{P}_{\mathrm{DC},q} = \eta_0 \sqrt{E\left\{ x_{q,n}^2 \right\}}, \; q = 0, 1, \cdots, N_t - 1, \tag{6.57}$$

and this scheme is named as the *minimum DC bias scheme*. Correspondingly, the emitted optical power of the qth LED unit is given by

$$P_{\mathrm{opt},q} = E\left\{ x_{q,n} + \bar{P}_{\mathrm{DC},q} \right\} = E\left\{ x_{q,n} \right\} + \bar{P}_{\mathrm{DC},q} = \bar{P}_{\mathrm{DC},q}, \tag{6.58}$$

where the equality holds since the expectation of $x_{q,n}$ is zero according to (6.50).

When the average optical power of all the LED units is constrained to P for illumination requirement, the biased signal should be scaled and the transmitted signal for the qth LED unit is written as

$$y_{q,n} = \alpha \left(x_{q,n} + \bar{P}_{\mathrm{DC},q} \right), \tag{6.59}$$

where the scaling factor can be calculated as

$$\alpha = \frac{N_t P}{\sum_{q=1}^{N_t} P_{\mathrm{DC},q}} = \frac{N_t P}{\eta_0 \sum_{q=1}^{N_t} \sqrt{E\left\{ x_{q,n}^2 \right\}}}. \tag{6.60}$$

Therefore, the actual DC bias for the qth LED unit is given by

$$P_{\mathrm{DC},q} = \alpha \bar{P}_{\mathrm{DC},q} = \frac{\sqrt{E\left\{x_{q,n}^2\right\}} N_t P}{\sum\limits_{q=1}^{N_t} \sqrt{E\left\{x_{q,n}^2\right\}}}, \; q = 0, 1, \cdots, N_t - 1. \tag{6.61}$$

It is shown in (6.61) that the emitted optical power varies for the N_t LED units and it also changes with the topology of users, which is not suitable for the illumination function of LEDs. To maintain data transmission and high quality illumination simultaneously, the *unified DC bias scheme* is considered, where the same DC bias is applied to all the LED units as

$$P_{\mathrm{DC},q} = P. \tag{6.62}$$

When all the N_t transmitters are considered, the DC bias should make sure the clipping distortion of the transmitter with maximum electric power is acceptable. Therefore, the scaling factor is calculated as

$$\alpha = \frac{P}{\eta_0 \sqrt{\max\limits_{1 \leqslant q \leqslant N_t} \left(E\left\{x_{q,n}^2\right\}\right)}}. \tag{6.63}$$

The unified DC bias scheme improves the illumination performance at the cost of energy efficiency since larger DC biases are imposed on most of the LED units.

6.4.2
ACO-OFDM-based MU-MIMO VLC

Since DCO-OFDM requires a high DC bias, asymmetrically clipped optical OFDM (ACO-OFDM) can be used to improve the energy efficiency. In ACO-OFDM, the even subcarriers are set to zero, while the symbols are only transmitted by the odd subcarriers [29]. After IFFT, the time-domain signals of ACO-OFDM are antisymmetric, where

$$x_{q,n} = -x_{q,n+N/2}, \; n = 0, 1, \cdots, N/2 - 1, \tag{6.64}$$

which can be clipped at zero without information loss. At the receiver, the symbols on the odd subcarriers can be directly detected by FFT since the clipping distortion only falls on the even subcarriers. However, scaling is still required to fulfill the illumination requirement of P. Since $x_{q,n}$ follows a Gaussian distribution when $N \geq 64$, the optical power of the clipped ACO-OFDM signals $x_{q,n}^{(c)}$ can be given by [28]

$$P_{\mathrm{opt},q} = E\left\{x_{q,n}^{(c)}\right\} = \sqrt{E\left\{x_{q,n}^2\right\}/2\pi}, \tag{6.65}$$

and the scaling factor for the qth LED unit is calculated by

$$\alpha = \frac{N_t P}{\sum_{q=1}^{N_t} P_{\mathrm{opt},q}} = \frac{N_t P}{\sum_{q=1}^{N_t} \sqrt{E\left\{x_{q,n}^2\right\}/2\pi}}. \tag{6.66}$$

When the same optical power is given, ACO-OFDM-based scheme can use higher-order constellations to improve the spectral efficiency since DC bias is not required. However, only the odd subcarriers are employed in ACO-OFDM, leading to the loss of spectral efficiency by half when the same constellation orders are used. Therefore, ACO-OFDM is not always better than DCO-OFDM especially when high spectral efficiency is required.

6.4.3 Performance evaluation

In the simulations, four LED units and two users are assumed in one room, while the other parameters are listed in Table 6.5. The performance of the MU-MIMO-OFDM VLC is validated in terms of the aggregate achievable spectral efficiency, which is calculated with all the SINRs at the receiver by $\sum_{p=1}^{N_r} \log_2\left(1 + \mathrm{SINR}_p\right)$. Two cases of users' locations are considered. In Case 1, User 1 is in the middle of the room with the coordinate of $[2.5\, 2.5\, 0.85]$, while the coordinate of User 2 is $[3.2\, 3.9\, 0.85]$. The channel matrices for Case 1 are in good conditions since the two users are not close. In Case 2, the two users' coordinates are $[2.05\, 1.6\, 0.85]$ and $[2.05\, 1.4\, 0.85]$, which are very close and their corresponding channel matrix is ill-conditioned. The minimum DC bias factor η_0 is set to 3 to achieve a tradeoff between clipping distortion and effective power.

Figure 6.14 depicts the average spectral efficiency with different average emitted optical power, where DCO-OFDM is utilized with minimum DC bias and unified DC bias according to Section 6.4.1. It can be seen that MMSE achieves higher spectral efficiency when the optical power is low and the noise is the dominant part of interference-plus-noise. Moreover, higher performance gain is observed in Case 2 since the channel matrices are more ill-conditioned. Besides, the system with minimum DC bias according to (6.57) and (6.60) outperforms that with unified DC bias according to (6.61) and (6.62) since more power are used for data transmission. However, the latter provides better illumination performance.

Figure 6.15 provides the performance comparison of ACO-OFDM-based and DCO-OFDM-based schemes, where zero-forcing precoding is employed and the minimum DC bias scheme is used after precoding for DCO-OFDM-based system. It is shown that ACO-OFDM-based schemes achieve higher spectral efficiency than DCO-OFDM-based schemes when the emitted optical power is low. However, when larger optical power is employed and the achievable spectral efficiency is above 6 bit/s/Hz, the performance of DCO-OFDM-based scheme is better, which is consistent with the previous analysis.

Table 6.5 Simulation parameters for MU-MIMO-OFDM VLC system configuration.

Room Size (length × width × height)	5 m × 5 m × 3 m
LED 1 coordinate	[1.25 1.25 3]
LED 2 coordinate	[1.25 3.75 3]
LED 3 coordinate	[3.75 1.25 3]
LED 4 coordinate	[3.75 3.75 3]
LED emission angle ϕ_q	60°
PD area $A_{\text{PD},p}$	1 cm^2
PD responsivity coefficient ρ_p	0.4 A/W
PD concentrator refractive index γ	1.5
Lambertian emission mode number m	1
Receiver FOV angle $\Psi_{c,p}$	62°
Pre-amplifier noise density i_{amp}	5 pA/Hz$^{-1/2}$
Ambient light photocurrent χ_{amb}	10.93 A/m^2/Sr
System bandwidth B	1 GHz
OFDM subcarrier number N	64
Cyclic prefix length N_{CP}	3

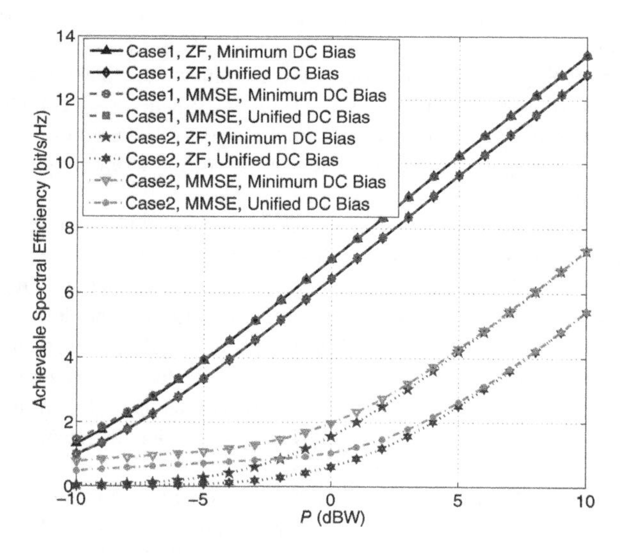

Figure 6.14 Average spectral efficiency of DCO-OFDM-based MU-MIMO VLC system with different average emitted optical power and DC bias.

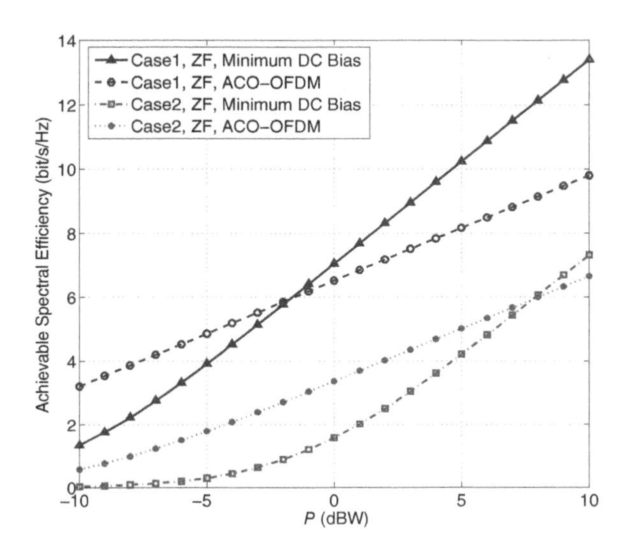

Figure 6.15 Average spectral efficiency comparison of ACO-OFDM-based and DCO-OFDM-based MU-MIMO VLC systems with different average emitted optical power.

6.5 Conclusion

In this chapter, imaging and non-imaging optical MIMO techniques have been introduced for VLC systems. In non-imaging MIMO systems, each receiver collects light with its own optical concentrator, and optical MIMO, OSM, and OSSK can be used to boost the data rate. For imaging MIMO systems, an imaging diversity receiver is utilized to distinguish the light from different transmitters. Meanwhile, in order to support data transmission for multiple users simultaneously, precoding techniques are employed to eliminate the inter-user interference, which consider the lighting constraints in VLC systems. Finally, MIMO-OFDM is discussed for single-user and multiuser VLC systems, which combines the MIMO with OFDM to provide high spectral efficiency and robust transmission.

References

1 L. Zeng, D. O'Brien, H. Minh, G. Faulkner, K. Lee, D. Jung, Y. J. Oh, and E. T. Won, "High data rate multiple input multiple output (MIMO) optical wireless communications using white LED lighting," *IEEE J. Sel. Areas Commun.*, vol. 27, no. 9, pp. 1654–1662, Dec. 2009.

2 T. Fath and H. Haas, "Performance comparison of MIMO techniques for optical wireless communications in indoor environments," *IEEE Trans. Commun.*, vol. 61, no. 2, pp. 733–742, Feb. 2013.

3 L. Zheng and D. N. C. Tse, "Diversity and multiplexing: A fundamental tradeoff in multiple-antenna channels," *IEEE Trans. Inf. Theory*, vol. 49, no. 5, pp. 1073–1096, May 2003.

4 R. Y. Mesleh, H. Haas,, S. Sinanovic, C. W. Ahn, and S. Yun, "Spatial modulation," *IEEE Trans. Veh. Technol.*, vol. 57, no. 7, pp. 2228–2241, Jul. 2008.

5 R. Mesleh, R. Mehmood, H. Elgala, and H. Haas, "Indoor MIMO optical wireless communication using spatial modulation," in *Proc. IEEE International Conference on Communications (ICC) 2010* (Cape Town, South Africa), May 23–27, 2010, pp. 1–5.

6 R. Mesleh, H. Elgala, and H. Haas, "Optical spatial modulation," *J. Opt. Commun. Netw.*, vol. 3, no. 3, pp. 234–244, Mar. 2011.

7 N. Ishikawa and S. Sugiura, "Maximizing constrained capacity of power-imbalanced optical wireless MIMO communications using spatial modulation," *J. Lightw. Technol.*, vol. 33, no. 2, pp. 519–527, Jan. 2015.

8 Q. H. Spencer, A. L. Swindlehurst, and M. Haardt, "Zero-forcing methods for downlink spatial multiplexing in multiuser MIMO channels," *IEEE Trans. Signal Process.*, vol. 52, no. 2, pp. 461–471, Feb. 2004.

9 Q. H. Spencer, C. B. Peel, A. L. Swindlehurs, and M. Haardt, "An introduction to the multi-user MIMO downlink," *IEEE Commun. Mag.*, vol. 42, no. 10, pp. 60–67, Oct. 2004.

10 Z. Yu, R. J. Baxley, and G. T. Zhou, "Multi-user MISO broadcasting for indoor visible light communication," in *Proc. IEEE International Conference on Acoustics, Speech and Signal Processing (ICASSP) 2013* (Vancouver, Canada), May 26–31, 2013, pp. 4849–4853.

11 H. Ma, L. Lampe, and S. Hranilovic, "Robust MMSE linear precoding for visible light communication broadcasting systems," in *Proc. IEEE Global Communications Conference (GLOBECOM) Workshops 2013* (Atlanta, GA), Dec. 9–13, 2013, pp. 1081–1086.

12 Y. Hong, J. Chen, Z. Wang, and C. Yu, "Performance of a precoding MIMO system for decentralized multiuser indoor visible light communications," *IEEE Photon. J.*, vol. 5, no. 4, p. 7800211, Aug. 2013.

13 B. Li, J. Wang, R. Zhang, S. Hong, C. Zhao, and L. Hanzo, "Multiuser MISO transceiver design for indoor downlink visible light communication under per-LED optical power constraints," *IEEE Photon. J.*, vol. 7, no. 4, p. 7201415, Aug. 2015.

14 J. Chen, Q. Wang, and Z. Wang, "Leakage-based precoding for MU-MIMO VLC systems under optical power constraint," *Opt. Commun.*, vol. 382, pp. 348–353, Jan. 2017.

15 H. Elgala, R. Mesleh, and H. Haas, "An

LED model for intensity-modulated optical communication systems," *IEEE Photon. Technol. Lett.*, vol. 22, no. 11, pp. 835–837, Jun. 2010.

16 M. Sadek, A. Tarighat, and A. H. Sayed, "A leakage-based precoding scheme for downlink multi-user MIMO channels," *IEEE Trans. Wirel. Commun.*, vol. 6, no. 5, pp. 1711–1721, May 2007.

17 Inc. CVX Research, "CVX: Matlab software for disciplined convex programming, version 2.0 beta," Sep. 2012.

18 Y. Wang and N. Chi, Demonstration of high-speed 2 × 2 non-imaging MIMO Nyquist single carrier visible light communications with frequency domain equalization, *J. Lightw. Technol.*, vol. 32, no. 11, pp. 2087–2093, Jun. 2014.

19 G. L. Stuber, J. R. Barry, S. W. Mclaughlin, Y. Li, M. A. Ingram, and T. G. Pratt, "Broadband MIMO-OFDM wireless communications," *Proc. IEEE*, vol. 92, no. 2, pp. 271–294, Feb. 2004.

20 M. Jiang and L. Hanzo, "Multiuser MIMO-OFDM for next-generation wireless systems," *Proc. IEEE*, vol. 95, no. 7, pp. 1430–1469, Jul. 2007.

21 A. H. Azhar, T. Tran, and D. O'Brien, "A gigabit/s indoor wireless transmission using MIMO-OFDM visible-light communications," *IEEE Photon. Technol. Lett.*, vol. 25, no. 2, pp. 171–174, Jan. 2013.

22 Q. Wang, Z. Wang, C. Qian, J. Quan, and L. Dai, "Multi-user MIMO-OFDM for indoor visible light communication systems," in *Proc. IEEE Global Conference on Signal and Information Processing (GlobalSIP) 2015* (Orlando, FL), Dec. 14–16, 2015, pp. 1170–1174.

23 Q. Wang, Z. Wang, and L. Dai, "Multiuser MIMO-OFDM for visible light communications," *IEEE Photon. J.*, vol. 7, no. 6, p. 7904911, Dec. 2015.

24 A. C. Boucouvalas, K. Yiannopoulos and Z. Ghassemlooy, "100 Gbit/s optical wireless communication system link throughput," *Electron. Lett.*, vol. 50, no. 17, pp. 1220–1222, Aug. 2014.

25 D. Tsonev, S. Videv, and H. Haas, "Towards a 100 Gb/s visible light wireless access network," *Opt. Exp.*, vol. 23, no. 2, pp. 1627–1637, Jan. 2015.

26 A. Gomez, K. Shi, C. Quintana, M. Sato, G. Faulkner, B. C. Thomsen, D. O'Brien, "Beyond 100-Gb/s indoor wide field-of-view optical wireless communications," *IEEE Photon. Technol. Lett.*, vol. 27, no. 4, pp. 367–370, Feb. 2015.

27 Q. Wang, Z. Wang, and L. Dai, "Iterative receiver for hybrid asymmetrically clipped optical OFDM," *J. Lightw. Technol.*, vol. 32, no. 22, pp. 3869–3875, Nov. 2014.

28 S. D. Dissanayake and J. Armstrong, "Comparison of ACO-OFDM, DCO-OFDM and ADO-OFDM in IM/DD systems," *J. Lightw. Technol.*, vol. 31, no. 7, pp. 1063–1072, Apr. 2013.

29 J. Armstrong and A. J. Lowery, "Power efficient optical OFDM," *Electron. Lett.*, vol. 42, no. 6, pp. 370–372, Mar. 2006.

7
Signal Processing and Optimization

In this chapter, we address several signal processing and optimization issues for visible light communications (VLCs). For multi-chip-based multi-input single-output VLC systems, an electrical and optical power allocation scheme is introduced in Section 7.1 to maximize the multiuser sum-rate in consideration of the luminance, chromaticity, amplitude, and bit error rate constraints. From the perspective of human color vision, the chromaticity constraint is defined within a MacAdam ellipse. As a result, the degree of freedom can be achieved by relaxing the chromaticity constraint from a fixed color point to an elliptic region.

Heterogeneous VLC and wireless fidelity (VLC-WiFi) systems offer a solution for future indoor communications that combines VLC to support high-data-rate transmission and radio frequency (RF) to support reliable connectivity. In such heterogeneous systems, vertical handover (VHO) is critical for improving the system performance. In Section 7.2, the VHO is formulated as a Markov decision process (MDP) problem and a dynamic approach is adopted to obtain a tradeoff between the switching cost and the delay requirement.

In VLC systems with narrow field of view (FOV), the photodiode (PD) shot noise modeled by Poisson statistics is signal-dependent since it originates from the quantum nature of the received optical energy rather than any external sources of noise, which changes the nature of efficient signaling problem from that of the conventional signal-independent additive white Gaussian noise (AWGN) model in RF systems. Complex modifications should be made to the signal processing and estimation techniques to maintain the transmission performance, which is introduced in Section 7.3.

7.1
Sum-rate maximization for the multi-chip-based VLC system

In VLC systems, white light emitting diodes (LEDs) are used as illumination sources and transmitters simultaneously. Compared to the phosphor-converted LEDs, multi-chip LEDs have higher modulation bandwidth. Since each chip of the multi-chip LEDs can be modulated independently, parallel communication channels are viable for information transmission. Therefore, multi-chip-based VLC systems have

Visible Light Communications: Modulation and Signal Processing. First edition. Zhaocheng Wang, Qi Wang, Wei Huang, and Zhengyuan Xu.
Published 2017 by John Wiley & Sons, Inc.

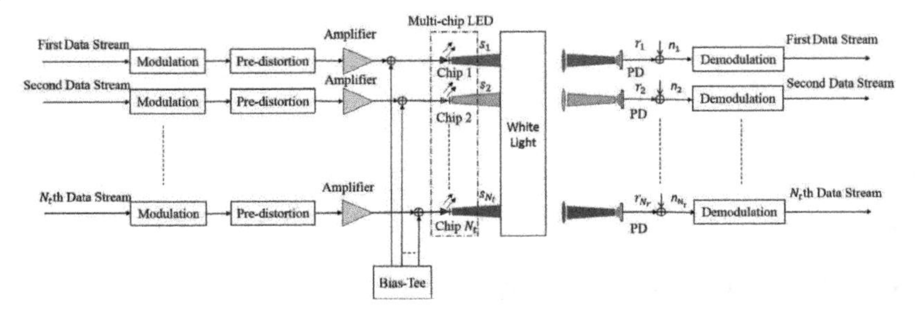

Figure 7.1 Block diagram of the multi-chip LED-based VLC system.

great potential for high-data-rate transmission. In this section, a multi-chip LED-based VLC system is introduced, where multiple independent parallel information bit streams are modulated onto the monochromatic light emitted by the LED chips. As brightness and chromaticity are the objective quality specifications of the color perceived by human eyes, luminance as well as chromaticity requirements should be satisfied in the design of the multi-chip-based VLC system. Specifically, chromaticity variation within MacAdam ellipse is considered since human eyes could not notice the small difference between two different colors in the MacAdam ellipse [1], which would be detailed in Section. 7.1.2.

7.1.1 System model

A multi-chip LED-based VLC system is shown in Fig. 7.1, where a multi-chip LED with N_t chips is used as the transmitter array. At the access point, each LED chip emits one monochromatic light and serves a single user. For different users, the user-specific optical filters and PDs are employed to concentrate their corresponding monochromatic light. For each link, pulse amplitude modulation (PAM) is employed to modulate the information bits due to its high spectral efficiency and low peak-to-average power ratio (PAPR).

For each transmitting chain i $(i = 1, 2, \cdots, N_t)$, the modulated signal d_i has zero mean and is normalized within $[-1, 1]$. To combat the nonlinearity of LEDs, time-domain predistortion is employed at the transmitter [2]. Moreover, amplifiers are used to fully exploit the dynamic range of LEDs and thus the electric signal power can be increased, where we denote the amplification coefficient as γ_i for transmitting chain i. Since intensity-modulated direct-detection (IM/DD) is adopted in VLC systems, a direct current (DC) bias $d_{\text{DC},i}$ is employed with the bias-tee circuits, and the resultant electric signal is given by

$$s_i = \gamma_i d_i + d_{\text{DC},i}, \quad i = 1, 2, \cdots, N_t. \tag{7.1}$$

At the jth receiver, the received signal r_j can be expressed as

$$r_j = h_{ij} s_i + w_j, \tag{7.2}$$

where h_{ij} is the channel gain and $w_j \sim N(0, \sigma^2)$ is the Gaussian noise.

7.1.2
Constraints on illumination and communication

In VLC, since illumination and communication should be maintained simultaneously, several constraints should be considered. For illumination, luminance and chromaticity constraints are imposed, while signal amplitude and bit error rate (BER) constraints are important for data communication. In the following, the various constraints are discussed.

1) Luminance constraint

When N_t chips are utilized for data transmission in multi-chip LEDs, indoor users should not perceive the luminance variation of the white light. That is

$$\mathbf{1}^{\mathrm{T}}\boldsymbol{\phi} = \sum_{i=1}^{N_t} \phi_i = P_t, \tag{7.3}$$

where P_t is the total luminous flux of the multi-chip LED.

Since the light intensity perceived by human eyes is determined by the average luminous flux, the average luminous flux instead of instantaneous luminous flux is optimized to improve the system performance. Besides, the optimization of the instantaneous luminous flux requires the knowledge of the DC bias for each signal at the receiver, which is impractical for implementation.

2) Chromaticity constraint

For the ith chip of the multi-chip LED, assuming the luminous flux emitted is Φ_i and the chromaticity coordinates of the emitted monochromatic light are (x_i, y_i). Based on Grassmann's laws of additive color mixture, the chromaticity coordinates of the mixed white light can be calculated by [3]

$$\begin{cases} \hat{x} = \dfrac{\boldsymbol{a}^{\mathrm{T}}\boldsymbol{\Phi}}{\boldsymbol{b}^{\mathrm{T}}\boldsymbol{\Phi}} = \dfrac{\sum_{i=1}^{N_t} \frac{x_i}{y_i}\Phi_i}{\sum_{i=1}^{N_t} \frac{1}{y_i}\Phi_i}, \\ \hat{y} = \dfrac{\mathbf{1}^{\mathrm{T}}\boldsymbol{\Phi}}{\boldsymbol{b}^{\mathrm{T}}\boldsymbol{\Phi}} = \dfrac{\sum_{i=1}^{N_t} \Phi_i}{\sum_{i=1}^{N_t} \frac{1}{y_i}\Phi_i}, \end{cases} \tag{7.4}$$

where $\boldsymbol{\Phi} = [\Phi_1\ \Phi_2\ \cdots\ \Phi_{N_t}]^{\mathrm{T}}$ is the luminous flux vector, $\boldsymbol{a} = [\frac{x_1}{y_1}\ \frac{x_2}{y_2}\ \cdots\ \frac{x_{N_t}}{y_{N_t}}]^{\mathrm{T}}$ and $\boldsymbol{b} = [\frac{1}{y_1}\ \frac{1}{y_2}\ \cdots\ \frac{1}{y_{N_t}}]^{\mathrm{T}}$ are the coefficient vectors.

As human eyes have certain limitation on color discrimination, quantifying chromaticity difference is subjective. MacAdam ellipse is used as a statistical measurement tool to describe the small chromaticity difference between two colors in the chromaticity diagram at the same luminance [1]. When two colors are located inside the same MacAdam ellipse, average human observers could not discern the color difference. The original ellipse in MacAdam's experiment is named as one-step MacAdam ellipse. In practice, larger ξ-step ($\xi > 1$) MacAdam ellipses are usually

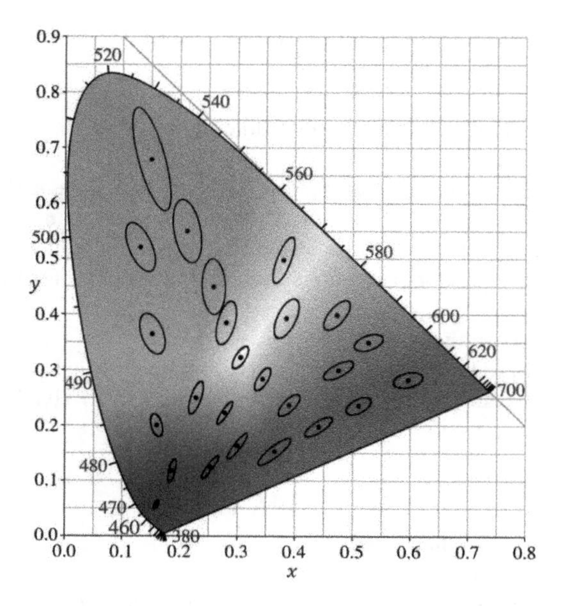

Figure 7.2 MacAdam ellipses in CIE 1931 color space chromaticity diagram with 10 times the actual size in MacAdam's experiment.

used, where the lengths of its major and minor axes are ξ times the lengths of the original ellipse's major and minor axes in MacAdam's experiment. As shown in Fig. 7.2, there are twenty-five 10-step MacAdam ellipses with different center points in the chromaticity diagram, which are 10 times the size of the 1-step MacAdam ellipses to make them observed clearly. Each of them varies in size and orientation. It can be observed that the blue region is much smaller than the green region. This is because human eyes have different sensitivities to different colors. The ξ-step MacAdam ellipse can be expressed as

$$g_{11}dx^2 + 2g_{12}dxdy + g_{22}dy^2 = \xi^2, \tag{7.5}$$

where g_{11}, g_{12}, and g_{22} are constant coefficients to describe the orientation and size of each ellipse; dx and dy are the differences of x and y coordinates between the color points on the ellipse and the center point of the ellipse.

Therefore, if the chromaticity coordinate of the white light moves within the MacAdam ellipse, small chromaticity change can be tolerated in practical VLC systems. Based on (7.4) and (7.5), the chromaticity constraint can be expressed as

$$\begin{aligned} g_{11}(\frac{\boldsymbol{a}^{\mathrm{T}}\boldsymbol{\phi}}{\boldsymbol{b}^{\mathrm{T}}\boldsymbol{\phi}} - x_0)^2 + 2g_{12}(\frac{\boldsymbol{a}^{\mathrm{T}}\boldsymbol{\phi}}{\boldsymbol{b}^{\mathrm{T}}\boldsymbol{\phi}} - x_0)(\frac{\mathbf{1}^{\mathrm{T}}\boldsymbol{\phi}}{\boldsymbol{b}^{\mathrm{T}}\boldsymbol{\phi}} - y_0) \\ + g_{22}(\frac{\mathbf{1}^{\mathrm{T}}\boldsymbol{\phi}}{\boldsymbol{b}^{\mathrm{T}}\boldsymbol{\phi}} - y_0)^2 \leqslant \xi^2, \end{aligned} \tag{7.6}$$

where x_0 and y_0 are the chromaticity coordinates of the center point inside the MacAdam ellipse.

3) Amplitude constraint

In VLC systems, the electric signal should be nonnegative, i.e., $s_i \geqslant 0$. Therefore, the signals should satisfy $\gamma_i \leqslant d_{\mathrm{DC},i} = c_i\phi_i$, where c_i is the constant coefficient to convert the luminous flux to the forward current. Moreover, since LEDs have a maximum forward current, the electric signal exceeding the maximum permissible amplitude A_i suffers from clipping distortion. As a result, the maximum amplitude constraint needs to be considered as well (i.e., $s_i \leqslant A_i$), and we have $\gamma_i \leqslant A_i - d_{\mathrm{DC},i} = A_i - c_i\phi_i$. Therefore, the amplitude constraint can be summarized as

$$0 \leqslant \gamma_i \leqslant \min(A_i - c_i\phi_i, c_i\phi_i), \quad i = 1, \cdots, N_t. \tag{7.7}$$

4) BER constraint

For the communication function, low BER at the receiver is required. For the ith user, we have $\mathrm{BER}_i \leqslant \mathrm{BER}_t$, where BER_t denotes the BER threshold to ensure the quality of service (QoS).

7.1.3 Sum-rate maximization

In the following, a typical point-to-point transmission scheme for the multi-chip LED-based VLC system is investigated, where the transmitter employs all its N_t colors (i.e., N_t parallel channels) to send the signals to the N_t receivers with individual optical filters (i.e., $N_r = N_t$). Moreover, an electric and optical power allocation scheme is employed to maximize the sum-rate for all users under the aforementioned constraints. With the constraint of BER (i.e., $\mathrm{BER}_i \leqslant \mathrm{BER}_t$), the achievable data rate for the ith user can be given by [4, 5]

$$v_i = \log(1 + \frac{\rho h_i \gamma_i}{\sigma_i}), \quad i = 1, \cdots, N_t, \tag{7.8}$$

where $\rho = \sqrt{(-\log(5\mathrm{BER}_t))^{-1}}$.

Afterwards, the sum-rate maximization problem in terms of the amplification coefficient γ_i and luminous flux ϕ_i can be formulated as

$$\begin{aligned}
\mathbf{P1}: \max_{\boldsymbol{\phi},\boldsymbol{\gamma}} & \sum_{i=1}^{N_t} \log(1 + \frac{\rho h_i \gamma_i}{\sigma_i}) \\
s.t.\quad & g_{11}(\frac{\boldsymbol{a}^{\mathrm{T}}\boldsymbol{\phi}}{\boldsymbol{b}^{\mathrm{T}}\boldsymbol{\phi}} - x_0)^2 + 2g_{12}(\frac{\boldsymbol{a}^{\mathrm{T}}\boldsymbol{\phi}}{\boldsymbol{b}^{\mathrm{T}}\boldsymbol{\phi}} - x_0)(\frac{\mathbf{1}^{\mathrm{T}}\boldsymbol{\phi}}{\boldsymbol{b}^{\mathrm{T}}\boldsymbol{\phi}} - y_0) \\
& + g_{22}(\frac{\mathbf{1}^{\mathrm{T}}\boldsymbol{\phi}}{\boldsymbol{b}^{\mathrm{T}}\boldsymbol{\phi}} - y_0)^2 \leqslant \xi^2, \\
& 0 \preceq \boldsymbol{\gamma} \preceq \min(\boldsymbol{A} - \boldsymbol{c} \circ \boldsymbol{\phi}, \boldsymbol{c} \circ \boldsymbol{\phi}), \\
& \boldsymbol{\phi} \succeq 0, \\
& \mathbf{1}^{\mathrm{T}}\boldsymbol{\phi} = P_t,
\end{aligned} \tag{7.9}$$

where $\boldsymbol{\gamma} = [\gamma_1\, \gamma_2\, \cdots\, \gamma_{N_t}]$ is the amplification coefficient vector, $\boldsymbol{A} = [A_1\, A_2\, \cdots\, A_{N_t}]$ is the maximum amplitude vector, and $\boldsymbol{c} = [c_1\, c_2\, \cdots\, c_{N_t}]$ is the luminous flux–forward current conversion coefficient vector. The notation $\preceq$ denotes the generalized inequality, and the notation $\circ$ denotes Hadamard product (i.e., for two matrices E and F, $(E \circ F)_{ij} = E_{ij} * F_{ij}$).

Since the compound function of the first inequality constraint in the optimization problem $\mathbf{P1}$ is non-convex, $\mathbf{P1}$ should be cast into a convex problem in order to find a global optimization solution. First, polar coordinates are used to transform the rotated MacAdam ellipse expressed by (7.5) to the ellipse with a standard form given by

$$\frac{p^2}{\alpha^2} + \frac{q^2}{\beta^2} = \xi^2, \tag{7.10}$$

where α is the length of the semi-major axis and β is the length of the semi-minor axis. From (7.5), α and β can be given by [6]

$$\begin{cases} \alpha = \sqrt{\dfrac{2}{(g_{11} + g_{22}) - \sqrt{(g_{11} - g_{22})^2 + 4g_{12}^2}}}, \\ \beta = \sqrt{\dfrac{2}{(g_{11} + g_{22}) + \sqrt{(g_{11} - g_{22})^2 + 4g_{12}^2}}}. \end{cases} \tag{7.11}$$

Then, the chromaticity point (p, q) on the standard ellipse can be written as

$$\begin{cases} p = \bar{x}\cos\theta + \bar{y}\sin\theta, \\ q = -\bar{x}\sin\theta + \bar{y}\cos\theta, \end{cases} \tag{7.12}$$

where we have $\bar{x} = \frac{\boldsymbol{a}^{\mathrm{T}}\boldsymbol{\phi}}{\boldsymbol{b}^{\mathrm{T}}\boldsymbol{\phi}} - x_0$, $\bar{y} = \frac{\mathbf{1}^{\mathrm{T}}\boldsymbol{\phi}}{\boldsymbol{b}^{\mathrm{T}}\boldsymbol{\phi}} - y_0$, and the rotated angle is donated as θ, given by

$$\theta = \begin{cases} 0, & \text{for } g_{12} = 0 \text{ and } g_{11} < g_{22}; \\ \frac{\pi}{2}, & \text{for } g_{12} = 0 \text{ and } g_{11} > g_{22}; \\ \frac{1}{2}\cot^{-1}\left(\frac{g_{11} - g_{22}}{2g_{12}}\right), & \text{for } g_{12} \neq 0 \text{ and } g_{11} < g_{22}; \\ \frac{\pi}{2} + \frac{1}{2}\cot^{-1}\left(\frac{g_{11} - g_{22}}{2g_{12}}\right), & \text{for } g_{12} \neq 0 \text{ and } g_{11} > g_{22}. \end{cases} \tag{7.13}$$

Correspondingly, the first constraint in $\mathbf{P1}$ is equivalent to the following expressions

$$\begin{cases} \dfrac{p^2}{\alpha^2} + \dfrac{q^2}{\beta^2} \leqslant \xi^2, \\ p = (\dfrac{\boldsymbol{a}^{\mathrm{T}}\boldsymbol{\phi}}{\boldsymbol{b}^{\mathrm{T}}\boldsymbol{\phi}} - x_0)\cos\theta + (\dfrac{\mathbf{1}^{\mathrm{T}}\boldsymbol{\phi}}{\boldsymbol{b}^{\mathrm{T}}\boldsymbol{\phi}} - y_0)\sin\theta, \\ q = (\dfrac{\mathbf{1}^{\mathrm{T}}\boldsymbol{\phi}}{\boldsymbol{b}^{\mathrm{T}}\boldsymbol{\phi}} - y_0)\cos\theta - (\dfrac{\boldsymbol{a}^{\mathrm{T}}\boldsymbol{\phi}}{\boldsymbol{b}^{\mathrm{T}}\boldsymbol{\phi}} - x_0)\sin\theta. \end{cases} \tag{7.14}$$

Moreover, due to the monotonicity of logarithmic function, the objective function is equivalent to

$$\det(\mathbf{I} + \mathbf{diag}(\boldsymbol{\kappa} \circ \boldsymbol{\gamma})) = \prod_{i=1}^{N_t}(1 + \frac{\rho h_i \gamma_i}{\sigma_i}), \tag{7.15}$$

where $\boldsymbol{\kappa} = [\frac{\rho h_1}{\sigma_1}\ \frac{\rho h_2}{\sigma_2}\ \cdots\ \frac{\rho h_{N_t}}{\sigma_{N_t}}]$. $\det(\cdot)$ denotes the determinant of a matrix, and $\mathbf{diag}(\cdot)$ is a diagonal matrix.

Therefore, the optimization problem **P1** can be reformulated as

$$\begin{aligned}
\mathbf{P2} : \max_{\boldsymbol{\phi},\boldsymbol{\gamma}}\ & \det(\mathbf{I} + \mathbf{diag}(\boldsymbol{\kappa} \circ \boldsymbol{\gamma})) \\
\textit{s.t.}\ \ & \frac{p^2}{\alpha^2} + \frac{q^2}{\beta^2} \leqslant \xi^2, \\
& 0 \preceq \boldsymbol{\gamma} \preceq \min(\boldsymbol{A} - \boldsymbol{c} \circ \boldsymbol{\phi}, \boldsymbol{c} \circ \boldsymbol{\phi}), \\
& \boldsymbol{\phi} \succeq 0, \\
& p = (\frac{\boldsymbol{a}^{\mathrm{T}}\boldsymbol{\phi}}{\boldsymbol{b}^{\mathrm{T}}\boldsymbol{\phi}} - x_0)\cos\theta + (\frac{\mathbf{1}^{\mathrm{T}}\boldsymbol{\phi}}{\boldsymbol{b}^{\mathrm{T}}\boldsymbol{\phi}} - y_0)\sin\theta, \\
& q = (\frac{\mathbf{1}^{\mathrm{T}}\boldsymbol{\phi}}{\boldsymbol{b}^{\mathrm{T}}\boldsymbol{\phi}} - y_0)\cos\theta - (\frac{\boldsymbol{a}^{\mathrm{T}}\boldsymbol{\phi}}{\boldsymbol{b}^{\mathrm{T}}\boldsymbol{\phi}} - x_0)\sin\theta, \\
& \mathbf{1}^{\mathrm{T}}\boldsymbol{\phi} = P_t.
\end{aligned} \tag{7.16}$$

Since the fourth and fifth constraints (i.e., the two equality constraints involving p, q, and $\boldsymbol{\phi}$) in **P2** are not affine or linear, two intermediate variables $m = \frac{p}{\alpha}\boldsymbol{b}^{\mathrm{T}}\boldsymbol{\phi}$ and $n = \frac{q}{\beta}\boldsymbol{b}^{\mathrm{T}}\boldsymbol{\phi}$ are introduced, and the sum-rate maximization problem can be transformed into [7]

$$\begin{aligned}
\mathbf{P3} : \max_{\boldsymbol{\phi},\boldsymbol{\gamma}}\ & \det(\mathbf{I} + \mathbf{diag}(\boldsymbol{\kappa} \circ \boldsymbol{\gamma})) \\
\textit{s.t.}\ \ & m^2 + n^2 \leqslant (\xi \boldsymbol{b}^{\mathrm{T}}\boldsymbol{\phi})^2, \\
& 0 \preceq \boldsymbol{\gamma} \preceq \min(\boldsymbol{A} - \boldsymbol{c} \circ \boldsymbol{\phi}, \boldsymbol{c} \circ \boldsymbol{\phi}), \\
& \boldsymbol{\phi} \succeq 0, \\
& m = \frac{1}{\alpha}[(\boldsymbol{a}^{\mathrm{T}} - x_0\boldsymbol{b}^{\mathrm{T}})\cos\theta + (\mathbf{1}^{\mathrm{T}} - y_0\boldsymbol{b}^{\mathrm{T}})\sin\theta]\boldsymbol{\phi}, \\
& n = \frac{1}{\beta}[(\mathbf{1}^{\mathrm{T}} - y_0\boldsymbol{b}^{\mathrm{T}})\cos\theta - (\boldsymbol{a}^{\mathrm{T}} - x_0\boldsymbol{b}^{\mathrm{T}})\sin\theta]\boldsymbol{\phi}, \\
& \mathbf{1}^{\mathrm{T}}\boldsymbol{\phi} = P_t.
\end{aligned} \tag{7.17}$$

Given a real-valued slack variable t, the first inequality constraint in **P3** is equivalent to

$$\begin{cases} m^2 + n^2 \leqslant t^2, \\ 0 \leqslant t \leqslant \xi \boldsymbol{b}^{\mathrm{T}}\boldsymbol{\phi}. \end{cases} \tag{7.18}$$

In (7.18), the first inequality defines a second order or Lorentz cone $\left(\text{i.e., } \{(z,w) \in \boldsymbol{R}^2 \times \boldsymbol{R} \mid z^{\mathrm{T}}z \leqslant w^2, w \geqslant 0\}\right)$ [8]. Hence, the problem becomes a conic optimization problem represented by

$$\begin{aligned}
\mathbf{P4}: \max_{\boldsymbol{\phi},\boldsymbol{\gamma}} \ & \det(\mathbf{I} + \mathbf{diag}(\boldsymbol{\kappa} \circ \boldsymbol{\gamma})) \\
\text{s.t.} \ & m^2 + n^2 \leqslant t^2, \\
& 0 \leqslant t \leqslant \xi \boldsymbol{b}^{\mathrm{T}} \boldsymbol{\phi}, \\
& 0 \preceq \boldsymbol{\gamma} \preceq \min(\boldsymbol{A} - \boldsymbol{c} \circ \boldsymbol{\phi}, \boldsymbol{c} \circ \boldsymbol{\phi}), \\
& \boldsymbol{\phi} \succeq 0, \\
& m = \frac{1}{\alpha}[(\boldsymbol{a}^{\mathrm{T}} - x_0 \boldsymbol{b}^{\mathrm{T}}) \cos\theta + (\mathbf{1}^{\mathrm{T}} - y_0 \boldsymbol{b}^{\mathrm{T}}) \sin\theta] \boldsymbol{\phi}, \\
& n = \frac{1}{\beta}[(\mathbf{1}^{\mathrm{T}} - y_0 \boldsymbol{b}^{\mathrm{T}}) \cos\theta - (\boldsymbol{a}^{\mathrm{T}} - x_0 \boldsymbol{b}^{\mathrm{T}}) \sin\theta] \boldsymbol{\phi}, \\
& \mathbf{1}^{\mathrm{T}} \boldsymbol{\phi} = P_t,
\end{aligned} \tag{7.19}$$

which can be solved by several convex optimization algorithms such as infeasible path-following algorithms [9]. In MATLAB, it can be solved by the optimization software package CVX [10].

7.1.4 Performance evaluation

The sum-rate performances of the multi-chip LED-based VLC system investigated on the assumption that three indoor users are at the same distance from the transmitter and use perfect optical filters to receive their own information, while Table 7.1 lists the parameters used in the simulations. The target BER is set to 10^{-3} for each user.

Figure 7.3 illustrates the maximum sum-rate under different luminous fluxes, where the step of MacAdam ellipse is set to 7 and the correlated color temperature (CCT) value is defined as 5000 K. Unlike conventional RF system where higher electric power results in higher achievable data rate, higher optical power for LED in VLC might even diminish the data rate. It is shown that the curves of the achievable data rate for red and green chips present the open-down parabolic shapes and have a peak point at 100 lm. The reason is that optical power is determined by the DC bias and narrower dynamic range would be available to amplify the modulated signal in case of clipping distortion when DC bias deviates from the middle position of the dynamic range of LED chips. Moreover, the variations of the data rate for red and green chips further cause the change of the sum-rate according to the same tendency. Due to the small portion of the blue component in the white light, the DC bias for the blue chip is always below the middle position of the dynamic range of blue LED chip in the luminous flux range. Consequently, the data rate for the blue chip increases slowly when DC bias approaches to the middle position.

The percentages of three light components to achieve the maximum sum rate are compared in Fig. 7.4 when the CCT is defined as 5000 K and chromaticity tolerance

Table 7.1 Simulation parameters for multi-chip LED-based VLC system.

Parameters		Values
Type of white LED		RGB LED
		(Cree Xlamp MC-E)
Number of LED chips N_t		3
Target BER for each user		10^{-3}
Maximum forward current for each chip		700 mA
Distance between LED and users r_i		2 m
Detect area A		1 cm^2
Lambertian emission order m		1
FOV of the receiver		60°
Angle of the irradiance		30°
Angle of the incidence		40°
Concentrator refractive index		1.5
Receiver responsivity		0.5 A/W
Noise variance		0.013 mA^2
Peak wavelength of each chip	Red	625 nm
	Green	530 nm
	Blue	460 nm
Chromaticity coordinates	Red	(0.7006, 0.2993)
	Green	(0.1547, 0.8059)
	Blue	(0.1440, 0.0297)
Luminance flux–forward current conversion coefficient	Red	0.0114 A/lm
	Green	0.0052 A/lm
	Blue	0.0427 A/lm

Table 7.2 Chromaticity coordinates for the MacAdam ellipse [11].

CCT	Center point	g_{11}	$2g_{12}$	g_{22}
2700 K	(0.459 0.412)	40×10^4	-39×10^4	28×10^4
3000 K	(0.440 0.403)	39×10^4	-39×10^4	27.5×10^4
3500 K	(0.411 0.393)	38×10^4	-40×10^4	25×10^4
4000 K	(0.380 0.380)	39.5×10^4	-43×10^4	26×10^4
5000 K	(0.346 0.359)	56×10^4	-50×10^4	28×10^4
6500 K	(0.313 0.337)	86×10^4	-80×10^4	45×10^4

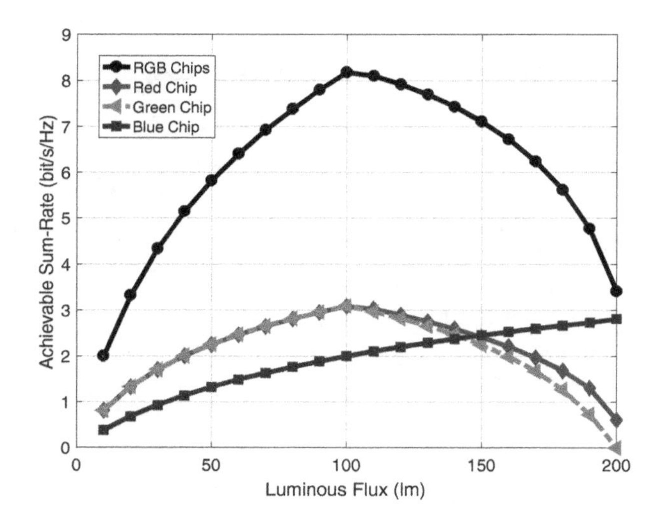

Figure 7.3 Maximum sum-rate for different luminous fluxes when CCT = 5000 K.

is within a 7-step MacAdam ellipse. In order to satisfy the white light constraint, the chromaticity variation is still restricted in a small range. Therefore, the percentages of all components almost remain unchanged, which are 0.3, 0.675, and 0.025 respectively, while the fluctuation of each component is very small.

In Fig. 7.5, different steps of the MacAdam ellipse are compared with CCT = 5000 K. It is shown that multiuser sum-rate is affected by the steps of the MacAdam ellipse. When 10-step MacAdam ellipse is considered, it achieves over 0.5 bit/s/Hz performance gain over the system with 1-step MacAdam ellipse.

The maximum sum-rates are also investigated in Fig. 7.6 under different CCT values with the chromaticity constraint based on MacAdam ellipse. Six CCT values are investigated according to ANSI C78.376–2001 [11, 12], which are 2700 K, 3000 K (warm white), 3500 K(white), 4000 K (cool white), 5000 K, and 6500 K (daylight), respectively. A seven-step MacAdam ellipse is chosen for every CCT value in the simulation. The parameters of the corresponding MacAdam ellipse are listed in Table 7.2. It can be seen that the curve of the sum-rate for each CCT value has an open-down parabolic tendency. When larger CCT value is used, higher sum-rate is achieved. The reason is that when a larger CCT value is considered, the proportion of the blue component becomes higher. The increase of the data rate in the blue light link has the main impact on the sum-rate compared to the other two links for a larger CCT. Besides, for a lower CCT value, the decrease of the electric power as well as the data rate in the red light link is much severer under large total luminous flux (above 100 lm). Thus, the difference of the sum-rates among all CCT values becomes more obvious.

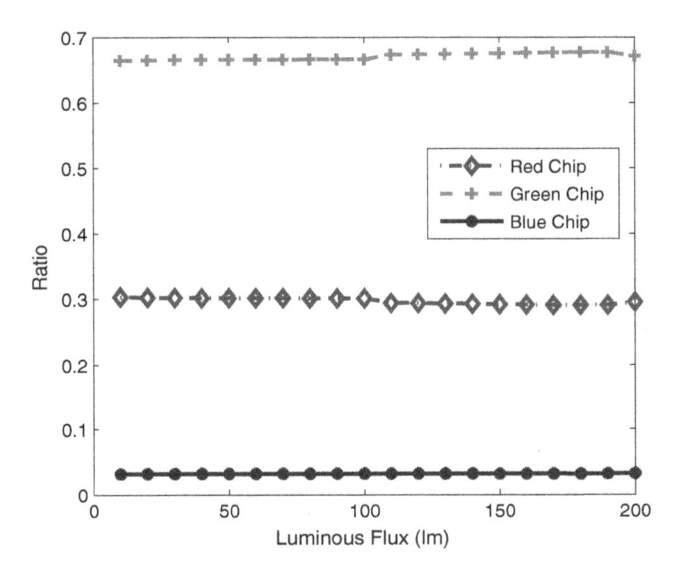

Figure 7.4 Percentages of three light components to achieve the maximum sum-rate for different luminous fluxes when CCT = 5000 K.

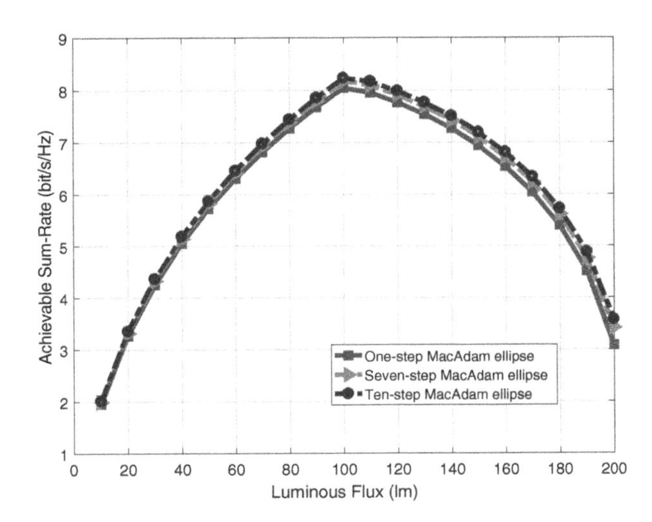

Figure 7.5 Comparison between different steps of MacAdam ellipse under different luminous fluxes when CCT = 5000 K.

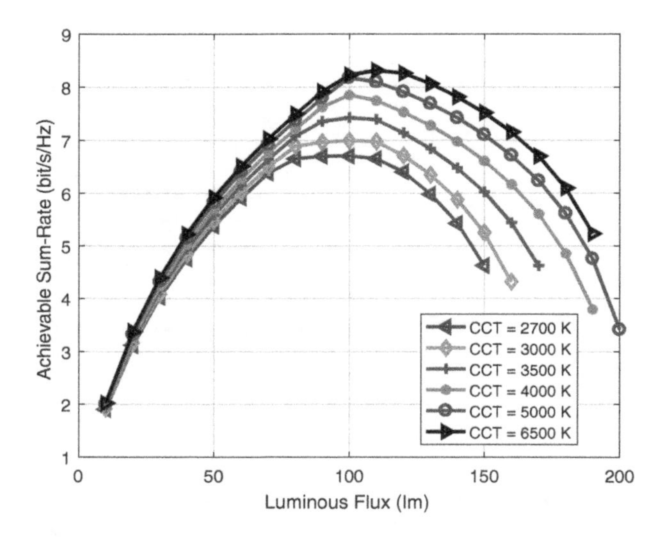

Figure 7.6 Maximum achievable sum-rate comparison for different CCT values under chromaticity constraint based on MacAdam ellipse.

7.2 Heterogeneous VLC-WiFi optimization

In VLC systems, line-of-sight (LOS) transmission is usually assumed, which cannot be always guaranteed especially in indoor applications due to random movements of mobile terminals (MTs). Therefore, heterogeneous network is considered to exploit the advantages of both VLC and conventional RF systems. When the vulnerable VLC LOS link is blocked, a vertical handover (VHO) from VLC to RF communications, for example, wireless local area network (WLAN), can be utilized to maintain the continuity of wireless data transmission.

In a heterogeneous VLC-RF network, it is preferred to exploit VLC LOS link for content distribution since higher data rate can be supported. When the LOS link is unavailable, however, MTs will suffer from severe quality of experience (QoE) degradation, and the system should switch to the RF connection to provide continuous data transmission with low QoE. When the VLC LOS link is recovered, the system should switch back to the VLC connection again to provide high QoE. However, LOS blockage is usually a temporary phenomenon lasting for a very short period. Therefore, it might be more efficient to wait for a while rather than switch immediately in some occasions, since switching back and forth will cause additional signaling overhead and latency, especially for non-real-time services. Therefore, "waiting" action might avoid unnecessary switching if the LOS link is recovered rapidly with an acceptable short blockage duration. Otherwise, "switching" action is preferred to avoid the possible communication breakdown. The main challenge lies in the efficient strategy to decide whether performing a VHO or not when LOS blockage happens.

VHO is a common problem in heterogeneous networks, such as integrated 3G/4G and WLAN networks [13]. Due to human activities, LOS link is intermittently interrupted and resumed. In general, LOS blockage is caused by random movements of MTs either in the shadow area or out of the coverage [14]. The link interruption from the first type of movements is usually short and temporary, while that from the second type usually lasts for a relatively long time. Therefore, VLC channels are quite different from conventional RF counterparts, and the conventional VHO schemes of all-radio heterogeneous systems could not be directly adopted in VLC-RF heterogeneous networks. Furthermore, VLC interruption mainly depends on the behaviors of MTs, which always follow certain patterns and can be depicted by factors such as location, group, time-of-day, and duration [15]. Hence, the movements of MTs can be forecasted by the record of behavior history, which can be used to model and predict the interruption of VLC links.

7.2.1 System model

Since both VLC and WiFi systems are commonly adopted for indoor scenarios, the heterogeneous VLC-WiFi system is investigated, which is illustrated in Fig. 7.7. Both VLC cell and WiFi radio cell are overlapped in the room, where the central area is covered by VLC to provide a relatively higher data rate, while a wider area is covered by WiFi. All MTs are assumed to support multi-mode transmission, which can download multimedia content either through VLC or WiFi network. Since the uplink traffic is usually much less than the downlink [16], only WiFi is utilized for uplink transmission.

As shown in Fig. 7.7, data packets from remote content providers are queued at the local access point (AP) for first-in-first-out (FIFO) transmission. At the AP, a controller is used to select the transmission mode of each packet, i.e., either VLC or WiFi. Since the coverage of WiFi network is larger than its VLC counterpart and WiFi works well under non-line-of-sight (NLOS) scenarios, it is reasonable to assume that WiFi link is always available in the integrated VLC-WiFi heterogeneous network. When VLC LOS link is blocked during the transmission of a packet in VLC mode, the packet has to be retransmitted, either through the recovered VLC link or WiFi channel. When the buffer of the AP overflows, new incoming packets have to be dropped.

The objective of an efficient VHO scheme is to find the balance between switching cost and traffic congestion. The design of an effective VHO decision-making algorithm is challenging due to the following reasons: 1) the best tradeoff point is determined by switching cost and delay requirement; 2) both signaling overhead and handover latency are induced during VHO process. For the first reason, since multimedia services might have different requirements, the controller may prefer to stay in VLC mode and wait for link recovery during LOS blockage if the switching cost is high, whereas it may immediately switch to WiFi mode if low transmission delay is required. For the second reason, the controller may stay in VLC mode if the VLC

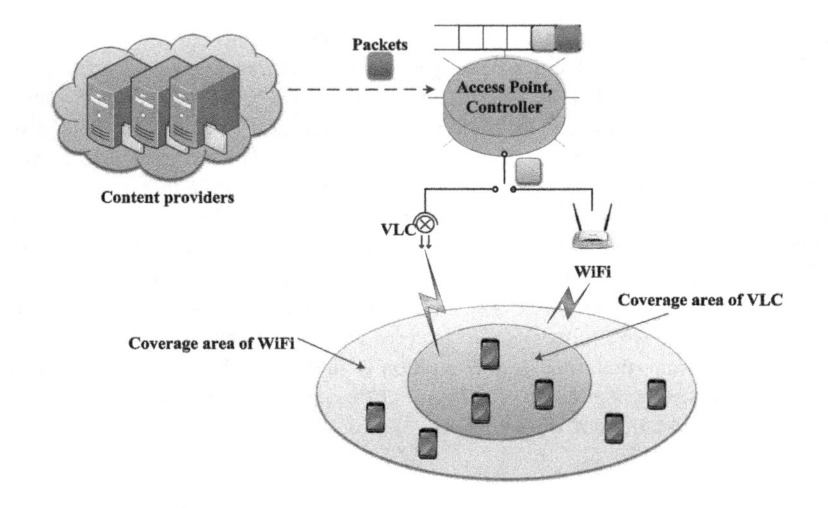

Figure 7.7 Illustration of the heterogeneous VLC-WiFi system.

link is unavailable when there is a high probability that it will be recovered soon, or stay in WiFi mode even if the VLC link is currently available when it is frequently blocked.

It is evident that the arrival process of packets and the service process for these two transmission modes follow Poisson distribution [17]. λ (packets/s), μ_1 (packets/s), and μ_2 (packets/s) are defined as the arrival rate of packets, the service rate of VLC, and the service rate of WiFi, respectively. An ON-OFF model is used to describe the intermittent interruption of VLC link, where ON represents that VLC link is available while OFF represents the opposite. γ_1 (s^{-1}) and γ_2 (s^{-1}) are described as two different rates at which the VLC channel changes from unblocking to blocking and vise versa. Thus, $1/\gamma_1$ and $1/\gamma_2$ represent the average duration of VLC link staying in ON or OFF, respectively. The packet delivery process of the heterogeneous VLC-WiFi system can be modeled by a Markov chain, where the VHO decision-making problem can be formulated as a Markov decision process and solved by a dynamic programming approach.

7.2.2 Efficient VHO scheme

For heterogeneous VLC-WiFi networks, we formulate the VHO decision-making problem as a continuous time MDP [18–21]. A two-state Markov process is used to model the intermittent VLC LOS blockage. Based on MDP, a VHO scheme is proposed in [22], which could achieve a tradeoff between switching cost and traffic congestion.

In the heterogeneous VLC-WiFi system, the state space of data delivery is defined

as

$$\Omega = \{(s, b, w), s \in \mathcal{S}, b \in \mathcal{B}, w \in \mathcal{W}\}, \tag{7.20}$$

where $\mathcal{S} = \{\text{ON}, \text{OFF}\}$ denotes the status of the VLC link, $\mathcal{B} = \{0, 1, ..., B\}$ represents the number of packets in the queue with size of B. $\mathcal{W} = \{1, 0\}$ is the action space for the transmission mode, where 1 means the packets are transmitted over VLC link, while 0 means the packets are transmitted over WiFi channel. The system stays in state $i = (s, b, w)$ when the condition of the optical channel is s, the number of packets in the system is b and the current transmission mode is w. The possible events include packet arrivals and departures, and the optical channel changes between ON and OFF. The event of packet departure includes two cases where the packet is transmitted either over VLC link or over WiFi channel. When any one of those events occurs, the system state is updated.

Let a stage be the period between two consecutive events, and the period is so short that any two events do not occur simultaneously. At the beginning of each stage, an action is taken to determine which state to enter. The solution of this MDP is a scheme denoted by p, indicating which action $p(i)$ should be taken for each state i. A two-dimensional (2-D) Markov process is used to describe the transition rates between different states, as shown in Fig. 7.8. Note that each state (s, b) in the figure actually includes two states $(s, b, 1)$ and $(s, b, 0)$. Let $Pr^{p(i)}_{i \to i'}$ denote the transition probability from state i to state i' with action $p(i)$, the transition probability for $s = \text{ON}$ is given by

$$Pr^{p(i)}_{i \to i'} = \begin{cases} \dfrac{\lambda}{\lambda + \eta\gamma_1 + \mu_{\text{ON}}}, & \text{if } i' = (\text{ON}, b+1, p(i)); \\ \dfrac{\eta\gamma_1}{\lambda + \eta\gamma_1 + \mu_{\text{ON}}}, & \text{if } i' = (\text{OFF}, b, p(i)); \\ \dfrac{\mu_{\text{ON}}}{\lambda + \eta\gamma_1 + \mu_{\text{ON}}}, & \text{if } i' = (\text{ON}, b-1, p(i)); \\ 0, \text{ else}, \end{cases} \tag{7.21}$$

where $\mu_{\text{ON}} = p(i)\mu_1 + (1 - p(i))\mu_2$, $\lambda + \eta\gamma_1 + \mu_{\text{ON}}$ is the sum of weighted transition rates, and η is the weighting factor with unit of packets. Similarly, the transition probability for $s = \text{OFF}$ is given by

$$Pr^{p(i)}_{i \to i'} = \begin{cases} \dfrac{\lambda}{\lambda + \eta\gamma_2 + \mu_{\text{OFF}}}, & \text{if } i' = (\text{OFF}, b+1, p(i)); \\ \dfrac{\eta\gamma_2}{\lambda + \eta\gamma_2 + \mu_{\text{OFF}}}, & \text{if } i' = (\text{ON}, b, p(i)); \\ \dfrac{\mu_{\text{OFF}}}{\lambda + \eta\gamma_2 + \mu_{\text{OFF}}}, & \text{if } i' = (\text{OFF}, b-1, p(i)); \\ 0, \text{ else}, \end{cases} \tag{7.22}$$

where $\mu_{\text{OFF}} = (1 - p(i))\mu_2$. Note that packet departure happens in situation $s =$ OFF only if the transmission mode is WiFi. When $b = 0$, no packet departs, while no packet arrives when $b = B$.

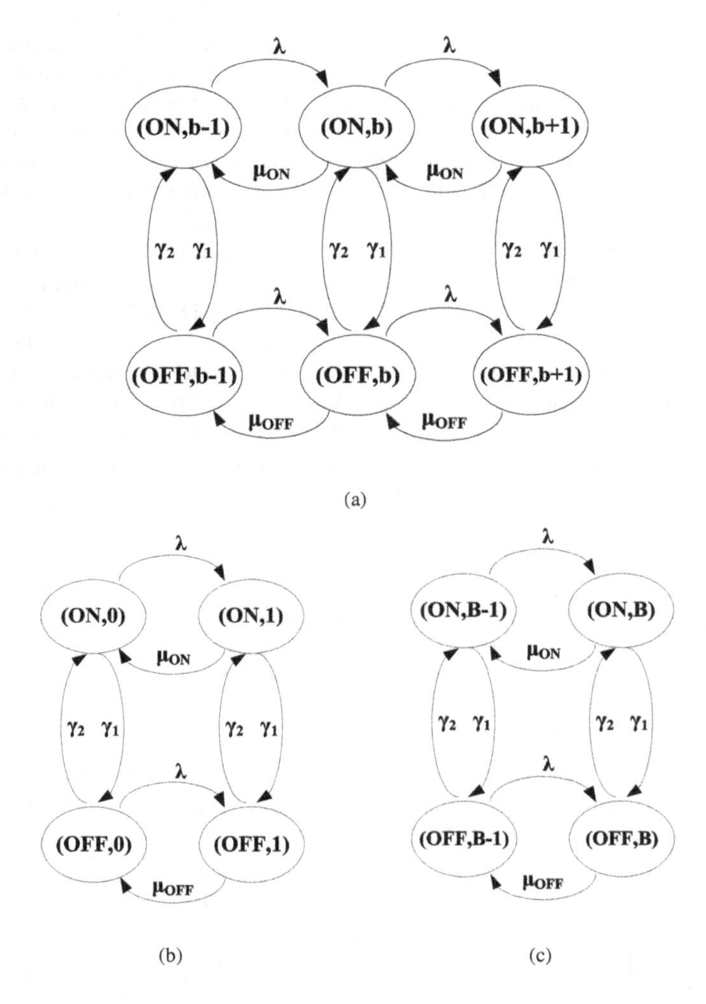

Figure 7.8 Illustration of transition rates for (a) $0 < b < B$, (b) $b = 0$ and (c) $b = B$.

In the VHO decision problem, the objective is to minimize a weighted sum of the traffic consumption, switching cost, and traffic congestion. Let $g(i, p(i))$ denote the cost incurred at state $i = (s, b, w)$ under action $p(i)$, we have

$$g(i,p(i)) = p(i)E_O + (1-p(i))E_W + \zeta b + |p(i) - w|(E_{sw} + \theta\lambda\tau), \quad (7.23)$$

which consists of two parts, denoted as $g_1(i,p(i)) = p(i)E_O + (1-p(i))E_W + \zeta b$ and $g_2(i,p(i)) = |p(i) - w|(E_{sw} + \theta\lambda\tau)$, respectively. $g_1(i,p(i))$ is a weighted sum of the energy cost for packet transmission and the delay cost induced by the packets queuing, where E_O and E_W are the energy consumption of staying in VLC and WiFi modes, and ζ is the tradeoff parameter. Compared with E_O and E_W, the energy consumption for the transceiver of either VLC or WiFi in the idle state is negligible, which is not considered in this problem. $g_2(i,p(i))$ denotes the switching cost, including the signaling overhead for switching the transmission mode, denoted as E_{sw}, and the handover latency, denoted as $\lambda\tau$, which represents $\lambda\tau$ packets accumulating in the buffer during the VHO process. θ is the scaling parameter which measures the importance of the latency.

To make the continuous time problem tractable, uniformization is required to transform it into a discrete one [23]. Define v_m as the uniform transition rate, which is no smaller than all the transition rates $v_i(p(i))$ of any state i and action $p(i)$, we have

$$v_m = \max\{\lambda + \eta\gamma_1 + \mu_1, \lambda + \eta\gamma_1 + \mu_2, \lambda + \eta\gamma_2 + \mu_2\}. \quad (7.24)$$

Denote $\hat{Pr}_{i\to i'}^{p(i)}$ as the new transition probability of the discrete-time Markov chain from state i to state i' under v_m, which is given by [23]

$$\hat{Pr}_{i\to i'}^{p(i)} = \begin{cases} \dfrac{v_i(p(i))}{v_m} Pr_{i\to i'}^{p(i)}, & \text{if } i \neq i'; \\ \dfrac{v_i(p(i))}{v_m} Pr_{i\to i'}^{p(i)} + 1 - \dfrac{v_i(p(i))}{v_m}, & \text{if } i = i'. \end{cases} \quad (7.25)$$

Afterwards, the cost per stage after uniformization is calculated as

$$\hat{g}(i,p(i)) = \frac{1}{\beta + v_m} g_1(i,p(i)) + g_2(i,p(i)), \quad (7.26)$$

where β is the parameter related to the discounted factor α [23]. α is calculated by $\alpha = \frac{v_m}{\beta + v_m}$ and β is chosen so that $\alpha \approx 1$.

To describe the VHO scheme, the discounted model is adopted [23]. Instead of minimizing the total cost, the cost of current stage and the discounted cost of future stages are minimized. According to Bellman's function [23], the average discounted sum of costs defined as the cost of current stage plus the expected cost of all future stages, is minimized by

$$V(i) = \min_{p(i)}\{\hat{g}(i,p(i)) + \alpha \sum_{i' \in \Omega} \hat{Pr}_{i\to i'}^{(p(i))} V(i')\}, \quad (7.27)$$

where $V(i)$ is the cost of state i.

The VHO solution is calculated by the value iteration algorithm demonstrated in Algorithm 7.1 [22]. Since all the components in (7.26) and $\hat{g}(i, p(i))$ are bounded, the iteration algorithm is convergent [23]. $Q_k(i, p(i))$ and $V_k^*(i)$ are denoted as the average cost for each state i of iteration k under action $p(i)$ and the optimal average cost for each state i of iteration k, respectively. In the algorithm, with the knowledge of $V_{k-1}^*(i)$, the optimal action $p_k^*(i)$ for each state i of iteration k is determined by selecting the action that minimizes $Q_k(i, p(i))$ in line 10 and the corresponding optimal cost $V_k^*(i)$ is obtained from line 11. Since the updated average costs $V_k^*(i)$ are also the input for iteration $k+1$, such procedure is repeated until the convergence criterion $||V_k - V_{k-1}|| \leq \epsilon$ is satisfied.

Algorithm 7.1 Value Iteration Algorithm

Input:

1: $\Omega, \mathcal{W}, \alpha$, transition probability $\hat{P}$, and cost function $\hat{g}$ after uniformization, convergence criterion parameter ϵ;

Output:

2: Action p;
3: **for** each $i \in \Omega$ **do**
4: $\quad V_0(i) = 0$;
5: **end for**
6: $\Delta = inf$;
7: **while** $\Delta > \epsilon$ **do**
8: $\quad$ **for** each $i \in \Omega$ **do**
9: $\quad\quad$ **for** each $p(i) \in \mathcal{W}$ **do**
10: $\quad\quad\quad Q_k(i, p(i)) = \hat{g}(i, p(i)) + \alpha \sum_{i' \in \Omega} \hat{Pr}_{i \to i'}^{p(i)} V_{k-1}(i')$;
11: $\quad\quad$ **end for**
12: $\quad\quad p_k^*(i) = \arg\min_{p(i)}\{Q_k(i, p(i))\}$;
13: $\quad\quad V_k^*(i) = Q_k(i, p_k^*(i))$;
14: $\quad$ **end for**
15: $\quad \Delta = ||V_k - V_{k-1}||$;
16: **end while**

With the VHO solution, the transition rate matrix $\mathbf{Q}$ is also derived. Let $\pi^*(s, b, w)$ and $\vec{\pi}^* = [\cdots \pi^*(s, b, w) \cdots]$ denote the stationary state probability of state (s, b, w) and the stationary probability vector, $\vec{\pi}^*$ can be obtained by solving the following set of equations [24]

$$\vec{\pi}^* \mathbf{Q} = \mathbf{0}, \tag{7.28}$$

$$\vec{\pi}^* \mathbf{1}^{\mathrm{T}} = 1, \tag{7.29}$$

where $\mathbf{1}$ is a vector with all unity entries. Packet blockage occurs when a new packet arrives at a time when the buffer overflows. Let P_b denote the blockage probability of packets, which is defined as the probability that the number of packets in the queue is B, we have

$$P_b = \sum_{s \in \mathcal{S}} \sum_{w \in \mathcal{W}} \pi^*(s, B, w). \tag{7.30}$$

7.2.3 Performance evaluation

The performance of the aforementioned VHO scheme, named as O-VHO, is compared with two benchmark schemes including immediate-VHO (I-VHO) and dwell-VHO (D-VHO) [14]. In I-VHO scheme, the controller immediately performs VHO to switch the transmission mode from VLC to WiFi once the optical channel is interrupted and from WiFi to VLC once the optical channel is restored. In D-VHO scheme, the controller waits for a dwell period of t_0 when the LOS blockage is detected. When the dwell time expires, if the optical link is still unavailable, the controller switches the transmission mode to WiFi, otherwise it stays in VLC mode. When the system detects that the VLC link is recovered, the controller immediately performs VHO from WiFi back to VLC. D-VHO can avoid potential ping-pong effects. To facilitate the simulations, we set t_0 to 0.5 s and 1 s, respectively.

Four metrics are used for performance evaluation. First, ρ denotes the *packet loss rate*, which is defined as the ratio of the number of dropped packets to the total number of arrived packets. Second, d denotes the *average delay* representing the average waiting time for a packet to be served. Third, l is the *average queue length*, which is defined as the average number of waiting packets. Finally, c is the *number of VHO*, which is proportional to signaling cost. When B packets are waiting in the queue, if there are packets that can be delivered successfully before new packet's arrival, then newly arrived packets will not be dropped. Therefore, P_b is the upper bound of ρ and $\rho \leq P_b$.

Simulations are carried out based on stochastic process over a period of continuous time. As mentioned previously, packet arrival and packet departure through two transmission modes are modeled as Poisson process with rates λ, μ_1, and μ_2, respectively [17]. The VLC LOS link blockage is modeled as an independent blocking event, whose duration has negative and exponential distribution with rate γ_2. Similarly, non-blocking duration of the LOS link is modeled by γ_1 [25]. Hereby, we mainly evaluate the switching cost during VHO process, thus the energy consumption of VLC and WiFi transmission is assumed to be the same. The simulation parameters are summarized in Table 7.3.

First, τ is set to 0.2 s and γ_2 is varied from 0.3 to 1.1 s^{-1}. Larger γ_2 means shorter average time when the system stays in the OFF state and the VLC link is more likely to be available. As shown in Fig. 7.9, when γ_2 increases from 0.3 to 1.1 s^{-1}, ρ attained from I-VHO, O-VHO, D-VHO with $t_0 = 0.5$ s and D-VHO with $t_0 = 1$ s is decreased by 68.08%, 61.56%, 59.82%, and 59.25% respectively, while d attained from those four schemes is decreased by 30.63%, 22.14%, 28.39%, and 30.71%, respectively. In addition, l attained from those four schemes is decreased by 29.86%, 21.25%, 27.24%, and 28.97%, respectively, and c attained from those four schemes

Table 7.3 Simulation parameters.

Parameter	Value
Packet arrival rate λ	1 packet/s
Handover delay τ	0–0.5 s
Packet departure rate of VLC μ_1	2 packets/s
Packet departure rate of WiFi μ_2	1.1 packets/s
Rate of optical channel changing from ON to OFF γ_1	0.4 s^{-1}
Rate of optical channel changing from OFF to ON γ_2	0.3–1.1 s^{-1}
Signaling cost for VHO E_{sw}	418
Energy consumption for transmission E_O and E_W	5×10^{-4} kwh
Buffer size B	10 packets
Scaling parameter θ	3000 J
Tradeoff parameter ζ	500 J
Weighting factor η	1 packet
Dwell time t_O	1, 0.5 s
Simulation period	30,000 s

is increased by 71.11%, 24.97%, 44.84%, and 31.09%, respectively. Although I-VHO achieves the lowest packet loss rate, shortest average delay, and average queue length, the number of VHO obtained by I-VHO is the highest. Besides, the gaps of the number of VHO between I-VHO and the other three handover schemes become larger with the increase of γ_2. Consequently, I-VHO is not preferred in practical scenarios.

As for D-VHO, its performance depends highly on the value of t_O. For example, when $t_O = 0.5$ s, it is similar to the proposed O-VHO in terms of d and l. However, when $t_O = 1$ s, the performance of D-VHO is worse than O-VHO. Furthermore, the values of c attained from D-VHO with $t_O = 0.5$ s and D-VHO with $t_O = 1$ s are much higher than O-VHO. As shown in Fig. 7.9, O-VHO always achieves the lowest number of VHO and has similar performance compared with I-VHO in terms of ρ, d, and l. Take $\gamma_2 = 0.8$ s^{-1} as an example, ρ, d, and l attained by O-VHO are 1.1903, 1.147, and 1.1454 times of those attained from I-VHO, while c attained by O-VHO is 54.94% of that attained from I-VHO. Hence, the proposed O-VHO can achieve a balance between switching cost and QoS requirement.

In Fig. 7.10, γ_2 is fixed to 1 s^{-1} and handover delay τ is changed. When τ increases from 0 to 0.5 s, ρ attained from I-VHO, O-VHO, and D-VHO with $t_O = 0.5$ s and D-VHO with $t_O = 1$ s is increased from 0.219%, 0.226%, 0.459%, and 0.798% to 2.228%, 3.378%, 3.685%, and 4.984%, respectively, while d attained from those four schemes is increased from 1.388, 1.389, 1.703, and 1.994 to 2.793, 3.51, 3.344, and 3.739, respectively, and l is increased from 1.384, 1.387, 1.695, and 1.979 to 2.733, 3.389, 3.216, and 3.552, respectively. c attained from O-VHO decreases with the increase of τ, while c attained from the other three schemes changes slightly. As

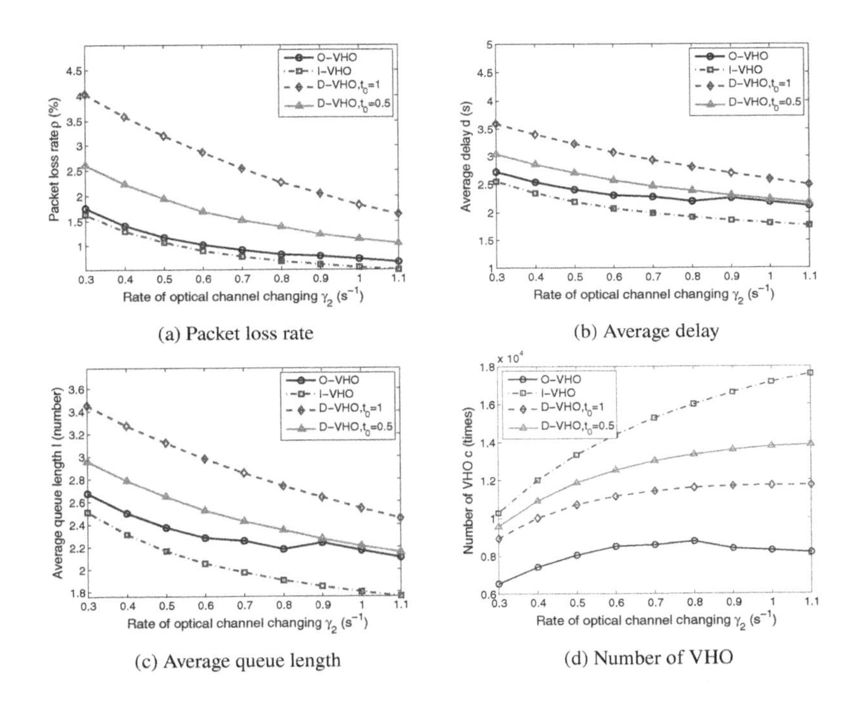

(a) Packet loss rate

(b) Average delay

(c) Average queue length

(d) Number of VHO

Figure 7.9 Impact of γ_2 on the system performance for four schemes.

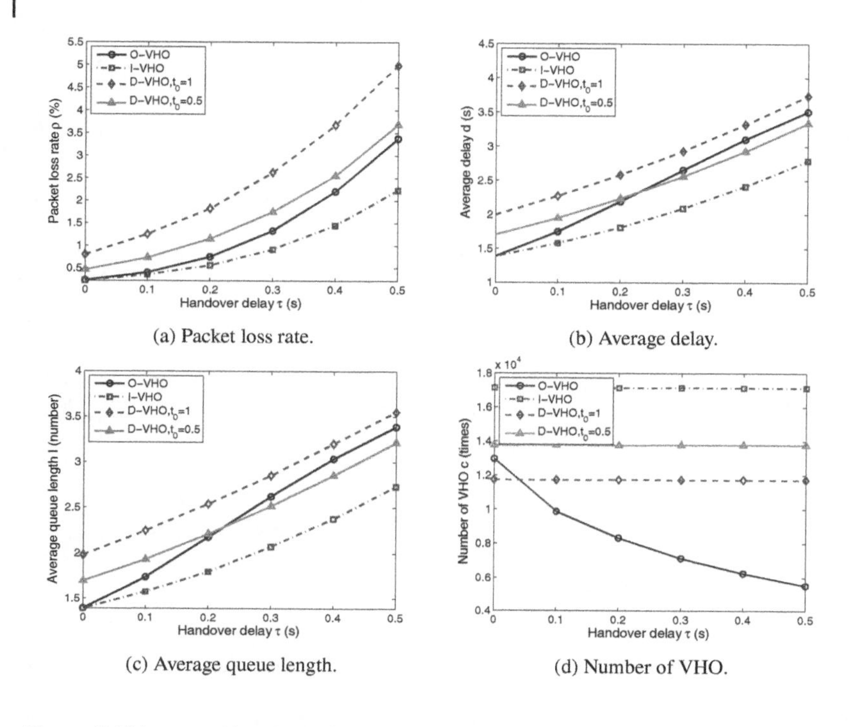

(a) Packet loss rate.

(b) Average delay.

(c) Average queue length.

(d) Number of VHO.

Figure 7.10 Impact of handover delay τ on the system performance for four schemes.

shown in (7.23), the value of the cost function grows for VHO action with the increase of τ. Therefore, more stages keep their transmission mode unchanged, leading to fewer VHO processes. I-VHO and D-VHO , on the other hand, do not consider the handover delay, and thus there is almost no fluctuation in the number of VHO with the increase of τ. It coincides with the phenomenon that the slope of the curves obtained by O-VHO is steeper than those obtained by the others in Fig. 7.10 (a)–(c). The performance degradation is not only a reflection of the increase of handover delay, but also the balance of cost and delay. If we compare the proposed O-VHO to D-VHO with $t_0 = 0.5$ s, d and l attained from O-VHO are lower than those attained from D-VHO when $\tau \leq 0.2$ s, while the opposite result is observed when $\tau > 0.2$ s. For all cases, O-VHO performs better than D-VHO with $t_0 = 0.5$ s in terms of ρ and c. D-VHO with $t_0 = 1$ s achieves a lower c than O-VHO when $\tau = 0$, but it does not perform well in other scenarios. Therefore, although D-VHO performs better than I-VHO and O-VHO in some occasions, it requires adaptive t_0 when the condition of the system changes. The proposed O-VHO, however, is able to make the near optimal VHO decision for all the cases dynamically and adaptively.

7.3 Signal estimation and modulation design for VLC with SDGN

Conventionally, the research about VLC is mainly focused on the system with signal-independent Gaussian noise (SIDGN), whilst much less effort is made to cope with the system degradation due to signal-dependent Gaussian noise (SDGN). In optical communication, the photodetector shot noise is originally modeled by Poisson statistics dependent on signal. As the number of received photons increases, the shot noise process can be well approximated by Gaussian statistics. Thus, the channel can be described by an AWGN model, including electrical thermal noise, signal-dependent shot noise, and signal-independent shot noise. When a single element photodiode is used as the reception device, the system will suffer from serious interference in direct background light, especially in the wide FOV case. This is because the background light is typically strong and can often be received at an average power much higher than the desired signal. Furthermore, it is difficult to reduce the enormous amount of noise signal from the background light in a wide FOV, even if an optical band-pass filter is used. Hence, the noise in VLC is typically assumed to be signal-independent. However, the signal-dependent distortion is observed in shot noise limited optical communication systems where the signal-dependent shot noise is comparable with background/thermal noise. Some examples of such systems are VLC links with high signal-to-noise ratios (SNRs) employing high-gain narrow FOV avalanche photodiode (APD) receivers [26], and free-space optical communication systems with high-sensitive receivers such as single photon avalanche diode (SPAD) array [27]. The SDGN channel model fundamentally changes the problem of efficient signaling, optimal signal detection/estimation, and system performance. This section discusses the optimal signal detection/estimation and efficient signal design in the presence of SDGN.

7.3.1 Signal estimation for VLC with SDGN

The general IM/DD channel with SDGN can be modeled as

$$Y = hX + kg(hX)Z_1 + Z_0, \tag{7.31}$$

where X is the nonnegative input signal and is proportional to the light intensity, which is generally characterized by the probability function $f_X(X)$. Z_1 and Z_0 are signal-independent random noise processes, which are assumed to be statistically independent of X. $g(hX)$ is the function of the input signal, k and h are scalar constant and channel state respectively, and Y is the noisy measurement. The term $kg(hX)Z_1$ is called signal-dependent noise. The commonly used signal-dependent model in VLC typically sets $g(hX)$ to $(hX)^p$, and the value of p used for characterizing the signal-dependent noise to $1/2$. Moreover, for a determined channel or a channel with channel-state-information-at-the-transmitter (CSIT), the scaling factor h represents the product of all the path losses and photoelectric conversion factors. Since h scales the SNR only, without loss of generality, we set $h = 1$ to simplify the analysis. Therefore, the general signal-dependent measurement model in (7.31) can be simplified as

$$Y = X + kX^p Z_1 + Z_0. \tag{7.32}$$

For ease of signal estimation, the observed value can be used for estimation, which results in a minimum-variance unbiased estimate. That is, when the estimation of X, denoted as $\hat{X}$, is equal to Y, the average error is given by

$$E(\hat{X} - X) = E(Y - X) = E(kX^p Z_1 + Z_0) = 0. \tag{7.33}$$

The estimator is unbiased since the average error is 0. The mean squared estimation error (MSEE) for this mismatched case is given by

$$E[(\hat{X} - X)^2] = k^2\sigma_1^2 E(X^{2p}) + \sigma_0^2. \tag{7.34}$$

According to the arguments in Jensen's inequality, for a convex function $d(X)$, if $d''(X) \geq 0$ for all X, then $d(E[X]) \leq E[d(X)]$, while for a concave function $d(X)$, the inequality is reversed. Thus, if $p = \frac{1}{2}$, equality holds in Jensen's inequality, and the mismatched MSEE is given by

$$E[(\hat{X} - X)^2] = k^2\sigma_1^2 E(X) + \sigma_0^2. \tag{7.35}$$

For $p > \frac{1}{2}$ and $p < \frac{1}{2}$, the MSEE is lower-bounded and upper-bounded by $k^2\sigma_1^2[E(X)]^{2p} + \sigma_0^2$, respectively.

Cramer–Rao lower bound (CRLB) is a well-known lower bound on the variance of any unbiased estimate for a fixed but unknown X. Given the conditional probability density function (PDF) $f(Y|X)$, the CRLB is given by

$$E\{(\hat{X} - X)^2|X\} \geq \left[-E\left(\frac{\partial^2 \ln f(Y|X)}{\partial X^2}\right)\right]^{-1}. \tag{7.36}$$

For the channel model given in (7.32), where Y is Gaussian with mean X and with variance $\sigma(X)$, the conditional PDF is given by

$$f(Y|X) = \frac{1}{\sqrt{2\pi\sigma(X)}} e^{-\frac{(Y-X)^2}{2\sigma(X)}}, \tag{7.37}$$

where

$$\sigma(X) = k^2\sigma_1^2 X^{2p} + \sigma_0^2. \tag{7.38}$$

Then the CRLB can be expressed as

$$\begin{aligned} E\{(\hat{X} - X)^2|X\} &\geq \frac{2[\sigma(X)]^2}{2\sigma(X) + [\sigma'(X)]^2} \\ &= \frac{k^4\sigma_1^4 X^{4p} + 2k^2\sigma_1^2\sigma_0^2 X^{2p} + \sigma_0^4}{k^2\sigma_1^2 X^{2p} + 2p^2k^4\sigma_1^4 X^{4p-1} + \sigma_0^2}, \end{aligned} \tag{7.39}$$

where $\sigma'(X)$ is the derivative of $\sigma(X)$ with respect to X. Although it is not obvious by inspection, the bound given by (7.39) may actually be smaller than the bound given by σ_0^2 for the SDGN channel case and the bound given in (7.34) for the MSEE of the mismatched case in the SDGN channel. There are cases that a properly matched signal-dependent estimator may potentially outperform the corresponding properly matched signal-independent estimator. Some optimal estimators are presented in the next to achieve the CRLB.

7.3.1.1 **MMSE estimation**

The minimum mean square error (MMSE) estimator is a commonly studied optimal estimator, whose criterion is to minimize the mean square estimation error, given by

$$\hat{X}_{\text{MMSE}} = \min_{\hat{X}(Y)} E\{[X - \hat{X}(Y)]^2\}. \tag{7.40}$$

Define estimation error as $e(X, Y) \triangleq X - \hat{X}(Y)$. Then MMSE is given by

$$\begin{aligned} E\{e^2(X,Y)\} &= \int_{-\infty}^{\infty} \int_0^A e^2(X,Y) f(X,Y) dX dY \\ &= \int_{-\infty}^{\infty} \left[\int_0^A e^2(X,Y) f(X|Y) dX \right] f(Y) dY, \end{aligned} \tag{7.41}$$

where A is the peak-intensity constraint for the nonnegative input signal X. Since $f(Y)$ is nonnegative, $E\{e^2(X,Y)\}$ can be minimized if the term $\int_0^A e^2(X,Y) f(X|Y) dX$ is minimized for all values of Y. Then, the conditional MMSE to estimate $\hat{X}$ can be minimized as

$$\hat{X}_{\text{MMSE}} = \min_{\hat{X}(Y)} E\{e^2(X,Y)|Y\} = \min_{\hat{X}(Y)} \int_0^A \left[X - \hat{X}(Y)\right]^2 f(X|Y) dX. \tag{7.42}$$

The $\hat{X}_{\text{MMSE}}$ is the value of $\hat{X}$ when the partial derivative is 0 with respect to $\hat{X}$. Thus,

$$\frac{\partial^2 E\{e^2(X,Y)|Y\}}{\partial \hat{X}} |_{X=\hat{X}_{\text{MMSE}}} = 0 = -2 \int_0^A (X - \hat{X}_{\text{MMSE}}) f(X|Y) dX. \tag{7.43}$$

After some mathematic manipulations, the $\hat{X}_{\text{MMSE}}$ can be expressed as

$$\hat{X}_{\text{MMSE}} = \int_0^A X f(X|Y) dX. \tag{7.44}$$

Thus, the MMSE estimate is the conditional mean of X given the measurement Y. For this reason, the estimator is also referred to as the conditional mean (CM) estimator. The expected value of the estimation error $e(X,Y) = X - \hat{X}_{\text{MMSE}}$ is given by

$$E[e(X,Y)] = E\{E[e(X,Y)|Y]\} = E\{E[X|Y] - E[\hat{X}_{\text{MMSE}}|Y]\}, \tag{7.45}$$

where the first term on the right side of (7.45) is simply the conditional mean of X given the measurement Y, or equivalently $\hat{X}_{\text{MMSE}}$ as given in (7.44). The second term on the right side is deterministic once Y is given. Thus we have

$$E[\hat{X}_{\text{MMSE}}|Y] = \hat{X}_{\text{MMSE}}. \tag{7.46}$$

Combining (7.44) and (7.46) in (7.45), we have

$$E[e(X,Y)] = 0. \tag{7.47}$$

Since the estimation error $e(X,Y)$ has zero-mean and variance $E[e^2(X,Y)]$, $\hat{X}_{\text{MMSE}}$ is also known as the minimum-variance unbiased (MVU) estimate.

Noteworthily, (7.44) is computationally inconvenient due to the difficulty of deriving $f(X|Y)$. To alleviate this problem, Bayes' rule can be used and $\hat{X}_{\text{MMSE}}$ can be expressed as

$$\hat{X}_{\text{MMSE}} = \frac{\int_0^A Xf(Y|X)f(X)dX}{\int_0^A f(Y|X)f(X)dX}. \tag{7.48}$$

In this form, it is only necessary to know the PDF of $f_X(X)$ and $f(Y|X)$, and generally, these are either given, assumed known, or easily derived. Due to high complexity and suboptimal performance of the MMSE estimator, it is worth investigating alternative estimators.

7.3.1.2 MAP estimation

The maximum *a posteriori* probability (MAP) estimator is another optimum estimator. The conditional probability $f(X|Y)$ is known as the *a posteriori* probability, which represents the probability distribution of the variable X, given that the measurement Y is known. Its mean value is the MMSE estimation shown in the previous section. On the other hand, its most probable value is the one maximizing $f(X|Y)$. This MAP estimator can be expressed as $\hat{X}_{\text{MAP}}$, which satisfies

$$f(\hat{X}_{\text{MAP}}|Y) \geq f(\hat{X}|Y), \hat{X}_{\text{MAP}} \neq \hat{X}. \tag{7.49}$$

Thus, MAP estimation is a conditional mean of X given Y, just like the MMSE estimation.

For MAP, the estimator is attained by taking the partial derivative of the *a posteriori* probability and equating the derivative to 0, provided that the derivative is valid. Thus $\hat{X}_{\text{MAP}}$ satisfies

$$\frac{\partial}{\partial X} f(X|Y)|_{X=\hat{X}_{\text{MAP}}} = 0. \tag{7.50}$$

Similar to MMSE estimation, the *a posteriori* probability $f(X|Y)$ is difficult to compute, thus Bayes' rule can be used again to rewrite (7.50) as

$$\frac{\partial}{\partial X}[\frac{f(Y|X)f(X)}{f(Y)}]|_{X=\hat{X}_{\text{MAP}}} = 0 \Leftrightarrow \frac{\partial}{\partial X}[f(Y|X)f(X)]|_{X=\hat{X}_{\text{MAP}}} = 0,$$

(7.51)

where the right equality holds based on the assumption that $f(Y)$ is independent of X. Moreover, since monotonic transformations preserve maxima and minima, the MAP estimation $\hat{X}_{\text{MAP}}$ equivalently satisfies

$$\frac{\partial}{\partial X}\ln f(X|Y)|_{X=\hat{X}_{\text{MAP}}} = \frac{\partial}{\partial X}\ln f(Y|X) + \frac{\partial}{\partial X}\ln f(X)|_{X=\hat{X}_{\text{MAP}}} = 0. \quad (7.52)$$

Actually, when $f(X|Y)$ satisfies the following two conditions: (1) the *a posteriori* probability $f(X|Y)$ is symmetrical; (2) $\lim_{X\to\infty} C(X,\hat{X})f(X|Y) = 0$, where $C(X,\hat{X}) = (X-\hat{X})^2$ is the quadratic cost function, MMSE and MAP estimators are equivalent.

To investigate the effect of SDGN on the MAP estimator, the channel model given in (7.32) and the conditional probability $f(Y|X)$ in (7.36) are used. A straightforward substitution of $f(Y|X)$ into (7.52) yields the MAP equation,

$$\begin{aligned}(Y-\hat{X}_{\text{MAP}})^2\sigma'(\hat{X}_{\text{MAP}}) + 2(Y-\hat{X}_{\text{MAP}})\sigma(\hat{X}_{\text{MAP}}) - \sigma'(\hat{X}_{\text{MAP}})\sigma(\hat{X}_{\text{MAP}})\\ + 2[\sigma(\hat{X}_{\text{MAP}})]^2\frac{\partial}{\partial \hat{X}_{\text{MAP}}}\ln f(\hat{X}_{\text{MAP}}) = 0,\end{aligned} \quad (7.53)$$

where $\sigma'(X)$ denotes the partial derivative of $\sigma(X)$ with respect to X. Generally, compared with the optimal MMSE criterion, the mathematical complexity of the MAP estimator is reduced.

7.3.1.3 ML estimation

Another commonly used optimal estimator is the maximum likelihood (ML) estimator, which is employed when prior statistical knowledge of the signal is unknown, and obtained by maximizing $f(Y|X)$ over X. Thus, the ML estimation $\hat{X}_{\text{ML}}$ is the value which satisfies the inequality

$$f(Y|\hat{X}_{\text{ML}}) \geq f(Y|\hat{X}), \hat{X}_{\text{ML}} \neq \hat{X}. \quad (7.54)$$

Computationally, the ML estimator can be found by equating the partial derivative of $f(Y|X)$ to 0 with respect to X, provided that the derivative exists. Similar to the MAP estimator, maximizing the logarithm of $f(Y|X)$ can lead to the ML solution. Thus, either of the following equivalent equations can be used to find the ML estimator, where the ML estimation is the value $\hat{X}_{\text{ML}}$ which satisfies

$$\frac{\partial}{\partial X}f(Y|X)|_{X=\hat{X}_{\text{ML}}} = 0 \Leftrightarrow \frac{\partial}{\partial X}\ln f(Y|X)|_{X=\hat{X}_{\text{ML}}} = 0. \quad (7.55)$$

Comparing ML estimation in (7.55) with MAP in (7.52), it is found that the MAP estimation is equivalent to the ML estimation when $f(X)$ is constant, thus the MAP estimation can be viewed as a specific case of ML estimation, where all values of X in the range of interest are assumed to have equal probability. The last term in (7.53) which embodies the *a priori* knowledge of X can be eliminated. The ML estimator is the solution of

$$(Y - \hat{X}_{\text{ML}})^2\sigma'(\hat{X}_{\text{ML}}) + 2(Y - \hat{X}_{\text{ML}})\sigma(\hat{X}_{\text{ML}}) - \sigma'(\hat{X}_{\text{ML}})\sigma(\hat{X}_{\text{ML}}) = 0. \tag{7.56}$$

Substituting the expression of $\sigma(\hat{X})$ and $\sigma'(\hat{X})$ into (7.56), the ML estimator is obtained by the positive root solution of

$$\begin{aligned}&[2k^2\sigma_1^2(p-1)]\hat{X}_{\text{ML}}^{2p+1} + [2Yk^2\sigma_1^2(1-2p)]\hat{X}_{\text{ML}}^{2p}\\ &+ [2pk^2\sigma_1^2(Y^2-\sigma_0^2)]\hat{X}_{\text{ML}}^{2p-1} - [2\sigma_0^2]\hat{X}_{\text{ML}}\\ &- [2pk^4\sigma_1^4]\hat{X}_{\text{ML}}^{4p-1} + [2\sigma_0^2Y] = 0.\end{aligned} \tag{7.57}$$

Noteworthily, compared with the MAP estimator, the ML estimator has lower complexity.

7.3.2 Suboptimal estimation for VLC with SDGN

The motivation of the development of suboptimal estimators for the SDGN channel as in (7.32) stems from the obvious shortcomings in optimal estimators [28]: (1) the mathematical complexity, (2) difficulty of implementation, (3) general lack of closed-form solutions, and (4) sensitivity to the *a priori* function $f_X(X)$. To avoid these problems, several suboptimal estimators are exploited at the cost of performance degradation.

7.3.2.1 Linearization

Obviously, the term X^p in (7.32) leads to extra computational complexity. An obvious suboptimal estimation scheme is to modify the measurement model by simply ignoring the term X^p. By doing so, the problem becomes one with a classical signal-independent noise model and the entire analysis of this approach is on the basis of the mismatch case discussed in the previous section with degraded performance. Actually, by linearizing the term X^p, nonlinearity can be removed. Specifically, in (7.32), X^p is expanded in a Taylor series at $E(X) = \mu_X$, while the second and higher order terms are dropped. The linearized measurement model becomes

$$\begin{aligned}Y &\approx X + k\left[p\mu_X^{p-1}X + \mu_X^p - \mu_X p\mu_X^{p-1}\right]Z_1 + Z_0\\ &= \left[1 + kp\mu_X^{p-1}Z_1\right]X + k\left[\mu_X^p - \mu_X p\mu_X^{p-1}\right]Z_1 + Z_0.\end{aligned} \tag{7.58}$$

This is in the form of $Y = AX+B$, where $A = \left[1+kp\mu_X^{p-1}Z_1\right]$ and $B = k\left[\mu_X^p - \mu_X p\mu_X^{p-1}\right]Z_1 + Z_0$, and the accuracy of this linear approximation depends on the dropped second and higher order terms $\frac{1}{2}(X-\mu_X)^2p(p-1)\mu_X^{p-2} + o(\mu_X^n)$. Since X is a random variable, it is more meaningful to consider the statistical properties of the second and higher order terms. Recall that when $p = \frac{1}{2}$, the mean value is defined as

$$\frac{1}{2}\sigma_X^2 p(p-1)\mu_X^{p-2} + o(\mu_X^n) = -\frac{1}{8}\sigma_X^2\mu_X^{-\frac{3}{2}} + o(\mu_X^n). \tag{7.59}$$

Obviously, the value of σ_X^2 and the nonlinear feature determine the accuracy of the linear approximation of (7.58).

As mentioned above, the signal-independent noises $Z_1 \sim \mathcal{N}(0, \sigma_1^2)$ and $Z_0 \sim \mathcal{N}(0, \sigma_0^2)$ are assumed to be statistically independent with X. Then, the expectation of the measurement based on the linearized model in (7.58) can be calculated as

$$E(Y) = E(A)\mu_X + E(B) = \mu_X. \tag{7.60}$$

This demonstrates that $\hat{X} = Y$ is an unbiased estimator for X.

7.3.2.2 **Modified noise cheating**

To tackle the SDGN issue, one way is to average a small portion of the measurement and use the local average as the estimate. Define the jth sample mean as $\bar{Y}_j$, given by

$$\bar{Y}_j = \frac{1}{n}\sum_{i=1}^{n} Y_{i,j}, \tag{7.61}$$

and the estimate $\hat{X}$ is taken to be

$$\hat{X}_j = \bar{Y}_j. \tag{7.62}$$

This estimator is a modification of noise-cheating, which is simpler to implement and analyze. Obviously, the modified noise-cheating estimator is unbiased since

$$E[\bar{Y}_j] = \frac{1}{n}\sum_{i=1}^{n} E[Y_{i,j}] = \frac{1}{n}\sum_{i=1}^{n} \mu_X = \mu_X. \tag{7.63}$$

The advantages of the modified noise-cheating estimator are as follows. It is easy to implement and does not need *a priori* statistical knowledge of the measurement process. Moreover, the modified noise-cheating estimator is robust to the *a priori* probability function of the signal X. However, this method also has some drawbacks. Since the modified noise-cheating algorithm is actually a finite-window low-pass filter, it loses high-frequency signal information, which makes this method not robust to the noise power spectrum.

7.3.2.3 **Modified MAP and James–Stein estimation with SDGN**

Compared with the MMSE estimator, the computational complexity of the MAP estimator is greatly reduced. However, the complexity is still high for practical implementation. To retain the ease of implementation, a proper choice is using a modified signal-independent MAP estimator. Assuming that X is normally distributed with mean μ_X and variance σ_X^2, the signal-independent MAP estimator is given by

$$\hat{X}_{\text{MAP}} = \frac{\sigma_X^2}{\sigma_X^2 + \sigma_0^2} Y + \frac{\sigma_0^2}{\sigma_X^2 + \sigma_0^2}\mu_X. \tag{7.64}$$

To modify the signal-independent MAP estimation, so that it will accommodate the SDGN effects better, μ_X in (7.64) is replaced with the mean sample $\bar{Y}_j$, as given in (7.61). Moreover, the signal variance σ_X^2 is adaptively estimated with the local sampled variance. Thus the modified signal-independent MAP estimator is given by

$$\hat{X}_j = Q_j Y_j + (1 - Q_j)\bar{Y}_j, \tag{7.65}$$

where Q_j is given by

$$Q_j = \frac{u_j}{u_j + \sigma_0^2}. \tag{7.66}$$

The sample variance u_j is given by

$$u_j = \frac{\sum_{i=1}^{n}(Y_{i,j} - \bar{Y}_j)^2}{n-1}, \tag{7.67}$$

where the denominator $n - 1$ is to make an unbiased estimate of the variance. Although the modified signal-independent MAP estimator is robust due to its employment of sample means as parameters, this estimator is designed for the signal with Gaussian distribution.

The empirical Bayesian estimator, i.e., the James–Stein estimator, estimates the mean of a multivariate normal distribution with uniformly lower MSEE than the sample mean. By estimating the sample variance $\sigma_Y^2 = \sigma_X^2 + \sigma_0^0$ with its proper sample variance u_j, the James–Stein estimator is given by

$$\hat{X}_j = \bar{Y}_j + Q_j^+(Y_j - \bar{Y}_j), \tag{7.68}$$

where $Q_j^+ \triangleq \max(0, (1 - \frac{\sigma_0^2}{u_j}))$, with which a lower MSEE can be obtained, while $\bar{Y}_j$ is given by (7.61). To guarantee the unbiased estimate of the variance, the estimate, denoted by u_j, is given by

$$u_j = \frac{\sum_{i=1}^{n}(Y_{i,j} - \bar{Y}_j)^2}{n-3}. \tag{7.69}$$

Note that when $Q_j = 1$, the estimate is given by $\hat{X}_j = Y_j$; when $Q_j = 0$, the estimate is $\hat{X}_j = \bar{Y}_j$, which is the modified noise-cheating estimator.

7.3.3
Efficient signal design for VLC with SDGN

The SDGN model fundamentally changes the problem of efficient signaling compared to the conventional SIDGN model. Complex modifications to signal processing techniques may be required to ensure optimal data transmission. Efficient transmission with nonuniform signaling is possible, which is capacity-achieving optimal signaling, but such a signaling scheme would require complex optimization and additional signal processing for coding and decoding. However, via transforming signals in square-root or frequency domains, the strength of the SDGN can be normalized and essentially transmission with signal-independent distortions is allowed [27]. Moreover, this approach can identify efficient communication schemes which can take the advantages of conventional coding, signal processing, and estimation/detection techniques. Additionally, the signal in the presence of the SDGN can be optimally designed and the improved performance can be achieved.

7.3.3.1 Efficient signaling in the square-root domain

A simple scheme was developed based on signaling in the square-root (SQR) domain, which transforms the optimal channel originally distorted by SDGN into a stationary SIDGN channel [26, 27, 29], so that the conventional coding, signal processing, and detection/estimation techniques with lower complexity can be utilized. The concept of signaling in the square-root domain and the efficient communication with square-root signaling will be introduced here.

For a general square-root transformation, it can normalize the variance of random variable X, where its variance σ_X^2 is the function of the mean μ_X as

$$\sigma_X^2 = l^2(\mu_X). \tag{7.70}$$

The corresponding normalizing transformation $Y = T(X)$ can be expressed as

$$T(X) = \int \frac{dX}{l(X)}, \tag{7.71}$$

for which $\sigma_Y^2 \approx 1$ and $\mu_Y \approx T(\mu_X)$ when the value of μ_X is sufficiently large. That means the signal-dependent noise becomes signal-independent after the transformation.

Based on this, recall that the variance of signal is $\sigma(X) = k^2\sigma_1^2 X + \sigma_0^2$ as shown in (7.38) and $p = 1/2$. Define the function $l(X) = \sqrt{k^2\sigma_1^2 X + \sigma_0^2}$ for the Gaussian random variable Y, which falls within the general family of random variables described above. The corresponding normalizing transform for Y is given by

$$T(Y) = \int \frac{dY}{\sqrt{k^2\sigma_1^2 Y + \sigma_0^2}} = \frac{2}{k\sigma_1}\sqrt{Y + \sigma_0^2/k^2\sigma_1^2}. \tag{7.72}$$

Then, the resultant transform is in a square-root form as the received signal is originally distorted by a Poisson-distributed shot noise. Thus, the normalized form of (7.72) is called square-root transform. Considering the normalization as

$$\tilde{Y} = T_{\text{sqr}}(Y) = \sqrt{Y + \sigma_0^2/k^2\sigma_1^2}, \tag{7.73}$$

then the stabilized variance of the output of the square-root transform can be expressed as

$$\sigma_{\tilde{Y}}^2 = \frac{k^2\sigma_1^2}{4}. \tag{7.74}$$

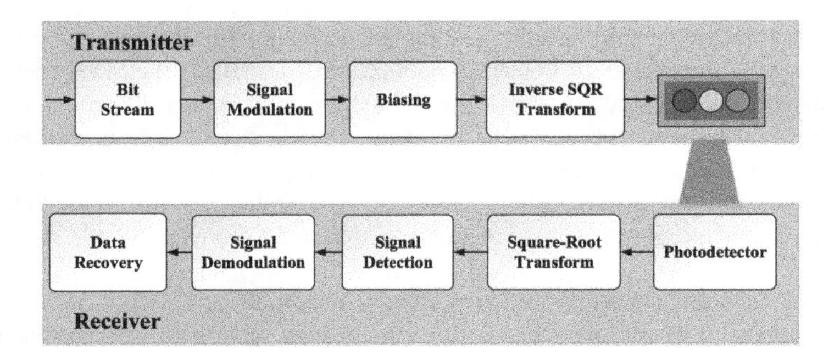

Figure 7.11 VLC system with square-root signaling.

Furthermore, the inverse transform can be expressed as

$$Y = T_{\text{sqr}}^{-1}(\tilde{Y}) = \tilde{Y}^2 - \sigma_0^2/k^2\sigma_1^2. \tag{7.75}$$

The system diagram of VLC with signaling in the SQR domain is shown in Fig. 7.11. Since the signal is generated in the SQR domain, the channel in Fig. 7.11 can be described as the conventional SIDGN channel in the SQR domain although it suffers from SDGN in the time domain. Then, conventional coding and signal processing techniques can be directly applied to the signal. After the biasing operation by $\sigma_0/k\sigma_1$ to generate non-negative signal, the modulated signal undergoes inverse square-root transform. At the receiver, the received signal is distorted by the SDGN. The distorted signal is transformed back into the SQR domain where the noise is signal-independent. Finally, using the conventional signal detection/estimation for SIDGN channel, the original data can be recovered.

7.3.3.2 **Optimal signal design for VLC with SDGN**

In the previous sections, linearization, modified noise cheating, and square-root transform are used to tackle the communication issues with SDGN, in which the conventional coding, signal processing, and detection/estimation techniques can be directly applied. However, these schemes are suboptimal in a SDGN channel. In the following, a joint signal and corresponding threshold design for an IM/DD system with SDGN under intensity constraints will be presented. To guarantee the optimal performance, ML-based estimation is applied.

For a general SDGN channel defined in (7.32), when we set $k = 1$ and $p = \frac{1}{2}$, we have

$$Y = X + \sqrt{X}Z_1 + Z_0. \tag{7.76}$$

As mentioned before, X is the nonnegative input signal and is proportional to the light intensity. Z_1 and Z_0 are the SIDGNs with $Z_1 \sim \mathcal{N}(0, \sigma_1^2)$ and $Z_0 \sim \mathcal{N}(0, \sigma_0^2)$, respectively. Moreover, Z_1, Z_0, and X are assumed to be statistically independent.

For a SDGN channel, the optimally designed signal is the one which has the minimum symbol error rate (SER). Assume that the collection of the intensity-modulated PAM signal is $\mathcal{X} = \{X_1, X_2, \ldots, X_M\}$ where M is the maximum intensity level. At the receiver, symbol X_m can be selected using an ML detector, thus

$$m = \arg\max_{n \in \mathcal{S}} f(Y|X_n), \tag{7.77}$$

where $\mathcal{S} = \{1, 2, \ldots, M\}$ contains the intensity levels, and $f(Y|X_m)$ is the conditional probability, given by

$$f(Y|X_m) = \frac{1}{\sqrt{2\pi(\sigma_1^2 X_m + \sigma_0^2)}} \exp\{-\frac{(Y - X_m)^2}{2(\sigma_1^2 X_m + \sigma_0^2)}\}. \tag{7.78}$$

In a SDGN channel, the ML detection rule is given by

$$\begin{aligned} \hat{X}_m &= \arg \max_{n \in \{1,\ldots,M\}} f(Y|X_n) \\ &= \arg \max_{n \in \{1,\ldots,M\}} \left[\ln\sqrt{\sigma_1^2 X_n + \sigma_0^2} + \frac{(Y - X_n)^2}{2(\sigma_1^2 X_n + \sigma_0^2)^2} \right]. \end{aligned} \tag{7.79}$$

The corresponding probability of detection error is given by [27]

$$P_e \approx \frac{1}{\log_2 M \cdot M} \sum_{m=1}^{M-1} Q\left(\frac{\tau_m - X_m}{2\sqrt{\sigma_1^2 X_m + \sigma_0^2}}\right) + Q\left(\frac{X_{m+1} - \tau_m}{2\sqrt{\sigma_1^2 X_{m+1} + \sigma_0^2}}\right), \tag{7.80}$$

where $\tau_m, m \in \mathcal{S}$ is the threshold, formulated as a function of the transmitted signal given by

$$\begin{aligned} &\ln\left(\sqrt{\sigma_1^2 X_m + \sigma_0^2}\right) + \frac{(\tau_m - X_m)^2}{2(\sigma_1^2 X_m + \sigma_0^2)} \\ &= \ln\left(\sqrt{\sigma_1^2 X_{m+1} + \sigma_0^2}\right) + \frac{(\tau_m - X_{m+1})^2}{2(\sigma_1^2 X_{m+1} + \sigma_0^2)}. \end{aligned} \tag{7.81}$$

For notational simplicity, the transmitted signals and corresponding thresholds are packed into vectors as

$$\mathbf{X}_M = [X_1\, X_2\, \cdots\, X_M]^{\mathrm{T}}, \text{ and } \boldsymbol{\tau}_M = [\tau_1\, \cdots\, \tau_{M-1}]^{\mathrm{T}}. \tag{7.82}$$

Further, due to safety and practical consideration, the input signal has peak and non-negative constraints as

$$P_r[0 \leq X_m \leq A] = 1,\; m \in \mathcal{S}, \tag{7.83}$$

and an average intensity constraint is given by

$$E[\mathbf{X}_M] \leq P. \tag{7.84}$$

We use ρ to denote the average to peak power ratio (APPR)

$$\rho \triangleq \frac{P}{A} \quad (0 < \rho \leq 1). \tag{7.85}$$

Note that if $\rho = 1$, the average intensity constraint is inactive, which corresponds to the case with only a peak-intensity constraint. Similarly, $\rho \ll 1$ corresponds to a very weak peak-intensity constraint. Then, the optimal signals and thresholds are found jointly by solving the following problem [30]

$$\begin{aligned} \min_{\mathbf{X}_M, \boldsymbol{\tau}_M} \quad & P_e(\mathbf{X}_M, \boldsymbol{\tau}_M) \\ \text{s.t.} \quad & [0 \ \boldsymbol{\tau}_M^{\mathrm{T}}]^{\mathrm{T}} \leq \mathbf{X}_M \leq [\boldsymbol{\tau}_M^{\mathrm{T}} \ A]^{\mathrm{T}}, \\ & \mathbf{1}^{\mathrm{T}}\mathbf{X}_M = \rho A \cdot M. \end{aligned} \tag{7.86}$$

Although the objective function is non-convex, the feasible region is convex for $\mathbf{X}_M, \boldsymbol{\tau}_M$. Such optimization can be solved via several non-convex optimization methods, such as the gradient projection procedure in [30].

The following simulation results illustrate the performance of the optimally designed signal. The intensity level M is set to 4, the peak intensity value A equals 8, and the APPR ρ is set 0.5. The optimally designed nonuniform signaling is compared with the conventional uniform signaling and signaling in the SQR domain. The thresholds τ_n for those scenarios are obtained by a similar gradient projection procedure but with fixed intensity.

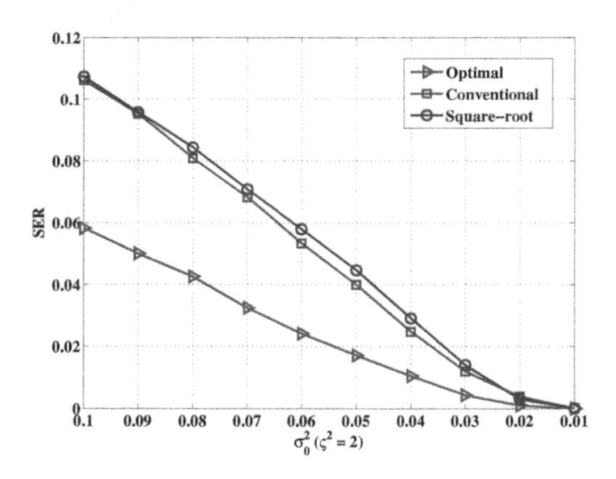

Figure 7.12 SER performance of the optimal signaling against thermal noise variance.

Figures 7.12–7.14 show the SER performance for optimal nonuniform signaling, uniform signaling, and signaling in the SQR domain, against different values of thermal noise variance, shot noise scaling factor $\varsigma^2 = \sigma_1^2/\sigma_0^2$, and APPR ρ. The simulation results indicate that the optimally designed signal achieves better SER performance than the conventional uniform signaling and signaling in the SQR domain under the intensity constraints. For the case with lower SNR, the uniform signaling has a superior performance to the signaling in the SQR domain. However, in the

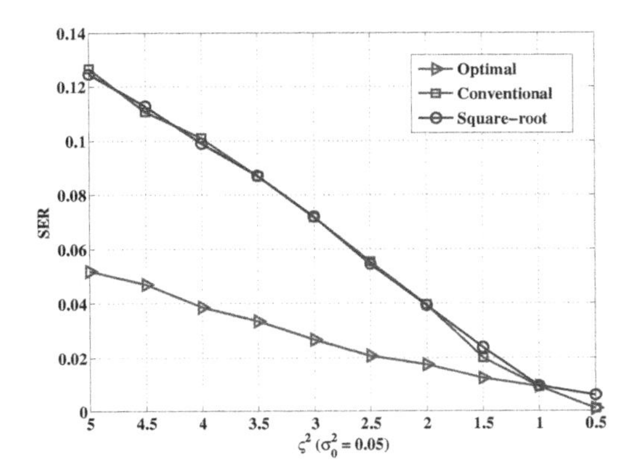

Figure 7.13 SER performance of the optimal signaling against shot noise variance.

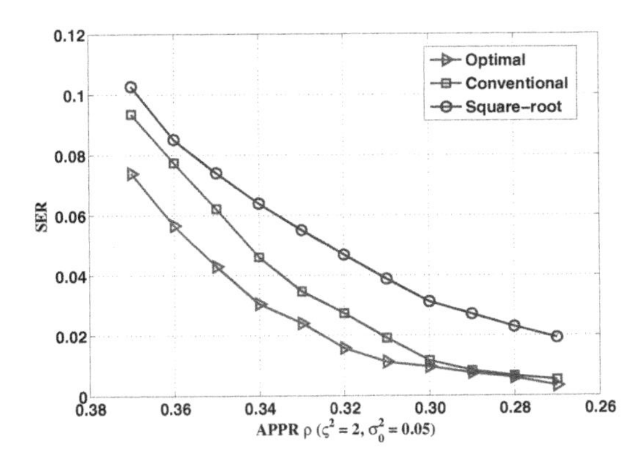

Figure 7.14 SER performance of the optimal signaling against the APPR.

cases with high SNR, signaling in the SQR domain performs better [30]. Furthermore, due to signal-dependent noise, the SER performance degrades as the APPR ρ increases.

7.4 Conclusion

In this chapter, signal processing and optimization issues are addressed for VLC systems. For multi-chip-based multi-input single-output VLC system, an electric and optical power allocation scheme is proposed to maximize the multiuser sum-rate in consideration of the luminance, chromaticity, amplitude, and bit error rate constraints. For heterogeneous VLC-WiFi systems, the vertical handover is formulated as a MDP problem and a dynamic approach is adopted to obtain a tradeoff between the switching cost and the delay requirement, thus offering a solution for future indoor communications that combines VLC to support high-data-rate transmission and RF to support reliable connectivity. For VLC with narrow FOV, since the PD shot noise is signal-dependent, novel signal processing and estimation techniques are proposed to maintain the transmission performance.

References

1 D. L. MacAdam, "Visual sensitivities to color differences in daylight," *J. Opt. Soc. Amer.*, vol. 32, no. 5, pp. 247–273, May 1942.

2 H. Elgala, R. Mesleh, and H. Haas, "Non-linearity effects and predistortion in optical OFDM wireless transmission using LEDs," *Inderscie. Int. J. Ultra Wideband Commun. Syst.*, vol. 1, no. 2, pp. 143–150, 2009.

3 G. Wyszecki and W. S. Stiles, *Color Science: Concepts and Methods, Quantitative Data and Formulae*, 2nd ed. New York: Wiley, 1982.

4 K. H. Park, Y. C. Ko, and M. S. Alouini, "On the power and offset allocation for rate adaption of spatial multiplexing in optical wireless MIMO channels," *IEEE Trans. Commun.*, vol. 61, no. 4, pp. 1535–1543, Apr. 2013.

5 Z. Yu, R. J. Baxley, and G. T. Zhou, "Multi-user MISO broadcasting for indoor visible light communication," in *Proc. IEEE International Conference on Acoustics, Speech and Signal Processing (ICASSP) 2013* (Vancouver, Canada), May 26–31, 2013, pp. 4849–4853.

6 H. Flanders and J. J. Price, *Calculus with Analytic Geometry*. New York: Academic Press, 1978.

7 R. Jiang, Z. Wang, Q. Wang, and L. Dai, "Multi-user sum-rate optimization for visible light communications with lighting constraints," *J. Lightw. Technol.*, vol. 34, no. 16, pp. 3943–3952, Aug. 2016.

8 S. Boyd and L. Vandenberghe, *Convex Optimization*. Cambridge: Cambridge University Press, 2004.

9 M. Hintermller and K. Kunisch. "Path-following methods for a class of constrained minimization problems in function space," *SIAM J. on Optim.*, vol. 17, no. 1, pp. 159–187, May 2006.

10 M. Grant, S. Boyd, and Y. Ye, "CVX: Matlab software for disciplined convex programming," Jun. 2015," [online], *http://cvxr.com/cvx/.*

11 *ANSI NEMA ANSLG C78.376-2001: Specications for the chromaticity of fluorescent lamps*, ANSI, 2001.

12 *ANSI NEMA ANSLG C78.377-2008: Specications for the chromaticity of solid state lighting products*, ANSI, 2008.

13 A. Ahmed, L. Boulahia, and D. Gaïti, "Enabling vertical handover decisions in heterogeneous wireless networks: A state-of-the-art and a classification," *IEEE Commun. Surv. Tuts.*, vol. 16, no. 2, pp. 776–811, 2014.

14 J. Hou and D. O'brien, "Vertical handover-decision-making algorithm using fuzzy logic for the integrated radio-and-OW system," *IEEE Trans. Wirel. Commun.*, vol. 5, no. 1, pp. 176–185, Jan. 2006.

15 W. Wanalertlak, B. Lee, C. Yu, M. Kim, S. M. Park, and W. T. Kim, "Behavior-based mobility prediction for seamless handoffs in mobile wireless networks," *Wirel. Netw.*, vol. 17, no. 3, pp. 645–658, 2011.

16 Z. Huang and Y. Ji, "Design and demonstration of room division multiplexing-based hybrid VLC network," *Chin. Opt. Lett.*, vol. 11, no. 6, p. 060603, Jun. 2013.

17 T. Nguyen, M. Chowdhury, and Y. M. Jang, "Flexible resource allocation scheme for link switching support in visible light

communication networks," in *Proc. International Conference on ICT Convergence (ICTC) 2012* (Jeju, South Korea), Oct. 15–17, 2012, pp. 145–148.

18 E. Stevens-Navarro, Y. Lin, and V. Wong, "An MDP-based vertical handoff decision algorithm for heterogeneous wireless networks," *IEEE Trans. Veh. Technol.*, vol. 57, no. 2, pp. 1243–1254, Mar. 2008.

19 B. J. Chang, J. F. Chen, C. H. Hsieh, and Y. H. Liang, "Markov decision process-based adaptive vertical handoff with RSS prediction in heterogeneous wireless networks," in *Proc. IEEE Wireless Communications and Networking Conference (WCNC) 2009* (Budapest, Hungary), Apr. 5–8, 2009, pp. 1–6.

20 C. Sun, E. Stevens-Navarro, V. Shah-Mansouri, and V. W. Wong, "A constrained mdp-based vertical handoff decision algorithm for 4G heterogeneous wireless networks," *Wirel. Netw.*, vol. 17, no. 4, pp. 1063–1081, May 2011.

21 J. Pan and W. Zhang, "An mdp-based handover decision algorithm in hierarchical LTE networks," in *Proc. IEEE Vehicular Technology Conference (VTC Fall) 2012* (Quebec City, Canada), Sep. 3–6, 2012, pp. 1–5.

22 F. Wang, Z. Wang, C. Qian, L. Dai, and Z. Yang, "Efficient vertical handover scheme for heterogeneous VLC-RF systems," *J. Opt. Commun. Netw.*, vol. 7, no. 12, pp. 1172–1180, Dec. 2015.

23 D. Bertsekas, *Dynamic Programming and Optimal Control*, 4th ed. Athena Scientific, Belmont, MA, USA, 2007.

24 S. Ross, *Stochastic Processes*. John Wiley and Sons, 1983.

25 J. Wang, R. V. Prasad, and I. Niemegeers, "Solving the uncertainty of vertical handovers in multi-radio home networks," *Comput. Commun.*, vol. 33, no. 9, pp. 1122–1132, Jun. 2010.

26 A. Tsiatmas, F. M. Willems, and C. P. Baggen, "The optical illumination channel," in *Proc. IEEE Symposium on Communications and Vehicular Technology in the Benelux (SCVT) 2012* (Eindhoven, Netherlands), Nov. 16, 2012, pp. 1–6.

27 M. Safari, "Efficient optical wireless communication in the presence of signal-dependent noise," in *Proc. IEEE International Conference on Communications Workshop (ICCW) 2015* (London, United Kingdom), Jun. 8-12, 2015, pp. 1387–1391.

28 G. K. Froehlich, "Estimation in signal-dependent noise," 1980.

29 P. R. Prucnal and B. E. Saleh, "Transformation of image-signal-dependent noise into image-signal-independent noise," *Opt. Lett.*, vol. 6, no. 7, pp. 316–318, Jul. 1981.

30 Q. Gao, S. Hu, C. Gong, and Z. Xu, "Modulation designs for visible light communications with signal-dependent noise," *J. Lightw. Technol.*, vol. 34, no. 23, pp. 5516–5525, Dec. 2016.

8
Optical Camera Communication: Fundamentals

In this chapter, the fundamentals of optical camera communication (OCC) are discussed. The OCC system employs the pervasive image sensor assembled in consumer electrons as the receiver. It occupies a wide spectrum, and can be easily built upon pervasive optical light sources and the pervasive consumer cameras, as discussed in Section 8.1. There are various potential applications, such as indoor localization, intelligent transportation, screen–camera communication, and privacy protection, which are introduced in Section 8.2.

To investigate the fundamental problems, including the channel characteristics and system performance, the pixel-sensor structure and the process during each phase of its operation for the complimentary metal-oxide-semiconductor (CMOS) image sensor are discussed in Section 8.3. Besides, the noise compositions, such as photon shot noise, dark current shot noise, source follower noise, sense node reset noise, and quantization noise at high illumination are analyzed. Based on the noise model, a unified communication model for OCC is derived. Moreover, the OCC channel capacity under an ideal pixel-matched channel with bounded inputs and intensity constraints is presented in Section 8.4.

8.1
Why OCC

Recently, OCC emerges as a new form of visible light communication. It employs an image sensor assembled in consumer electronic devices, such as smartphone, iPad as a receiver to serve as an alternative to the photodiode (PD) or avalanche photodiode (APD) based receiver [1, 2]. OCC allows easier implementation of various services into smart devices. A study group IEEE 802.15.7r1 is dedicated to revision of formerly established IEEE 802.15.7 visible light communication (VLC) standard, and incorporating new physical layers to support OCC functionalities and medium access control (MAC) modifications [3]. The technologies for smart devices have been developed greatly in the past few years, especially for the smartphones. According to the 2015 annual Mobility Report from Ericsson, there are 2.6 billion smartphone subscriptions in the world. That number will increase to 6.1 billion by 2020. Moreover,

Visible Light Communications: Modulation and Signal Processing. First edition. Zhaocheng Wang, Qi Wang, Wei Huang, and Zhengyuan Xu.
Published 2017 by John Wiley & Sons, Inc.

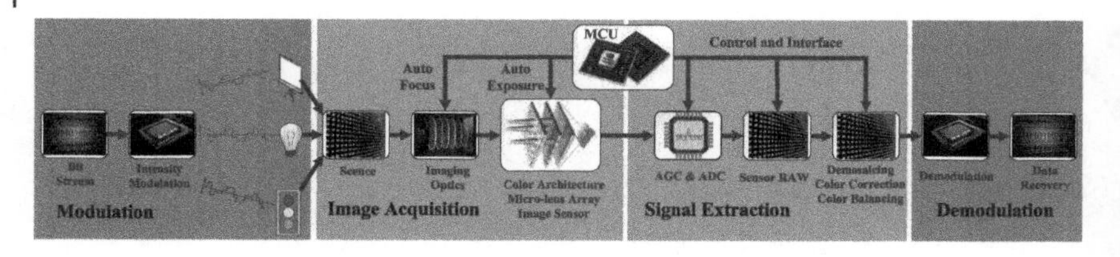

Figure 8.1 The diagram of the transceiver for optical camera communication (OCC) [4].

the majority of smart devices are equipped with light emitting diode (LED) flash, and front and/or rear cameras. This opens a possibility of pragmatic form of VLC implementation based on these devices as a transceiver pair, without additional hardware modifications. Therefore, there is a growing interest in VLC implementation using LEDs and displays for wireless data transmission, and cameras as receivers.

8.1.1 Wide spectrum

According to the proposal of IEEE 802.15.7r1, the wavelength of optical light for OCC is in the range of 10,000 nm, spanning from infrared (IR) to 190 nm [3], which is much wider than that for VLC or radio frequency (RF). Adoption of such an unregulated and unused spectral region will alleviate the mobile communication traffic congestion. However, only some spectral regions are suitable for optical wireless communications. For example, the wavelength in the range of 780–950 nm is currently the best choice for the IR-based communication, and the ultraviolet-C band is suitable for wide field of view (FOV) applications. In the visible light band, VLC performance may vary with wavelength. The frame rate of many consumer-grade image sensors is about 30–60 fps, which is suitable for low-rate applications. An industrial grade image sensor offers a high frame rate of up to millions of frames per second, which opens a new horizon for high-speed OCC.

8.1.2 Image-sensor-based receiver

A typical OCC system is shown in Fig. 8.1. The signals are modulated in intensity, color, or time-frequency domain for pervasive optical light sources (illumination LED, display or traffic light). The cameras, which are integrated in pervasive consumer electronic devices, are used as receivers to detect the signals beyond traditional imaging.

The major driving force of OCC applications stems from the availability of commercial visible light LEDs for data transmission and the possibility of utilizing the camera in the smart devices to decode signal received from LEDs. The commercial visible light LEDs are widely deployed in the indoor or outdoor scenarios, which include LED-based infrastructure lighting, LED flashes, LED tags, displays, image patterns, new generation projectors, traffic lights, rear and front lights of vehicles, roadside luminaries, and other outdoor LED signage. Some LEDs are modulated for lighting control by pulse-width modulation (PWM) and dimming circuit. They are probably first for fast migration to data transmission applications with slight modifications. The widespread LED displays are also capable of transmitting non-visual information, imperceptible to human eyes.

In addition to vastly existing transmitters, an OCC system aims to use camera as a convenient receiver, which consists of an imaging lens, an image sensor, and a readout circuit. The imaging lens projects light onto the image sensor, which is comprised of multiple PD-based pixels to detect the incident optical (photon) radiation. Each activated pixel generates a voltage proportional to the number of impinging photons. It is also connected to an external circuit to convert the pixel voltage into binary data. According to the readout circuit configuration and exposure mechanism, image sensor can be classified into two categories, namely, the global shutter type charge coupled device (CCD) image sensor and the rolling shutter type CMOS image sensor. In a CCD image sensor, the global shutter exposes all pixels per frame simultaneously. However, in a CMOS image sensor, the rolling shutter exposes one row/column of pixels at a time, and reads out the pixel voltages row-by-row or column-by-column. Most personal mobile devices or high-end professional camcorders use CMOS image sensors due to excellent performance/cost tradeoffs. According to the Grand View Research's report about image sensor market [5], CMOS image sensors take 83.8% of market share by 2013, while CCD image sensors take only 16.2%. By 2015, CMOS shipments amount to 3.6 billion units or 97% market share, compared to CCD shipments of just 95.2 million, or 3% market share. The majority of commercial CMOS image sensors utilize a rolling shutter, while only a few expensive high-end CMOS image sensors can support global shutter, since the global shutter is hard to be accomplished with the current CMOS technology. Although these image sensors have their own advantages and disadvantages, it is preferable to use the CMOS image sensor as the OCC receiver for its high frame rate.

8.1.3 Advantages of image sensor receiver

The image sensor has a high resolution and can classify multiple spatially separated light sources. It is thus a natural multielement optical receiver. Meanwhile, it has a multicolor filter layout and easily separates the blended multicolor lights. Therefore, it is also a natural multicolor receiver. Moreover, a front optical lens can help distinguish the light sources in the image sensor and reduce the channel correlation. These appealing features make the image-sensor-based receiver suitable for imaging optical multi-input multi-output (MIMO) settings, and are expected to support a huge number of parallel transmissions.

Pervasiveness: An OCC system uses widespread transmitters and receivers. The LED screens display arbitrary images of objects. They can be used as transmitters if they are illuminated by temporally modulated signals. Even the reflective surface illuminated by a signal-carrying light source can be used as the transmitter. The pervasive camera assembled in the smartphone and iPad, as well as the smartglass and tachographscan, can serve as an alternative receiver to the PD- or APD-based VLC receiver.

A multicolor receiver [7]: Usually, a color filter array (CFA), or color filter mosaic (CFM), is a mosaic of tiny color filters placed over the pixel sensors of an image sensor to capture color information. The color filters filter the light by wavelength range, such that the separate filtered intensities contain information about the color of light. Typical CFAs used in image sensor include Bayer filter, RGBE (red, green, blue, emerald-cyan) filter, CYYM (cyan, two yellow, magenta) filter, CYGM (cyan, yellow, green, magenta) filter, and RGBW (red, green, blue, white) Bayer filter. Due to the Bayer-pattern or Foveon X3 color filter layout, the image sensor is a natural multicolor receiver. Figure 8.2(a) shows the Bayer-pattern color filter, which contains 25%, 50%, and 25% of red, green, and blue filters, respectively. Using such a Bayer-pattern filter, each pixel produces signal components corresponding to three colors. One can demodulate the multicolor modulated signal for a pixel-based receiver. For a Bayer-pattern color filter sensor, interpolation is required for image-processing, during which the missing color data is estimated from neighboring pixel data. However, the pattern images are less sharp than they otherwise could be, due to undersampling, and optical blurring is introduced due to the lateral displacement of the color filters. The Foveon X3 color filter pattern image sensor can avoid the drawbacks of the traditional Bayer color filter pattern. As shown in Fig. 8.2(b), the Foveon X3 image sensor directly measures red, green, and blue colors at each location by stacking color pixels on top of one another, increasing the sampling density in the image plane. It also enhances sharpness in luminance and chrominance, as well as robustness to color aliasing artifacts.

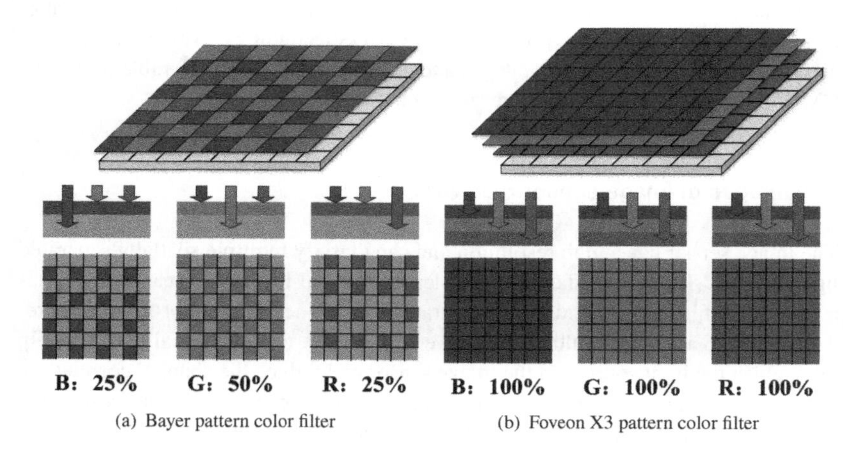

Figure 8.2 Color filter arrays in image sensor.

An optical MIMO receiver [8, 9]: Optical MIMO can help achieve high data rate by utilizing spatial multiplexing or spatial diversity. However, channel correla-

Table 8.1 Typical CFAs in image sensor.

Pattern	Name	Description	Pattern size
	GRGB Bayer filter	Bayer pattern (green, red, green, blue) CFA is the most commonly used CFA in image sensor	2 × 2
	RGBE filter	RGBE (red, green, blue, emerald-cyan) CFA can reduce the color reproduction errors and record natural images closer to the natural sigh	2 × 2
	RGBW filter	RGBW (red, green, blue, white) CFA allow the pixel to respond to all colors of light, and more of light is detected, rather than absorbed	2 × 2
	CYYM filter	CYYM (cyan, two yellow, magenta) CFA would result in a more sensitive imager (less light sapping dye over each pixel)	2 × 2
	CYGM filter	A CYGM (cyan, yellow, green, magenta) CFA uses mostly secondary colors, again to allow more of the incident light to be detected	2 × 2

tion is a major concern for optical MIMO. Thus, it is not practical to apply MIMO techniques directly in an optical wireless communication (OWC) system since the channel condition is very sensitive to transmitter spacing and receiver position. Intuitively, an ill-conditioned channel matrix may occur due to homogenous behaviors of different direct current (DC) channel gains. In such a case, the channel matrix may lose rank or its condition number may be very large. Some methods have been proposed to reduce the channel correlation. The non-imaging like blocked receiver, mirror diversity receiver, angle diversity receiver were proposed to reduce the channel correlation. However, the problem is that the complex receiver structures limit its practical application. A simple way of decorrelation is to apply channel matrix pseudo-inversion at the receiver side no matter the channel matrix is rank-deficient or not. However, it might result in noise amplification in commonly encountered sparse or low-correlation MIMO channels. Singular value decomposition (SVD)-based spatial multiplexing MIMO is another choice, where the pre-coding matrix and the combining matrix are adopted at the transmitter and receiver to form parallel channels respectively. For SVD-based optical MIMO, the transceiver design is based on the assumption that channel state information (CSI) is perfectly known at the transmitter and receiver. Otherwise, the performance will dramatically degrade due to imperfect CSI. However, perfect channel estimation is impossible in practical systems. Moreover, the challenges of uplink provision, CSI feedback, and the high computational complexity of designing the pre-coding and combining matrix for massive-MIMO, limit the application of the SVD-based decorrelation method.

For an image-sensor-based receiver, optical lens can help distinguish the images of

LEDs at the image sensor and reduce the channel correlation, which make the image-sensor-based receiver one of the most efficient ways for optical MIMO systems. Using a well-designed lens, the imaging receiver can clearly separate the signal from different light sources to achieve omnidirectional receiving and provide high spatial diversity for decoding of the MIMO signals. Moreover, due to the high resolution of the modern image sensor, there are millions or even more pixels integrated into a pixel array. Ideally, every pixel can be used as a receiver unit. Light is emitted from each light source and then projected onto a detector array, where it may strike any pixel or group of pixels on the array, and be in arbitrary alignment with them. In such a way, an image sensor still serves well as an efficient optical MIMO receiver.

An anti-interference receiver [10]: Generally speaking, if a conventional single PD is used as a receiver in VLC, the system cannot perform well under direct sunlight or other direct interference light sources, because the FOV of a conventional PD is wide and the direct interference light is typically strong to overwhelm the desired signal. Furthermore, it is very difficult to reduce the enormous amount of noise from background light to the desired optical signal level with a wide FOV, even if an optical band-pass filter is used. Therefore, when a single element photodiode is used outdoors, directed linkage with small optical beam divergence is required. On the contrary, an image sensor is composed of a PD/pixel array with massive number of available pixels. It is able to spatially separate different sources. A series of experiments have illustrated that ambient noise can be eliminated via an image sensor, due to the spatial filtering and separation. Furthermore, from the noise suppression perspective, a tracking algorithm based on image-processing techniques is considerably easier to implement than those based on mechanical techniques.

8.1.4 Challenges for OCC implementation

Despite various appealing features mentioned above, there are several factors that make OCC deployment challenging.

Limited frame rate [6]: The transmission speed of an OCC system is inherently limited by camera's (or image sensor) frame rate. Most of the consumer electronic cameras typically operate at 30 fps, which limits the achievable data rate. Exposure time and frame speed are mutually exclusive parameters affecting signal quality and data decoding rate. Exposure time has a direct relationship with the frequency of transmitted signal. This relationship can be understood based on the rolling shutter camera principle, in which the accumulated photons are collected and read out. In order to decode data efficiently, exposure time should be significantly smaller than transmission refresh frequency of the transmitter. Long exposure time results in a partial exposure problem in cameras using rolling shutter. Therefore, the received image will have a reduced contrast between bright and dark bands, which may lead to bit errors. According to the Nyquist sampling requirement, the receiver-side frame rate should be twice as fast as the transmitter-side signal rate.

Synchronization [6]: For most of commercial image sensors, the frame rate is unstable and limited by the manufacturing process and image post-processing.

Moreover, the frame rate of commercial image sensor is diverse. If the frame rate mismatches between transmitter and camera receiver, the synchronization problem arises for the image-sensor-based receiver. For example, liquid crystal display (LCD) has a frame rate of 30–60 fps, whist a maximum frame rate of 30 fps is for typical smart device cameras. Meanwhile, the frame rate of a camera is not constant during signal capture due to variable exposure period and software issues relevant to firmware and application programming interface (API). During the exposure period, image sensor pixels get bombarded by light signal, and eventually become saturated. Saturation of a pixel requires sufficient light exposure, which depends on the quantum efficiency, fill factor of the image sensor, and the illumination intensity. From the illumination perspective, the synchronization of the LED-based transmitters differs from the display-based transmitters. Therefore, different exposure time settings in a camera for two types of transmitters should be controlled manually. The exposure time also depends on the frequency of the OCC signal transmission. Meanwhile, both firmware and API constraints contribute to the variation of camera frame rate. The frame rate mismatch results in mixed frames or lost frames.

Shot noise [11]: In OCC, the shot noise in photodetectors, originally modeled by Poisson statistics, is signal-dependent. This is because it originates from the quantum nature of the received optical energy rather than any external sources of noise. As the number of received photons increases, the shot noise process can be well approximated by Gaussian statistics. The channel can be described by an additive white Gaussian noise (AWGN) model which includes electrical thermal noise as well as the signal-dependent (caused by the optical signal) and signal-independent (caused by background light) shot noise. In the image-sensor-based receiver, the signal-dependent distortions can be typically observed, because the signal-dependent shot noise is comparable with the background/thermal noise. Typically, an OCC system has a high signal-to-noise ratio (SNR) and highly narrow FOV receiver for each pixel or pixels block-based receiver. The signal-dependent noise model fundamentally changes the problem of efficient signaling compared to the conventional signal-independent AWGN model in VLC or RF communication. Thus, complex modifications to signal processing and estimation/detection techniques may be required to ensure optical data transmission and reception.

Perspective distortions [12]: Perspective distortions depend on the nature of the camera imaging mechanism and manifest as the deformation in size and shape of the captured object on the image. They lead to visual compression or magnification of the object's projection on the image. When the transmitter, such as the screen, is at an out-of-focus distance from the camera lens (or at an oblique angle), these distortions become prominent and create interference among adjacent screen pixels on the camera image, termed as inter-pixel interference (IPI). The combined effect of background noise and IPI degrades the received signal quality and hence reduces information capacity in camera channels.

Misalignment and blur [13]: Mismatch between the physical size of transmitter and camera pixels can cause the transmitter misaligned with a camera pixel, even if the transmitter is at the camera focus. Such misalignment will cause a deviation of the distortion factor for each pixel as the perspective changes. Meanwhile, some vibrations on the pixels are inevitable, especially when the camera is not stable, which means that the area of misalignment can keep changing with perspective. Such dynamic change in perspective arises primarily when the camera is hand-held, due to handshakes or lateral movements. Misalignment also applies to many more stationary scenarios. In such cases, the distortion effect is in the form of blurry and mixed frames due to motion blur and diffraction limited optical subsystem. The blur effect arises from movement within or between camera frames, and has been well studied in computer vision literature. The impact of misalignments and lens blur becomes smaller as one block covers more pixels on the camera and only affects pixels near the boundary.

Ambient light: Ambient light is a source of interference in OCC. It changes the luminance at the received pixels and causes decoding errors, resulting in information loss at the receiver. Especially, the main source of noise is flicker noise, which is caused by background lighting. If the peak-peak intensity value of flicker noise is larger than half of the modulation intensity gap, the performance will dramatically degrade.

The comparisons between OCC and the other short-range communication methods are concluded in Table 8.2.

8.2 OCC applications: beyond imaging

The image sensor receivers available on today's mobile devices prompt a new direction of research and applications where VLC can be combined with mobile computing to realize novel forms of sensing and communication applications, which are discussed in this section.

8.2.1 Indoor localization

Location-based services have been observed tremendous growth in last few years. Mobile device localization in outdoor scenarios largely depends on Global Positioning System (GPS) [14]. On the other hand, accurate indoor positioning can enable a wide range of location-based services across many sectors. Retailers, supermarkets, and shopping malls, for example, are interested in indoor positioning because it can provide improved navigation for customers to easily locate merchandise or for customer tracking. The overall demand for indoor positioning in the retail sector is predicted to grow to $5 billion by 2018.

Despite the strong demand forecast, indoor positioning remains a "grand challenge", and no existing system offers accurate location and orientation using unmodified smartphones. The widely used GPS is not suitable for indoor positioning because the measurement errors due to signal attenuation, radio disturbance, and multipath effect will degrade its positioning accuracy.

In the past decade, various indoor positioning approaches have been proposed and

Table 8.2 Characteristics of OCC [3].

Comparison Between VLC and OCC		
	VLC	OCC
Receiver	Photodetector (PD)	Image sensor (Camera)
Interference	High	Low
SNR	Low	High
MIMO Multiplexing	Easy to implement	Difficult to implement
Decoding	Signal processing (low complexity)	Image processing (high complexity)

The Benefits for OCC				
	Bluetooth	WiFi	VLC	OCC
Interference	Yes	Yes	Yes	Less
Security	High	High	Highest due to LOS	Highest due to LOS
Link Setup	Scan-and-link	Scan-and-link	LOS-Link	Look-and-Link
Protocol	IEEE.802.15.1	IEEE.802.11a & IEEE.802.11b	IEEE.802.15.7	IEEE.802.15.7r1
Frequency Band	2.4 GHz	2.4 /3.6 /5 GHz	Visible light (400–800 THz)	IR, Visible light, UV
Data Rate	800 kbps	11 Mbps	PHY I: 11.67–266.6 kbps PHY II: 1.25–96 Mbps PHY III: 12–96 Mbps	Lower than VLC Enhanced by • numbers of LEDs • camera's resolution • camera's frame rate
Cover Range	30 m	46–100 m	Near	Can be extended up to kms

developed, including radio frequency identification (RFID), Bluetooth, wireless fidelity (WiFi), ultra wide band (UWB), wireless local area network (WLAN), and ZigBee. Among those alternatives, WiFi-based and other RF-based indoor localization have been proven to be attractive where existing WiFi access point (AP) deployment is leveraged to identify the client's location. However, due to metope reflection and non-line-of-sight (NLOS) path, the positioning errors of these RF based systems are up to the order of decimeters or even higher. Although it is low cost, a WiFi-based indoor localization system offers low accuracy and no orientation information. Thus it is not ideal for navigation and shelf-level advertising. The complex multipath cancelation techniques are required to improve the accuracy. Moreover, RF-based positioning is impossible when RF is prohibited such as in a hospital, airplane, or coal mine. Those environments avoid interference to the mission-critical electronic equipment or explosion trigger.

Similar to WiFi-based localization, indoor visible light communication system can also be leveraged for indoor localization [15]. Typically, multiple LED luminaries are available each time for triangularization-based positioning. It has been surveyed that there are 10 times more LED luminaries than the number of WiFi APs in a typical indoor building. The higher density allows more accurate triangulation of the mobile device, resulting in higher accuracy. Compared with a conventional RF-based positioning system, a visible light positioning (VLP) system is free of the electromagnetic radiation and provides high positioning accuracy. The conventional PDs and the ubiquitous commercial cameras can be used as receivers. The PD-based VLP approaches rely on the received signal strength (RSS) or time difference of arrival (TDOA) [16]. However, the positioning precision of an RSS method is determined by the model of source radiation and the PD characteristic. Meanwhile, perfect synchronization among the LED transmitters is required for the TDOA-based positioning methods.

In recent years, many research groups focus on camera-based VLP, and several VLP systems have been implemented. In 2003, a positioning beacon system using digital camera and LEDs was proposed [17]. The LED array is divided into sub-arrays, and each one has its own visual pattern. High-frequency switching of the sub-arrays is not noticeable for the human eyes. A digital camera is used as a receiver to capture a sequence of images of the LED-positioning beacon transmitter, and then location codes are decoded correctly. Another practical VLP system is Luxapose [18]. It leverages the rolling shutter effect of the CMOS camera to decode location codes, and adopts angle-of-arrival (AOA) to estimate the position of the camera-based receiver. Luxapose has been shown to achieve localization accuracy about 0.1 m within 3° orientation error. Similar to Luxapose, an imaging sensor can be used to receive light signals from multiple luminaries, each of which creates a visual landmark. In 2014, a light-weight indoor positioning system called PIXEL took the polarizer to convert unpolarized light into polarized light [19]. It utilizes single pixel LCD and a diffuser in front of the light source in order to disperse light of different colors of different polarizations. Although PIXEL can solve the flickering problem, the polarizer will reduce the light intensity and degrade the illumination performance. Even though there is still much room for improvement in a camera-based VLP system, very high accuracy and ability to leverage the existing lighting infrastructure make it a good candidate for future indoor positioning applications.

8.2.2 Intelligent transportation

As new and revolutionary lighting sources, LEDs are superior to conventional incandescent lights due to the low power consumption, long lifetime and low heat generation. Since LEDs are semiconductor devices and have a short response time, they are possible to transmit data in conjunction with illumination. Such a feature is well suited for intelligent transport system (ITS) [20]. ITS can help control the traffic jams and traffic accidents with the development of information technology. Widespread use of LEDs in ITS presents an opportunity for VLC applications, for example, LED traffic light and signs broadcast driving assistance data to cars (road-to-vehicle communications), and LED car brake lights can be used to transmit warning data to a car behind (vehicle-to-vehicle communications). These systems contribute to exchange safety information between roadway infrastructures and vehicles.

In the area of ITS, VLC offers several advantages. Since VLC links are visible, installation of roadside equipment is much easier. Additionally, previously installed facilities, such as LED traffic lights or LED sign boards, can be used as transmitters or receivers. Furthermore, since the transmitters, or LED light sources, are designed for illumination, the SNR is high for VLC, and eye safety is maintained for dual-use lighting. Conventionally, if a single element photodiode is used as a VLC reception device, the VLC system cannot be used under direct sunlight, especially for wide FOV cases because the average received power is much higher than that of the desired signal [10]. Furthermore, it is very difficult to reduce the enormous amount of noise from background light to the optical signal with a wide FOV, even if an optical band-pass-filter is used. Therefore, when a single element photodiode is used outdoors, narrow optical beam and FOV are required. In some cases, a telephoto can help the receiver to track the light source, but requires complex mechanical tracking.

Image sensors emerge as new OCC receivers attributed to their multifunctionality, low manufacturing costs, low power consumption, and a massive number of pixels to spatially separate sources and filter out light noise [10]. These sensors also facilitate image-processing and accurate tracking, applicable for ITS. The communication pixels specialized for receiving high-speed optical signals are integrated with ordinary image pixels into a pixel array. The field trial results show the novel design of optical communication image sensor (OCI) [21] enables vehicle-to-vehicle VLC at a data rate of 10 Mbps.

It is possible that a camera can recognize objects, track their locations, and receive LED modulated data at the same time. Exploring millions of pixels of an image-sensor-based receiver, parallel data transmissions can be realized from multiple individually modulated LED traffic lights, traffic signs or car brake lights. Moreover, by using image or video processing to detect and recognize moving vehicles, safety applications can be integrated. For example, as methods of enhancing driving safety, collision warning, pedestrian detection, and range estimation for nearby vehicles are potential candidate functions to be incorporated into the VLC systems. A map of surrounding vehicles is drawn, showing clearly car-to-car distances, and broadcasting real-time speeds to the neighbors through the LED headlight and taillight. Other nearby vehicles receiving this information can adjust their speed accordingly to maximize the fuel efficiency and minimize the chances of collisions. Therefore, camera-based OCC approach is very attractive for ITS, particularly on congested highways where the traffic density and the resulting contention impose challenges on traditional RF communications.

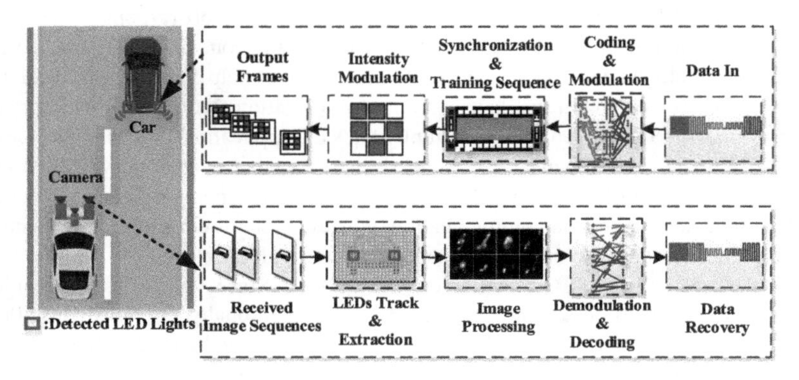

Figure 8.3 The application of OCC in intelligent transportation system.

8.2.3 Screen–camera communication

In many occasions, mobile users would like to have quick connections with neighboring users, and exchange messages using device-to-device communications without routing messages through a remote base station in a cellular network. VLC over a handset screen–camera link has become an attractive short-range wireless communication solution due to the high availability of camera-equipped smartphones over the past years [12, 22, 23], where information can be encoded as a stream of images and displayed in screens of smartphone, laptop, or advertisement boards. Due to the short wavelengths and narrow beams of visible light, screen–camera links are highly directional, less interfering, and secure as compared with the RF techniques such as Bluetooth and WiFi. By controlling the direction, FOV and distance, screen–camera communication simplifies the complicated authentication process for quickly setting up link connections. In addition, screen–camera communication is user-friendly. It is well suited for file transfer between smartphones when either no wireless connections are available or security is much concerned. Through analysis and experiments, such links have been shown capable of achieving hundreds of kbps to Mbps transmission.

Another unique scenario to make screen–camera communication attractive is service provision in the area of high user density [12]. For example, in a congested party, trade fair or business reception, many people would like to exchange brochures, photos or video demos. Traditional RF-based communication performs poorly because the available spectrum is congested under a large number of user contentions. In contrast, the high directionality of screen–camera communication enables a multitude of such communication links simultaneously without inter-user interference. Besides, in a public area, such as museum, gallery, or shopping mall, there is an interest in providing additional information about the displayed objects [12]. For example, in a gallery or museum, a user may access a video presentation about the artifact the user is viewing. The screen–camera communication provides additional details or side-information accompanying the primary video watching.

The performance and capabilities of digital camera and screens improve at a fast pace and consistently reduced price. Additionally, new technologies such as microlens arrays provide sophisticated but compact optical capabilities in a significantly smaller size. Moreover, with the advent of electronic shutters, industrial grade image sensors are now able to support up to thousands or even millions of frames per second. Although a consumer camera supports a range of frame rate from 10 fps to 1000 fps nowadays, the frame rate is increasing continually. Those advantages make the screen–camera communication a good candidate for future short-range and convenient communication.

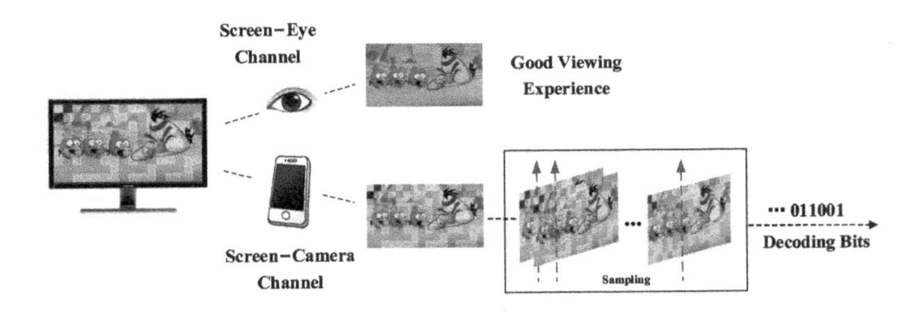

Figure 8.4 The application of OCC in screen–camera communication.

8.2.4 Privacy protection

The basic OCC systems have been developed and implemented, which improve the visual communication over screen–camera links. However, there exists a security issue on how to prevent unauthorized users from videotaping a video played on a screen or a projector [23], whilst not affecting the viewing experience of legitimate audiences, such as in a cinema or a lecture hall for high-quality display. According to a survey, over 90% of the illegal online contents are delivered from the pirate movies. Unauthorized videotaping during the exhibition could cause undesirable information leakage. To avoid the copyright piracy issues, many exhibitions have imposed no-camera policies. Moreover, videotaping a presentation or project demonstration could also cause the infringement of copyright and even plagiarism. Copyright protection is becoming increasingly important.

With the rapid spread of camera-enabled mobile devices or wearable devices, recording video in a cinema or a lecture hall becomes extremely easy and hard to detect. Traditionally, to protect the copyright, copyright is first filed for a digital property indicating that it is protected by law and unauthorized usage is illegal. Various technologies have been developed in the industry and research community for conveying this copyright protection information and/or protecting the digital property copyright from being violated, such as the watermarking. Film industry often implements expensive security strategies or equips guards with night vision goggles to prevent film piracy. Unfortunately, these technologies are ineffective in preventing attendees from taking pirate video for later redisplay. Thus, it is necessary to develop a universal technology that can be used to protect the video displayed in a variety of devices from pirate videotaping using typical mobile devices, such as smartphone or smartglasses, without introducing extra hardware.

In contrast to the existing techniques, which are used to maximize decodability of screen-camera communication, the copyright system needs to maximize the quality degradation of the display-camera channel while retaining the quality of the screen-eye channel [23]. Along this line, one can explore the fundamental differences among human vision, video encoding, screen display, and video-recording mechanisms. By taking advantage of the limited disparities (e.g., the spectral and temporal color mixture, the flicker effect, critical flicker frequency, and rolling shutter) between human vision and video-recoding, it is possible to develop the techniques for protecting the copy of the video, and preventing audience from taking a high-quality pirated copy of the video.

8.3 Fundamentals of OCC

In this section, based on the structure of active pixel sensor (APS) equivalent circuit and actual communication scenarios, the noise characteristic and channel model are investigated.

8.3.1 Optical imaging system

An image sensor is one of the most important building blocks in an optical camera based system [24, 25]. The block diagram of an imaging system architecture is shown in Fig. 8.5. First, the source radiation is focused on the image sensor using the imaging optics. The image sensor comprising a two-dimensional array of pixels converts the light incident at its sensitive area into an array of electrical signals. To perform color imaging, a CFA such as Bayer pattern color filter array or Foveon X3 pattern color filter array is typically deposited in a certain pattern on the top of the image sensor array. The converted electrical signals are read out and digitized by an analog-to-digital converter (ADC). To recover a full color image with Bayer pattern color filters, spatial interpolation process known as demosaicking is needed. Further digital signal processing is performed as white balancing and color correction to diminish the adverse effects of faulty pixels and imperfect optics. Other processing and control operations are also included for performing autofocus, auto-exposure, and general camera control. All these components of an imaging system determine the overall performance together. However, it is the image sensor that often limits the ultimate performance.

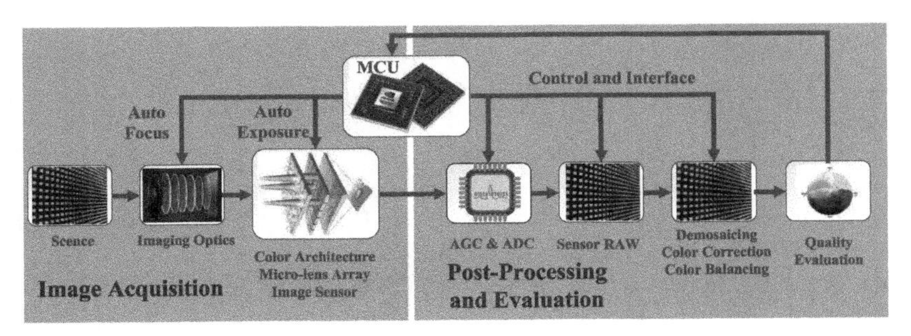

Figure 8.5 The imaging system pipeline.

8.3.2
Image sensor architecture

An image sensor consists of an array of pixels. Each pixel contains a photodiode that converts incident light into photocurrent and then the photocurrent is converted into electric charge or voltage by some readout circuits [26]. The percentage of area occupied by the effective photodiode in a pixel is known as fill factor. The rest of the readout circuits is located at the periphery of the array and multiplexed by the pixels. An array includes about tens of megapixels for high-end applications, where the size of each pixel is about 1.22 μm $\times$ 1.22 μm to 7.5 μm $\times$ 7.5 μm. A microlens array is typically deposited on the top of the pixel array to increase the fill factor and allow more light incident on each photodiode in certain exposure time. Figure 8.6 shows anatomy of the CMOS image sensor and APS photodiode, respectively.

Since the mid-1960, combinations of p-n or n-p-n junction have been used to convert optical light into electrons, and read the signal out from the arrays of pixels. The earliest solid-state image sensors were the bipolar and MOS photodiode arrays. Afterwards, CCDs quickly became the dominant ones. The research on CMOS image sensors started in the mid-1980s, and the passive pixel sensor (PPS) became the alternative technology for CMOS image sensor in the early 1990s [27]. However, since typical sizes of the available CMOS technologies were too large and it was impossible to accommodate more than a single transistor and three interconnect lines in a PPS pixel, thus, the fill-factor was lower for CMOS image sensor. Because of the need for transistors, the pixels read out, and amplification, the PPS CMOS

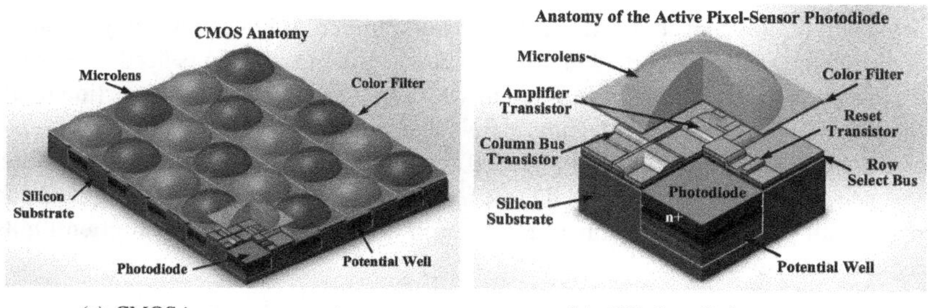

(a) CMOS image sensor anatomy (b) APS photodiode anatomy

Figure 8.6 The structure of a CMOS image sensor.

image sensors have much lower performance than CCDs, which limits their applicability to low-end machine-vision applications. The ability to produce CCD image sensor with the necessary number of pixels for application gave CCDs a big advantage over CMOS image sensor. However, improvements in CMOS fabrication technology and increasing pressure to reduce power consumption for battery-operated devices began the re-emergence of CMOS as a viable imaging device. It is generally regarded that the first all-CMOS sensor array to produce acceptable images is the APS image sensor [28]. The APS design uses the linear integration method for measuring light because of the large output signal generated. By adding an amplifier to each pixel, it can significantly increase sensor speed and improve SNR, thus overcoming the shortages of PPS. With the advent of deep submicron CMOS and integrated microlens technologies, APS has made CMOS image sensor a viable alternative to CCD sensor. While CMOSs and CCDs continue to compete for a share of the image sensor market, the ability to design customized integrated circuits with photodetectors to perform specific functions is an enormous advantage over CCD image sensor. Taking further advantage of technology scaling, the digital pixel sensor (DPS) integrates an ADC in each pixel [29]. The DPS architecture offers several advantages over an analog image sensor. These include better scaling with CMOS technology due to reduced analog circuit performance demands, and the elimination of read-related column fixed-pattern noise (FPN) and column readout noise. More significantly, massive parallel conversion and digital readout provide very high-speed readout, which makes it possible to enhance the sensor's dynamic range (DR) via programming and multiple sampling.

For these two widely used image sensors in cameras for microscopy, namely the CCD image sensor and CMOS image sensor, there are a number of similarities between these two technologies. But one major distinction is the way each sensor reads the signal accumulated at a given pixel. Figure 8.7 shows the readout architectures of CCD and CMOS image sensors. In a CCD image sensor, every pixel is exposed simultaneously at the same time, and the charge is shifted out of the array via vertical and horizontal CCDs, converted into a voltage via a simple follower amplifier, and then serially read out. In a CMOS image sensor, charge voltage signals are read out

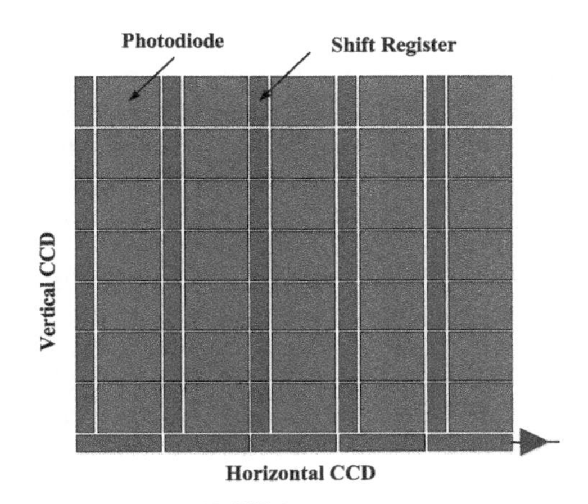

(a) CCD image sensors

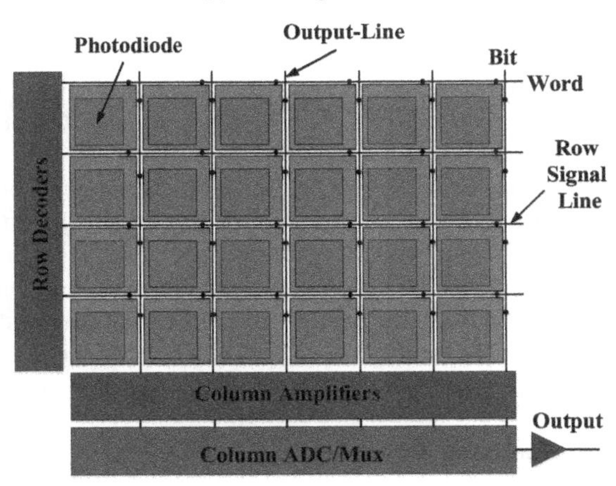

(b) CMOS image sensors

Figure 8.7 Readout architectures of interline transfer of different image sensors [26].

row-by-row similar to a random access memory using row and column selection circuits. Each readout architecture has its advantages and shortcomings, as presented in Table 8.3.

The main advantage of the CCD readout architecture is minimal pixel overhead, making it possible to design an image sensor with very small pixel sizes [26]. Moreover, the charge transfer is passive and therefore does not introduce temporal noise or pixel-to-pixel variations due to device mismatches, known as FPN. The CCD however has an inherent disadvantage when frame rate is of concern. When the exposure is complete, the signal from each pixel is serially transferred to a single ADC. The CCD's ultimate frame rate is limited by the rate that individual pixels can be transferred and then digitized. The more pixels to transfer in a sensor, the slower the total frame rate of the image sensor. It also has high power consumption due to the need of high-rate and high-voltage clocks to achieve near-perfect charge transfer efficiency. By comparison, the random access readout of CMOS image sensors provides the potential of high-speed readout and window-of-interest operations at low power consumption [26]. A CMOS chip enhances the frame rate performance by using an ADC for every column of pixels up to the thousands. The total number of pixels digitized by any converter is significantly reduced, enabling shorter readout times and consequently faster frame rates. Meanwhile, there are many parallel ADCs sharing the workload, the entire sensor array must still be converted one row at a time. This results in a small time delay between each row's readout. Nowadays, the frame rate reaches hundreds of frames per second for a commercial image sensor and even millions of frames per second for an industrial grade image sensor with high resolution. Other differences between CCD image sensors and CMOS image sensors arise from differences in their fabrication technologies. CCDs are fabricated by specialized technologies solely optimized for imaging and charge transfer. Control over the fabrication technology also makes it possible to scale pixel size down without significant degradation in performance. The disadvantage of using such specialized technologies, however, is the inability to integrate other camera functions on the same chip with the sensor. CMOS image sensors, on the other hand, are fabricated by mostly standard technologies, and thus are ready to integrate with other analog and digital processing and control circuits. Such integration further reduces imaging system power and size, and enables the implementation of new sensor functionalities.

8.3.2.1 Pixel-based photodetection

The core of the sensing element of a CMOS detector is the photosensitive element of the circuit. Photogates, phototransistors, and photodiodes all can be used as the sensing elements. The photodiode is simply a junction between a p-type and a n-type semiconductor, commonly known as a p-n junction. Although a simple p-n junction can be used for light detection, the more advanced p-i-n junction with an intrinsic region between the p-type and n-type is often used to improve the device efficiency. In an APS, a photodiode is usually made by forming an n-type region on a p-type semiconductor substrate, or vice versa. This can be realized by epitaxial growth, diffusion, or ion implantation.

Table 8.3 Comparison between CCD and CMOS image sensors.

Attributes	CCD Image Sensor	CMOS Image Sensor
Electronic shutter	Global shutter	Rolling shutter
Sensor noise	Low	High
Dark current	Low	High
SNR	High	Comparatively low
Fill factor	100%	< 100%
Integration capability	Low	High
Power consumption	High	Low
Cost	High cost	Low cost
Frame rate	Comparatively low speed	High speed

The two important metrics that are used to characterize the effectiveness of detection by a photodetector are external quantum efficiency (QE) and dark current. External QE is the fraction of incident photo flux that contributes to the photocurrent in a PD as a function of wavelength. It is typically combined with the transmittance of each color filter to determine its overall spectral response. External QE can be expressed as the product of internal QE and optical efficiency (OE). Internal QE is the fraction of photons incident on the PD surface that contributes to the photocurrent, which mainly depends on PD geometry and doping concentrations and is always less than 1 for silicon-photodetector. OE is the photo-to-photo efficiency from the pixel surface to the photodetector's surface. The geometric arrangement of the PD with respect to other elements of the pixel structure, i.e., shape and size of the aperture, length of the dielectric tunnel, position, shape and size of the PD, also affects OE [30]. Experimental evidence shows that OE can have a significant role in determining the resultant external QE, and eventually determining the pixel response function (PRF) or modulation transfer function (MTF) for image sensor. Figure 8.8 shows the 2-D MTF for pixel-based photodetector with different shapes of active area. It demonstrates that different geometrical shapes of the pixel active area will significantly influence the PRF or MTF.

The second important characteristic of the photodiode is its leakage or dark current. Dark current is the photodiode current when no illumination is present. One kind of the dark current under reverse bias is caused by saturation current. On the boundaries of the depletion region, minority carriers can diffuse into the depletion region, causing saturation current. Apart from the diffusion contribution to the dark current, carriers that are generated by thermal excitation inter-band trap (defect) states in the depletion region contribute to the dark current. The generated dark current is also contributed by other sources, including surface leakage current, Frankel–Poole current, and impact ionization current. It is noted that the dark current is detrimental to imaging performance under low illumination, as it introduces shot noise which is uniformly distributed over sensor array and cannot be corrected [26]. In CCDs, the

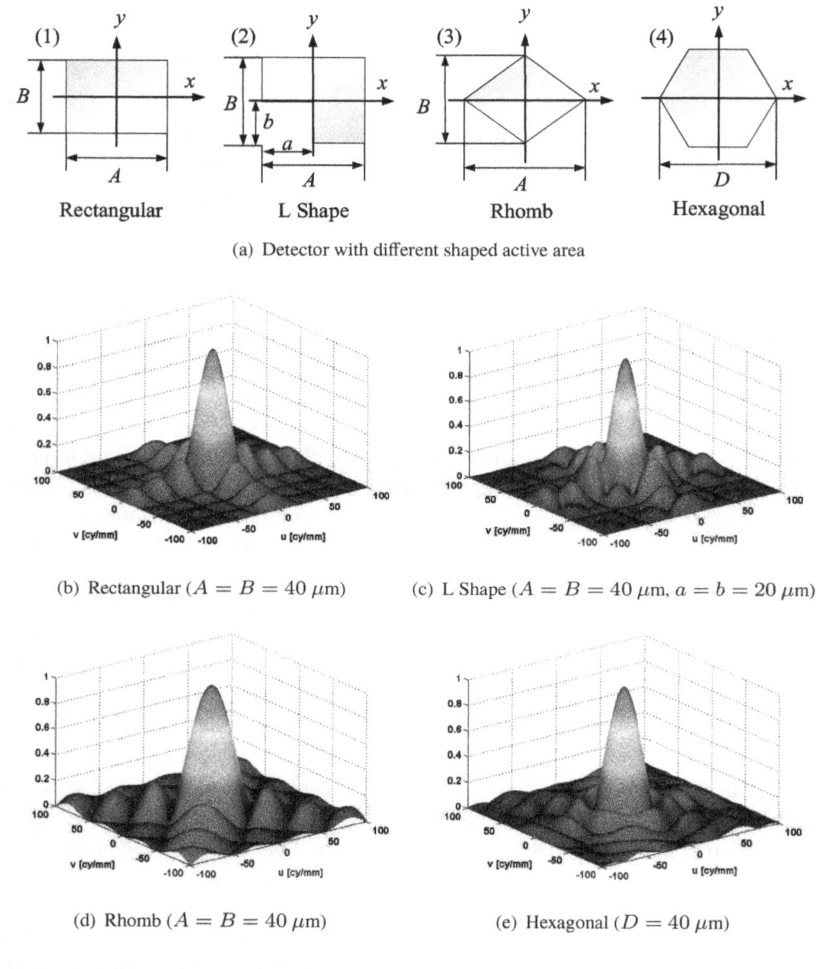

(a) Detector with different shaped active area

(b) Rectangular ($A = B = 40\ \mu$m)

(c) L Shape ($A = B = 40\ \mu$m, $a = b = 20\ \mu$m)

(d) Rhomb ($A = B = 40\ \mu$m)

(e) Hexagonal ($D = 40\ \mu$m)

Figure 8.8 2-D MTF for the image sensor detector with different shape active areas.

high-resistivity wafers can be used to minimize traps from metallic contamination as well as buried channels, and multiphase pinned operations can be used to minimize surface generated dark current. Dark current in the standard submicron CMOS process is the order of magnitude higher than that in a CCD. Fortunately, several process modifications are employed to reduce the dark current nowadays.

8.3.2.2 **PPS, APS, and DPS architectures**

There are different forms of readout architectures in CMOS image sensor, including the PPS, which is the earliest CMOS image sensor architecture, the three, four and five transistors (3T, 4T, and 5T) per pixel APS, which are the most popular architectures at present, and DPS, which includes a photodiode and an ADC for each pixel. The main advantage of PPS is its small pixel size. The column readout, however, is slow and vulnerable to noise and disturbances. The APS and DPS architectures can solve these problems with more transistors in each pixel.

Each of the APS architectures has its advantages and disadvantages. The 3-T APS pixel includes a reset transistor, a source follower transistor and a row select transistor. The current source component of the follower amplifier is shared by a column of pixels, and the readout is performed on one row at a time. A 4-T pixel is either larger or has a smaller fill factor than a 3-T pixel implemented with the same technology. The 4-T APS architecture employs a pinned diode, which adds a transfer gate and a floating diffusion (FD) node to decouple the read and reset operations from the integration period, enabling true correlated double sampling (CDS). Moreover, in a 4-T pixel, the capacitance of the FD node can be selected independently of the photodiode size, allowing conversion gain to be optimized for the sensor application. The 5-T APS pixel adds a transistor to tackle the blooming issue and provide reset operation for photodiode. By appropriately setting the gate voltage of the reset transistor in an APS pixel, the blooming effect, which is the overflow of charge from a saturated pixel into its neighboring pixels, can be mitigated. However, since more transistors and more complex control circuit are needed, the fill factor will be reduced.

DPS is the third and more advanced CMOS image sensor architecture, where ADC is performed locally at each pixel, and digital data is read out from the pixel array in a manner similar to a random access digital memory. The advantages of DPS over analog PPS and APS, include simplicity, scalability, on-chip processing, lower power consumption, wide dynamic range, the reduced read-related column FPN, and readout noise. More significantly, employing an ADC and memory at each pixel enables massive parallel analog-to-digital conversion and high-speed digital readout processes, providing a potential of high frame rate. The main drawback of DPS is that more transistors per pixel are needed, resulting in larger pixel size and lower fill factor [26].

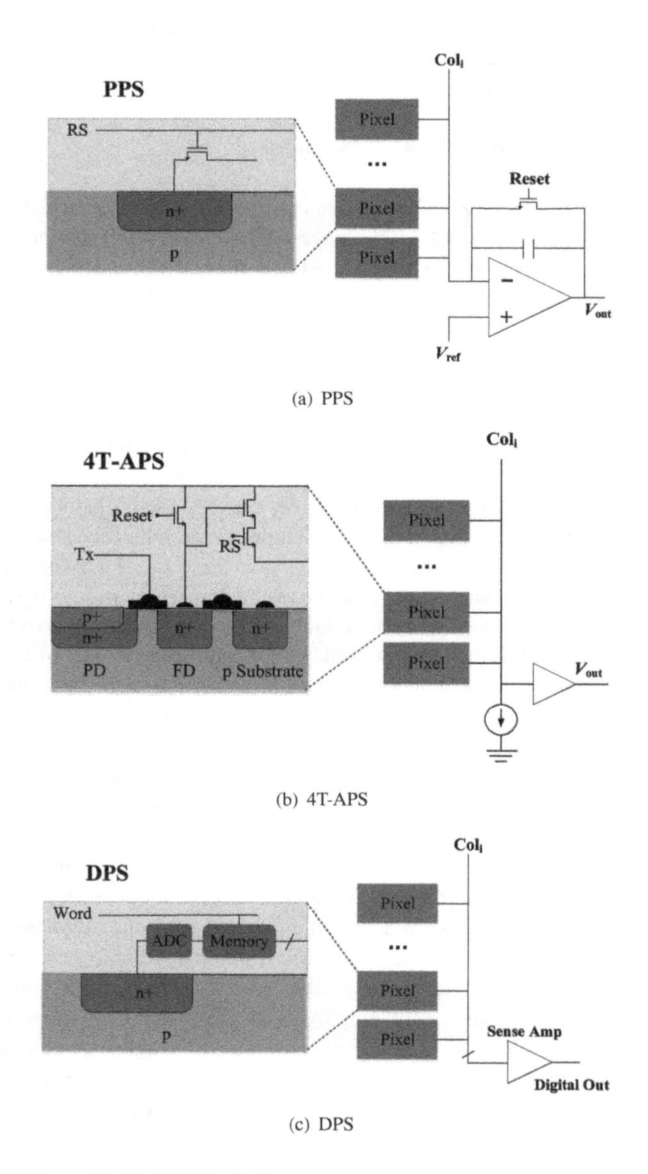

Figure 8.9 Schematic of a CMOS pixel sensor.

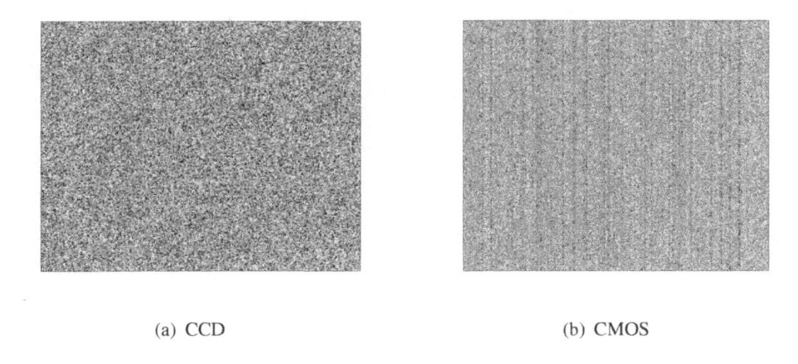

(a) CCD (b) CMOS

Figure 8.10 PFN noise for CCD and CMOS image sensors.

8.3.3 Noise characteristics in the image-sensor-based receiver

In general, ambient noise, temporal noise, and FPN are the main fundamental and technology related noises in CMOS image sensor. Temporal noise is the dominant noise in the image sensor, which is generated in pixel during each phase of its operation. Sources of temporal noise include photodetector shot noise, pixel reset circuit noise, readout circuit thermal and flicker noise, and quantization noise. In addition to temporal noise, CMOS image sensor also suffers from FPN, which is the pixel-to-pixel output variation under uniform illumination due to the device and interconnect mismatches across the image sensor array.

In this part, we analyze the sources of noise in a typical APS during each phase of its operation, including photon-to-charge operation, charge-to-voltage operation, and voltage-to-digital signal operation, as shown in Fig. 8.11. Moreover, the nonlinearities are also taken into account [4, 32].

From photon to charge

In pixel sensor, the photon-capturing process has an uncertainty that arises from random fluctuations when photons are collected by the photodiode, which will induce photon shot noise. The shot noise can be described by the Poisson distribution in a low light level, and can be approximated by a Gaussian distribution [33] in the case of a high light level. Denote the average number of photons I_{photon} collected by a single pixel during unit integration time as

$$I_{\text{photon}} = \text{round}\left(\frac{I_{\text{irrad}} \cdot A_{\text{p}}}{E_{\text{p}}}\right), \tag{8.1}$$

where I_{irrad} is sensor's irradiance with units of [W/m^2], A_{p} is the area of a pixel sensor, $E_{\text{p}} = \frac{h \cdot c}{\lambda}$ is the energy of a single photon at the wavelength λ, and h is Planck's constant. The photon shot noise corresponding to this pixel sensor can be

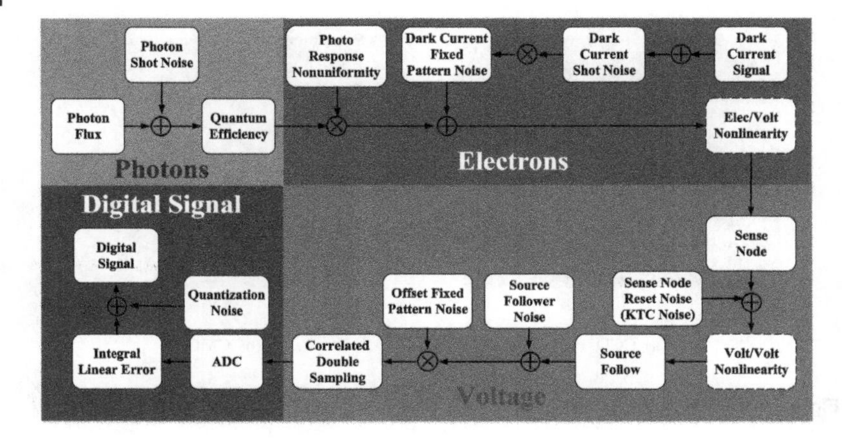

Figure 8.11 The diagram of the photosensor model.

modeled as a Poisson process $\mathcal{P}$ with mean λ_{ph}

$$I_{\text{photon,shot}} = \mathcal{P}(\lambda_{\text{ph}}), \quad \text{where } \lambda_{\text{ph}} = I_{\text{photon}}. \tag{8.2}$$

Then, the collected photons during exposure time t_{int}, including the shot noise, are converted to electrons I_{e^-} for each pixel as

$$I_{e^-} = QE \cdot I_{\text{photon}} \cdot t_{\text{int}} = I_{\text{ph}} \cdot t_{\text{int}}, \tag{8.3}$$

where QE is the quantum efficiency [e$^-$/incident photons] for the given wavelength, and it indicates the ability of a semiconductor to produce electrons from incident photons.

The photo response nonuniformity (PRNU) is the spatial variation in pixel output under uniform illumination mainly due to the variation in substrate material during the fabrication of the photodiodes. Moreover, the PRNU is signal-dependent and fixed-pattern (time-invariant) according to amounts of experimental measurements. The PRNU can be modeled as Gaussian distribution as

$$I_{\text{PRNU},e^-} = I_{e^-} + I_{e^-} \cdot \mathcal{N}(0, \sigma_{\text{PRNU}}^2), \tag{8.4}$$

where σ_{PRNU} is the PRNU factor value.

The dark current is inevitable in pixel sensor. It is induced by the thermally generated electrons that discharge the pixel, together with the surface defects and imperfections of the semiconductor manufacturing process. The average dark current I_{dc} [e$^-$/sec/pixel] can be characterized by

$$I_{\text{dc}} = A_{\text{p}} I_{\text{FM}} T^{3/2} \exp\left(\frac{-E_{\text{gap}}}{2k_{\text{B}}T}\right), \tag{8.5}$$

where I_{FM} [nA/cm^2] is the dark current figure-of-merit at 300 K, T is the temperature in K, k_{B} is the Boltzman's constant, and E_{gap} [eV] is the band gap energy of the semiconductor which varies with temperature.

When pixel exposure begins, a dark current is generated even if there is no light. The longer the integration time t_{int}, the stronger the dark signal S_{dark,e^-} (number of electrons per pixel), which can be modeled as

$$S_{\text{dark},e^-} = I_{\text{dc}} \cdot t_{\text{int}}. \tag{8.6}$$

The dark signal S_{dark,e^-} varies from pixel to pixel. It is linear with the integration time and doubles with every 6–8°C increase of temperature. Measurements are conducted at the room temperature +25°C. Moreover, since the electrons are generated randomly, the dark signal S_{dark,e^-} is a subject of dark current shot noise, which is due to the random arrival of the generated electrons and therefore described by a Poisson process as

$$I_{\text{dark,shot},e^-} = \mathcal{P}(\lambda_{\text{dark},e^-}), \text{ where } \lambda_{\text{dark},e^-} = S_{\text{dark},e^-}. \tag{8.7}$$

In a practical sensor, pixels cannot be manufactured exactly the same, and there will be the variations in the photodetector area that are spatially uncorrelated. Consequently, the average dark signal is not uniform but has a spatial-random and fixed pattern noise structure, which is called dark current fixed pattern noise. The dark current FPN can be modeled as log-normal distribution in the case of short integration time [32].

$$I_{\text{dark,FPN},e^-} = I_{\text{dark,shot},e^-}\left(1 + \ln\mathcal{N}(0, \sigma^2_{\text{dark,FPN},e^-})\right), \tag{8.8}$$

where

$$\sigma_{\text{dark,FPN},e^-} = \xi_{\text{dark,FPN}} \cdot I_{\text{dc}} \cdot t_{\text{int}}, \tag{8.9}$$

and $\xi_{\text{dark,FPN}}$ is the average dark current FPN factor that is typically $0.1, \cdots, 0.4$ for CMOS and CCD sensors.

In most of the commercial CMOS sensors, the source follower noise is significant and should be included in a photosensor model. The source follower noise generated due to the source follower amplifier has a resistance that generates thermal noise, imperfect contacts between two materials at the junction, moreover the random trapping and emission of mobile charge carriers resulting in discrete modulation of the channel current. The variance of source follower noise [e^- rms] can be expressed as

$$\sigma_{\text{SF}} \approx \frac{\sqrt{\sum_{f=1}^{f_{\text{clock}}} S_{\text{SF}}(f) \cdot H_{\text{CDS}}(f)}}{A_{\text{SN}} \cdot A_{\text{SF}}(1 - \exp[-t_s/\tau_{\text{D}}])}, \tag{8.10}$$

where $S_{\text{SF}}(f)$ is the power spectrum of the noise, $H_{\text{CDS}}(f)$ is the CDS transfer function, f_{clock} is the readout clock frequency, A_{SN} is a sense node conversion gain, A_{SF} is a source follower gain, t_s is the CDS sample-to-sampling time, and τ_{D} is the CDS dominant time constant and usually related to t_s as $\tau_{\text{D}} = 0.5t_s$. Using parameters provided in the specifications of the image sensor, we can model this Gaussian distribution source follower noise as

$$I_{\text{SF},e^-} = \text{round}\left[\mathcal{N}(0, \sigma^2_{\text{SF}})\right]. \tag{8.11}$$

From charge to voltage

The process of charge–voltage conversion is performed as follows. The light signal I_{light,e^-} contains photon shot noise and the PRNU

$$I_{\text{light},e^-} = I_{e^-}(1 + \mathcal{N}(0, \sigma^2_{\text{PRNU}})). \tag{8.12}$$

The dark signal I_{dark,e^-} consists of dark current shot noise and dark current FPN. It can be expressed as

$$I_{\text{dark},e^-} = I_{\text{dark,shot},e^-}\left(1 + \ln\mathcal{N}(0, \sigma^2_{\text{dark,FPN},e^-})\right). \tag{8.13}$$

Then total number of electrons I_{total,e^-} is a result of the light signal I_{light,e^-}, dark signal I_{dark,e^-} , and source follower noise I_{SF,e^-} which are summed together and rounded

$$I_{\text{total},e^-} = \text{round}(I_{\text{light},e^-} + I_{\text{dark},e^-} + I_{\text{SF},e^-}). \tag{8.14}$$

After that, the number of electrons I_{total,e^-} is truncated to the full well (the maximum number of electrons in the pixel) and rounded. Then electron–voltage conversion is applied to convert the electrons to voltages by multiplying the sense node conversion gain G_{SN}. Specifically, the charge–voltage conversion uses the sense node gain A_{SN} [V/e$^-$] as a parameter in the range of 1 μV/e$^-$ to 5 μV/e$^-$. The conversion from charge to voltage is performed in the sense node as follows

$$V_{\text{SN,V}} = V_{\text{ref}} - I_{\text{total},e^-} \cdot A_{\text{SN}}, \tag{8.15}$$

where $V_{\text{SN,V}}$ is the sense node voltage, and V_{ref} is the reference voltage.

Prior to measuring each pixel's charge packet, the sense node capacitor is reset to a reference voltage level. Noise is generated at sense node by an uncertainty in the reference voltage level, which is called reset noise and is a significant contributor to dark noise. The variance of reset noise can be expressed as

$$\sigma_{\text{reset}} = \sqrt{\frac{k_B T}{C_{\text{SN}}}}, \tag{8.16}$$

where C_{SN} is the sense node capacitance [F]. Moreover, the reset noise may be performed as an addition to non-symmetric distribution to the reference voltage V_{ref}, and its distribution depends on sensor's architecture and the reset technique. Here, log-normal distribution is used to model the reset noise for soft-reset techniques, as

$$V_{\text{SN,reset,V}} = \ln\mathcal{N}(0, \sigma^2_{\text{reset}}). \tag{8.17}$$

Noteworthily, when the reset noise exists, the reference voltage in (8.15) should be modified as

$$V_{\text{ref,new}} = V_{\text{ref}} + V_{\text{SN,reset,V}}. \tag{8.18}$$

After that, the sense node voltage is multiplied by the source follower gain A_{SF} [V/V] as

$$V_{SF,V} = V_{SN,V} \cdot A_{SF}. \tag{8.19}$$

In particular, for a CMOS image sensor, pixels in the same column of the photosensor share a column amplifier. Differences in the gain and offset of these column amplifiers contribute to a column-wise offset fixed pattern noise. The offset FPN appears as "stripes" in the received image, which will significantly degrade the system performance, but can be suppressed by the noise reduction circuits.

In nowadays image sensors, the noise reduction circuits such as CDS circuits are employed to eliminate or reduce the FPN and reset noise, as illustrated in Fig. 8.12. The CDS circuits consist of two sample-and-hold circuits. During the pixel read-out cycle, two samples are taken: the first when the pixel is in reset state and the second when the charge has been transferred to the read-out node [26]. During the reset stage, the photodiode capacitance is charged to a reset voltage. The reset voltage is read by the first sample-and-hold in a CDS circuit. Then the exposure begins: the photodiode capacitor is discharged during an integration time at a rate proportional to the incident illumination. This voltage is then read by the second sample-and-hold of the CDS. The CDS circuit subtracts the signal pixel value from the reset value. Although the dark current FPN and reset noise can be removed by CDS in CCD sensors, it is difficult to remove them in CMOS image sensors even after application of CDS. Moreover, CDS can suppress offset FPN and reset noise but increases the read noise power.

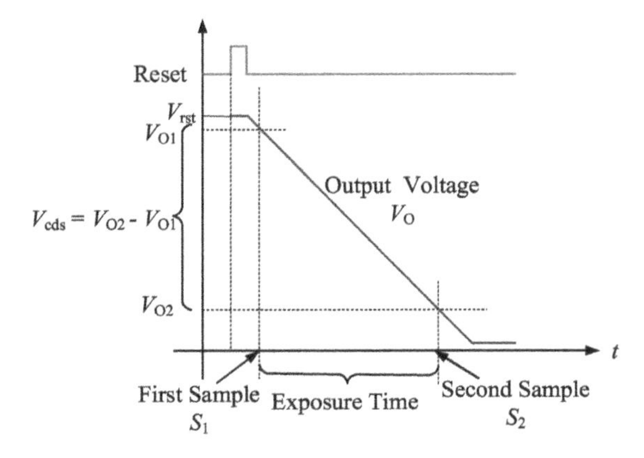

Figure 8.12 System diagram of correlated double sampling.

From voltage to digital numbers

ADC is employed to transform the voltage signal into discrete codes and the accuracy of output gray value corresponding to the voltage signal is determined by the resolution of ADC, which indicates the number of discrete values that can be produced over the range of analogue value as

$$K_{\mathrm{ADC}} = \frac{V_{\mathrm{ADC,ref}} - V_{\min}}{N_{\max}}, \tag{8.20}$$

where $V_{\mathrm{ADC,ref}}$ is the maximum voltage, $V_{\min}$ is the minimum quantifiable voltage, and $N_{\max} = 2^N$ is the number of voltage intervals. Therefore, the output of an ADC can be expressed as

$$ADC_{\mathrm{code}} = \mathrm{round}\left(\frac{V_{\mathrm{SF,V}} - V_{\min}}{K_{\mathrm{ADC}}}\right), \tag{8.21}$$

where $V_{\mathrm{SF,V}}$ is the total voltage signal accumulated by the end of the integration time and conversion, as in (8.19).

Due to the finite precision in an ADC, the quantization error is inevitable. The probability distribution of quantization noise is generally assumed to be uniform. Denote q_{ADC} as the quantizing step of the ADC, the variance of quantization noise can be expressed as

$$\sigma^2_{\mathrm{ADC}} = \frac{q^2_{\mathrm{ADC}}}{12}, \tag{8.22}$$

where $q = 2^{-b}$, and $(b+1)$ is the ADC bit.

Communication SNR for OCC

Different from the noise and SNR defined in image processing, in this part we analyze the noise from the communication perspective at a high illumination level. Generally speaking, intensity-modulated direct-detection (IM/DD) is commonly used in OCC due to its practical simplicity. The input signal modulates the optical intensity of the emitted light, which is transmitted to the receiver over a free space link. The input signal is proportional to the light intensity and is nonnegative. The pixel, being basically a power-detecting unit, responds to the instantaneous field count rate process produced from the receiver area. Its output appears as shot noise process whose count rate is proportional to the instantaneous received power. The noise in the received signal is caused by several effects. First, the exact number of arriving photons at the pixel-based receiver during a given integral time is a random process and is modeled by the mentioned Poisson distribution with a rate proportional to the input signal power. Second, the signal is corrupted by the noise from the background radiation. Third, the received signal is impaired by the signal-dependent noise. Noteworthily, other FPNs except PRNU are not damage factors for a fixed single pixel or pixel-group-based communication receiver, because these FPNs are time-invariant and signal-independent. Moreover, the read noise is another source noise in OCC, which is a combination of the remained noise generated between the photodiode and the output of the ADC circuitry.

The received intensity signal before being converted into electrons can be expressed as

$$R_y = hX + R_{\mathrm{back}}, \tag{8.23}$$

where X is the nonnegative input signal and is proportional to the light intensity, h is the channel gain, and R_{back} is the received intensity due to background light radiation. For an ideal pixel-matched case, if we only consider the OCC system suffers from aforementioned types of noise, but ignore the generalized optical or electrical interference, then the channel gain can be assumed as $h = 1$. The converted electrical signal can be expressed as

$$Y = X + Z_{\text{shot}} + Z_{\text{PRNU}} + Z_{\text{read}}, \tag{8.24}$$

where X is the desired input signal, Z_{shot} is the shot noise including the photon shot noise and dark current shot noise, and Z_{PRNU} is the PNUN noise. Z_{read} is the read noise, which contains any noise that is not a function of the signal, and is a combination of the remaining noise generated between the photodiode and the output of the ADC circuitry, consisting of sense node reset noise, source follower noise and ADC quantization noise. Assuming CDS is performed, part of the reset noise can be effectively cancelled. Then, we can quantify the communication SNR at the end of integration as

$$\text{SNR}_{\text{OCC}} = \frac{E\{X^2\}}{\sigma_{\text{shot}}^2 + \sigma_{\text{prnu}}^2 + \sigma_{\text{read}}^2}, \tag{8.25}$$

where the power of the desired signal is given by

$$E\{X^2\} = (I_{\text{ph}} t_{\text{int}})^2. \tag{8.26}$$

The shot noise variance is given by

$$\sigma_{\text{shot}}^2 = q[I_{\text{ph}} + I_{\text{back}} + I_{\text{dc}}]t_{\text{int}}, \tag{8.27}$$

where I_{back} is the current induced by background radiation. Moreover, only the PRNU is considered in pixel-based receiver with the variance of

$$\sigma_{\text{prnu}}^2 = \sigma_{\text{PRNU}}^2[(I_{\text{ph}} t_{\text{int}})^2 + (I_{\text{back}} t_{\text{int}})^2]. \tag{8.28}$$

The read noise, which consists of reset noise, source follower noise, and ADC quantization noise, has the variance of

$$\sigma_{\text{read}}^2 = q^2(\sigma_{\text{reset}}^2 + \sigma_{\text{SF}}^2 + \sigma_{\text{ADC}}^2). \tag{8.29}$$

As mentioned above, SNR is a function of photocurrent, pixel area, integration time, and so on. In addition, one can leverage the spatial diversity to improve SNR by grouping multiple pixels into a block to map one transmitter. Assuming B pixels in the block, the average output gray value for this block is $Y = \frac{1}{B}\sum_{i=1}^{B} Y_i$. By grouping multiple pixels as a block receiver, the noise will be reduced by spatial averaging. Thus, the pixels block receiver can support different SNR transmissions with different block sizes. In the following, we analyze SNR with different parameters. Table 8.4 lists the key specifications of the typical image sensor used in simulations.

Table 8.4 The key image sensor specifications.

Manufacture process	CMOS	PRNU factor	0.6%–1%
Pixel size	3.75 μm × 3.75 μm	Dark current FPN factor	1%
Number of pixels	1080 × 720	Column offset FPN factor	0.1%
Wavelength λ	550 nm	Dark current figure of merit	1.00 nA/cm^2
Fill factor	55%	Sense node gain	5.00 μV/e$^-$
Quantum efficiency	65%	Read nose σ_{read}	30 e$^-$
Full well	60,000 e$^-$	ADC bit	12 bit

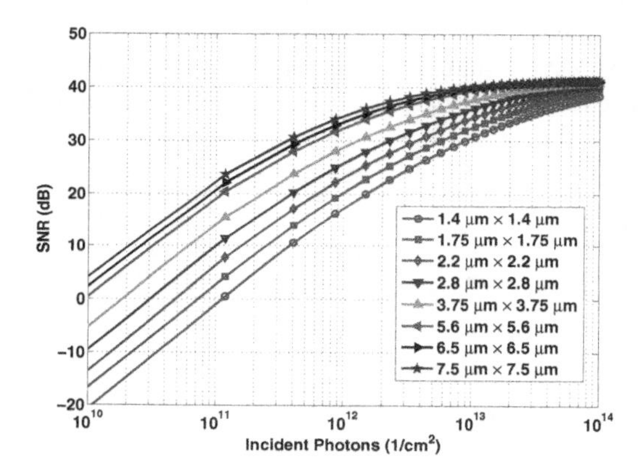

Figure 8.13 SNR in OCC with different pixel size areas.

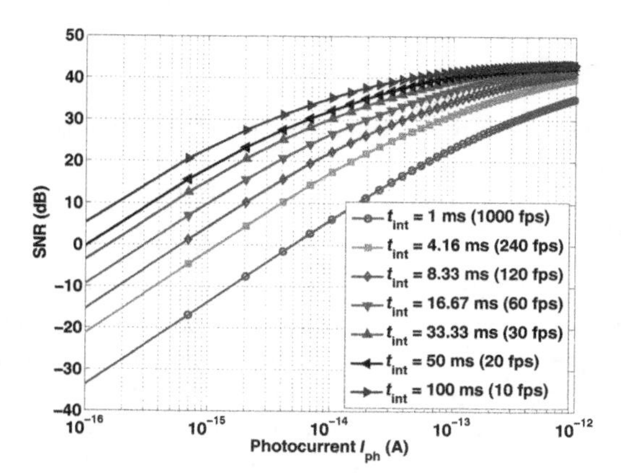

Figure 8.14 SNR in OCC with different integration time.

Figures 8.13–8.14 show the SNR in OCC with different areas of pixel and integration time. We set pixel area $A_p = 3.75\ \mu m \times 3.75\ \mu m$, and integration time $t_{int} =$ 33.3 ms in simulation if they are not specified. It can be seen that SNR increases with the photocurrent I_{ph}. First, SNR increases 20 dB as photocurrent increases 10 dB when readout noise and dark current shot noise dominate for small photocurrent. SNR increases 10 dB as photocurrent increases 10 dB when shot noise dominates. Further increase of photocurrent leads to significant PRNU, and SNR flattens out. The achieved maximum SNR roughly approaches the well capacity before saturation. Moreover, SNR increases with pixel area and integration time. However, even the pixel size is scaled down to $1.4\ \mu m \times 1.4\ \mu m$, SNR remains 17–37 dB when the average incident photon density is about 10^{12}–10^{14}/cm^2 as shown in Fig. 8.13. Most commercial CMOS image sensors provide frame rate exceeding 30 fps. For example, iPhone 7 supports 240 fps, and the frame rate of some special image sensor can be even up to 1000 fps. According to the simulation results shown in Fig. 8.14, SNR can range from 25 dB to 38 dB in OCC systems under a general parameter setting, even the frame rate exceeds 200 Hz.

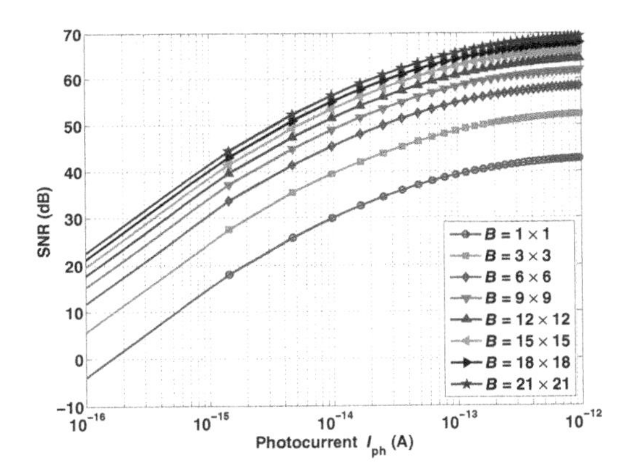

Figure 8.15 SNR in OCC with different block size.

To improve communication SNR in OCC, a potential solution is to leverage the diversity structure by grouping multiple pixels into a block receiver to map the same transmitter, as in RF-MIMO. Figure 8.15 shows SNR with different pixel block sizes. It can be seen that SNR can dramatically increase with the pixel block size or diversity order. When the pixel block size is 3×3, 6×6, or 9×9, SNR in these blocks can be improved by about 9.5 dB, 15.6 dB, and 19 dB compared with 1×1 pixel block when photocurrent is $I_{ph} = 10^{-13}$ A. However, as the pixel block size continues to increase, SNR improvement slows down. It is seen that even the block pixel-based detection with diversity structure can obtain higher SNR and support higher capacity, the effective data throughput for imaging MIMO system with spatial multiplexing is degraded since the number of parallel channels is reduced. Thus, the diversity gain is smaller than the multiplex gain in pixel-matched case for imaging MIMO system.

8.3.4 Channel model for OCC

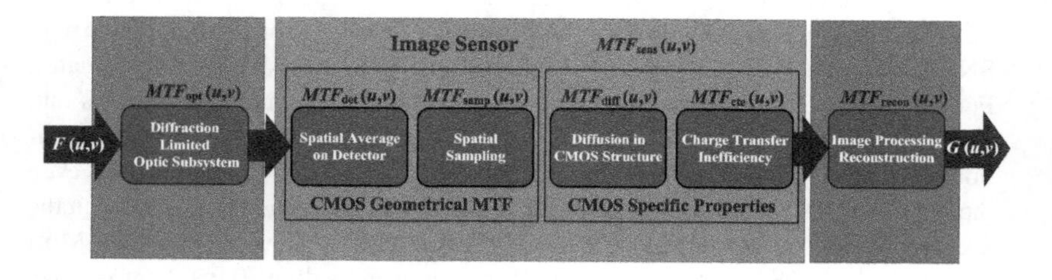

Figure 8.16 The equivalent system model for CMOS image-sensor-based OCC.

The system diagram for image-sensor-based OCC can be modeled as the space and linear shift-invariant system (LSI) [34]. As shown in Fig. 8.16, $f(x, y)$ is the intensity distribution for light sources in aperture plane of the optical lens group. $MTF_{\text{opt}}(u, v)$ is the modulation transfer function for the diffraction limited optical subsystem, $MTF_{\text{sensor}}(u, v)$ is the MTF for the CMOS image sensor, which is determined by the geometrical shape of the active area $MTF_{\text{det}}(u, v)$ and pixel sampling $MTF_{\text{samp}}(u, v)$, charge diffusion $MTF_{\text{diff}}(u, v)$, and charge collection efficiency $MTF_{\text{cte}}(u, v)$. $MTF_{\text{recon}}(u, v)$ is the MTF for the imaging processing subsystem such as interpolation, color transformations, and is usually assumed to be $MTF_{\text{recon}}(u, v) = 1$ to simplify the analysis. The equivalent model for the whole CMOS image-sensor-based receiver can be expressed as

$$\begin{aligned} MTF_{\text{sys}}(u, v) &= MTF_{\text{opt}}(u, v) \cdot MTF_{\text{det}}(u, v) \cdot \\ &\quad MTF_{\text{samp}}(u, v) \cdot MTF_{\text{diff}}(u, v) \cdot MTF_{\text{cte}}(u, v). \end{aligned} \tag{8.30}$$

In line-of-sight (LOS) OCC communication links, pointing accuracy is an important factor for link performance and reliability. However, the relative motions between transceiver result in random optical beam sways, which in turn cause pointing error and signal fading at the receiver. In the following, we will discuss the factors which contribute to the pointing error or pixel-mismatch, including the optical interference induced by the diffraction limited optical subsystem, electrical interference due to carrier diffusion, spatial nonideal condition such as linear misalignment or geometry perspective distortion, and the jitter variance [37]. However, the unified statistical channel model for OCC with pointing errors is still unavailable. Moreover, since the light sources are diverse, different light sources have different radiation patterns, such as Lambertian radiation, batwing-type radiation, and side-emitting radiation. For the widely used LEDs, to make a general and an accurate radiation pattern, the emitting surfaces (chip, chip array, or phosphor surface), the light redirected by both the reflecting cup and the encapsulating lens must be taken into account [38].

Diffraction limited optical subsystem (on-axis image irradiance)

Imperfect focusing or blur effect will degrade the OCC performance. The blur effect is attributed to the camera lens and more formally termed as lens-blur. Lens-blur causes the received light energy to spread to areas outside the pixel, where the spread range depends on the type of lens. Lens-blur can be viewed as a low-pass filter that suppresses the high-frequency components in the image, such as edges and high-contrast regions. To simplify the analysis, we assume that ideal lens is used in the image sensor and the radial distortion and eccentricity distortion need not be taken into account. Actually, a camera lens usually consists of 15 or more optical elements in sequence, which have different thicknesses and focal lengths and are unevenly spaced along the optical axis. The combination of these elements is typically chosen to make fabrication more cost-effective, the size of the optical system smaller, and some of the aberrations mitigated. However, there still lacks an optical model which consists of a variable number of thin lens with different distances from one another and different focal lengths. Fortunately, the equivalent single thick-lens model is nearly identical to the ideal thin-lens model, with a non-negligible lens thickness D_{lens} and a focal length as

$$\frac{1}{f_{\mathrm{c}}} = \frac{1}{D_{\text{foc-img}}} + \frac{1}{D_{\mathrm{obj}} - D_{\mathrm{lens}}}. \tag{8.31}$$

For example, if the aperture lens area is assumed to be circular with diameter d, the PSF for the equivalent thick-lens system can be calculated as [33]

$$f_{\mathrm{opt}} = \frac{2\pi}{\lambda f_{\mathrm{c}}} \int_0^{d/2} r J_0(\frac{\pi r \rho}{\lambda f_{\mathrm{c}}}) dr = (\frac{\pi d^2}{4\lambda f_{\mathrm{c}}})(\frac{2J_1(\pi d\rho/\lambda f_{\mathrm{c}})}{\pi d\rho/\lambda f_{\mathrm{c}}}), \tag{8.32}$$

where $\rho = \sqrt{x^2 + y^2}$ and $J_0(x)$ and $J_1(x)$ are Bessel functions.

Carrier diffusion induced crosstalk in CMOS image sensors

The optical beam or optical distribution is also affected by the carrier diffusion or electrical crosstalk. In [39], a unified MTF model for CMOS image sensor is proposed. The sensor is built on the epitaxial layer, which is deposited on a substrate layer and presents a doping gradient. As shown in Fig. 8.18, the charge diffusion and sampling aperture are taken into account in the unified model as

$$MTF_{\text{sensor}}(u, v) = MTF_{\text{diff}}(u, v) \cdot MTF_{\text{geom}}(u, v), \tag{8.33}$$

where MTF_{geom} is determined by the pixel sampling active area and geometrical shape, MTF_{diff} is derived by solving the steady-state diffusion equation based on the Fourier transform [39]. The MTF model due to sampling aperture geometry $MTF_{\text{geom}}(u, v) = MTF_{\text{det}}(u, v) \cdot MTF_{\text{samp}}(u, v)$ for a square shape pixel is given by

$$MTF_{\text{geom}}(u, v) = \text{sinc}(a\pi u) \cdot \text{sinc}(b\pi v), \tag{8.34}$$

where a, b are the sensitive widths of APS along the x direction and y direction. Then, the PSF in the spatial domain is the 2-D spatial inversion Fourier transform

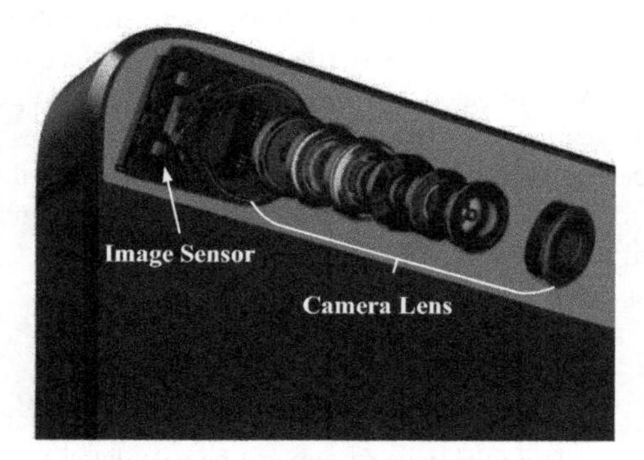

(a) The elements of an example camera lens

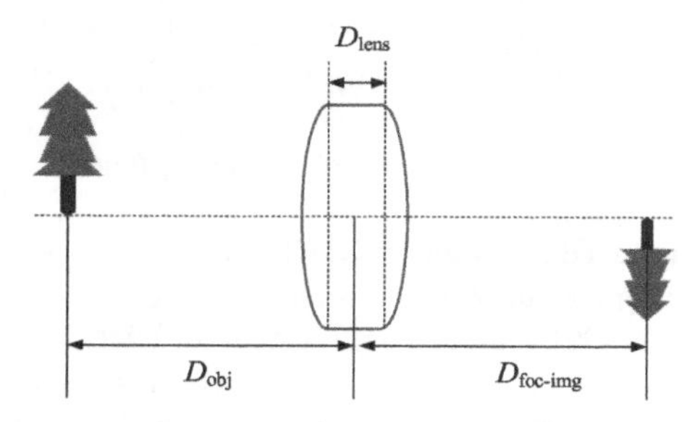

(b) The equivalent ideal thick-lens model

Figure 8.17 Diffraction limited optical subsystem.

of the MTF. Without loss of generality, $f_{\text{sensor}}(x, y)$ can be modeled as a Gaussian filter [11] as

$$f_{\text{sensor}}(x, y) = \frac{1}{2\pi\sigma_{\text{sensor}}^2} e^{-\frac{x^2+y^2}{2\sigma_{\text{sensor}}^2}}, \tag{8.35}$$

where σ_{sensor}^2 is determined by sensitive width, substrate, charge diffusion, and CMOS structure.

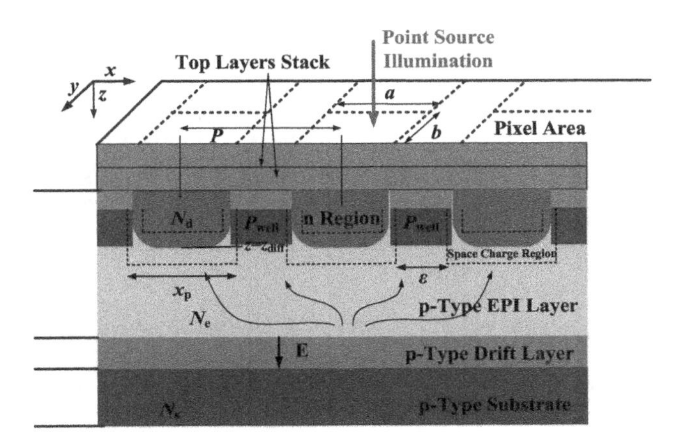

Figure 8.18 Geometrical model of CMOS image sensor by Ibrahima Djite [39].

Pointing error model

Taking both source radiation pattern, diffraction optical subsystem and charge diffusion into account, the spatial distribution of the optical intensity in the pixel-based receiver can be expressed as

$$I_{\text{beam}}(\boldsymbol{\zeta}) = I_{\text{source}}(\theta) * f_{\text{opt}} * f_{\text{sensor}}, \tag{8.36}$$

where $*$ denotes 2-D convolution, and $I_{\text{source}}(\theta)$ denotes the source radiation pattern with irradiance angle θ. Considering that most commercial image sensors adopt square detection, with a dimension of (c, d) as shown in Fig. 8.19(a), the attenuation of geometric spread with pointing vector $\boldsymbol{r}$ is expressed as

$$h_{\text{p}}(\boldsymbol{r}) = \int_{\mathcal{A}} I_{\text{beam}}(\boldsymbol{\zeta} - \boldsymbol{r}) d\boldsymbol{\zeta}, \tag{8.37}$$

where $h_{\text{p}}(\cdot)$ represents the fraction of the power collected by the detector and $\mathcal{A}$ is the detector area. When a pointing error of $\boldsymbol{r}$ is present, h_{p} is a function of the radial displacement and the angle [40, 41]. Assuming the radial displacement vector at the receiver aperture plane as $\boldsymbol{r} = [r_x \, r_y]^{\text{T}}$, where r_x and r_y denote the displacements along the horizontal and elevation axes at the detector plane, respectively. In general, the r_x and r_y, both follow a nonzero mean Gaussian distribution, i.e., $r_x \sim \mathcal{N}(\mu_x, \sigma_x^2)$, $r_y \sim \mathcal{N}(\mu_y, \sigma_y^2)$, then the radial displacement $r = \|\boldsymbol{r}\| = \sqrt{r_x^2 + r_y^2}$ follows the Beckmann distribution [41]:

$$f_r(r) = \frac{r}{2\pi\sigma_x\sigma_y} \times \int_0^{2\pi} \exp\left(- \frac{(r\cos\phi - \mu_x)^2}{2\sigma_x^2} - \frac{(r\sin\phi - \mu_y)^2}{2\sigma_y^2} \right) d\phi. \tag{8.38}$$

Note that, the Beckmann distribution is a versatile model to describe the point error, leading to a variety of models. For example, when $\mu_x = \mu_y = 0$, $\sigma_x = \sigma_y$, the Beckmann distribution is equivalent as Rayleigh distribution. When $\mu_x = \mu_y = 0$, $\sigma_x \neq \sigma_y$, then the Beckmann distbution is equivalent as Hoyt distribution.

Combining (8.37) and (8.38), the channel fading due to beam spread and point error can be expressed as

$$h_{\mathrm{p}} = \int_{-c/2}^{c/2} \int_{-d/2}^{d/2} \int_{r>0} I_{\mathrm{beam}}(x, y, r) f_r(r) dr dy dx. \tag{8.39}$$

As for the imaging optical MIMO case with a large single-aperture pixel-based receiver, the PDF of pointing error is similar to the single-input single-output (SISO) case. However, the difference is that the symmetrical light array should be assumed as an equivalent beam shape with uniform distribution. Figure 8.19 demonstrates the detector and beam footprint for transmitter and pixel-based detector plane for SISO and MIMO cases, respectively.

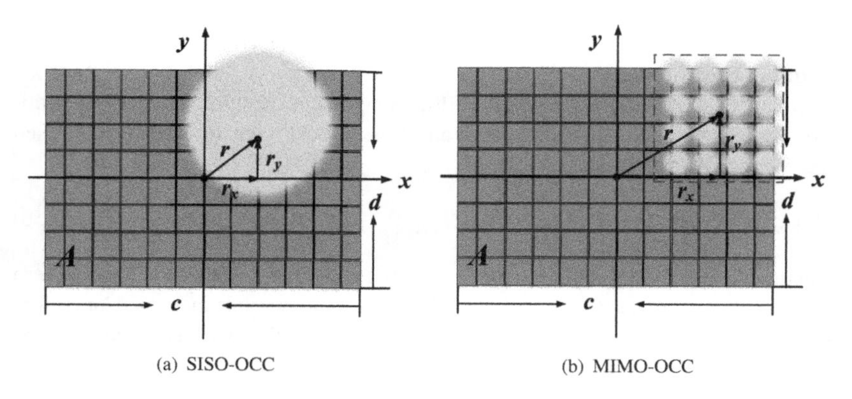

Figure 8.19 Detector and beam footprint for transmitters with misalignment and inter-pixel-interference in the pixel-based detector plane.

Channel statistical model with generalized point error

The probability distribution of channel state h can be expressed as

$$f_{\mathrm{h}}(h) = \int_{r>0} \frac{A_r I(\theta)}{d^2} \cdot T_s(\psi) \cdot g(D, f_c, r, \psi) \cdot \cos\psi \cdot f_r(r, I, D, f_c) dr,$$

$$(8.40)$$

where d denotes the distance between LED source and focal plane, $A_r\cos\psi$ is the effective collection area of the detector, ψ is the incident angle and smaller than the detector FOV Ψ_c, $T_s(\cdot)$ is the transmission of optical lens, $g(D, f_c, r)$ is the concentrator of gain with lens thickness D and focal length f_c, and $f_r(r, I, D, f_c)$ denotes the channel fading due to pointing error.

8.4 Capacity bounds for OCC

8.4.1 SISO-OCC channel capacity with M-SDGN

In the pixel-matched case, we only consider the mentioned front-end noise sources, but ignore the generalized optical and electrical interference and the alignment error. Then, the converted electrical signal in (8.24) from a pixel readout circuit can be rewritten as

$$Y = X + XZ_2 + \sqrt{X}Z_1 + Z_0. \tag{8.41}$$

Here, $X = I_{\text{ph}}t_{\text{int}}$ is the desired electrical signal and is proportional to the channel input. $Z_0 \sim \mathcal{N}_{\mathbb{R}}(0, \sigma^2)$ is the signal-independent Gaussian noise, including shot noise contributed by background light and dark current, background light induced PRNU, and read noise. $Z_1 \sim \mathcal{N}_{\mathbb{R}}(0, \varsigma_1^2\sigma^2)$ and $Z_2 \sim \mathcal{N}_{\mathbb{R}}(0, \varsigma_2^2\sigma^2)$ are both the signal-dependent noise, which is contributed by input signal induced shot noise and PRNU, respectively. The parameter

$$\begin{aligned}\sigma^2 =& q(I_{\text{dc}} + I_{\text{back}})t_{\text{int}} + \sigma_{\text{PRNU}}^2(I_{\text{back}}t_{\text{int}})^2 \\ &+ q^2(\sigma_{\text{reset}}^2 + \sigma_{\text{SF}}^2 + \sigma_{\text{ADC}}^2)\end{aligned} \tag{8.42}$$

describes the strength of the signal-independent noise, where I_{back} is the current induced by background radiation. Parameters $\varsigma_2^2, \varsigma_1^2$ are the ratio of the input-dependent noise variance $\varsigma_2^2\sigma^2, \varsigma_1^2\sigma^2$ to the input-independent noise variance σ^2, respectively. Here, Z_2, Z_1 are assumed to be independent of Z_0.

The conditional PDF of this mixed-signal-dependent Gaussian noise (M-SDGN) channel is given by

$$f_{Y|X}(y|x) = \frac{1}{\sqrt{2\pi\sigma^2(1 + \varsigma_2^2x^2 + \varsigma_1^2x)}}e^{-\frac{(y-x)^2}{2\sigma^2(1+\varsigma_2^2x^2+\varsigma_1^2x)}}, \quad x \in \mathbb{R}^+, y \in \mathbb{R}. \tag{8.43}$$

Further, due to safety and practical consideration, the input signal has a peak intensity (amplitude) and nonnegative constraint as

$$P_r[0 \leq X \leq A] = 1, \tag{8.44}$$

and an average intensity constraint as

$$E[X] \leq P. \tag{8.45}$$

We use ρ to denote the average-to-peak-power ratio (APPR) as

$$\rho \triangleq \frac{P}{A}. \tag{8.46}$$

Note that the input signal is proportional to the light intensity. Thus the power constraint is imposed on the signal itself, instead of its square as in RF communications.

The capacity bound can better reflect the physical channel properties and ultimate communication performance. In the following, we will focus on this Gaussian channel with M-SDGN, and present the capacity with bounded inputs and intensity constraints. The channel model (8.41) is a special case of the general signal-dependent Gaussian noise (SDGN) channel model. Unfortunately, the channel capacity of this M-SDGN channel is unknown, and it is difficult to apply the sphere-packing method or the dual expression approach. However, following similar arguments as in [42–46], it has been proven that the mutual information function is a concave, continuous, and weakly differentiable function over a compact and convex space of input distribution [4]. Thus, invoking the Karush–Kuhn–Tucker (KKT) Theorem results in sufficient and necessary conditions for the capacity-achieving input distribution. There is a unique optimal input distribution F_0, which achieves maximum mutual information. Finally, it can be proven that the capacity-achieving distribution for this M-SDGN channel is discrete with finite number of mass points from the complex analysis. The result is given in the following theorem.

Theorem 1. *$\mathcal{C}$ is achieved by a random variable, denoted by X_0 with probability distribution function $F_0 \in \mathcal{F}_X$, i.e.,*

$$\mathcal{C} = \max_{F_X \in \mathcal{F}_X} I(F_X) = I(F_0), \tag{8.47}$$

for some $F_0 \in \mathcal{F}_X$. A sufficient and necessary condition for F_0 to achieve capacity is

$$i_{F_0}(x) \leq I(F_0), \quad \forall x \in [0, A]. \tag{8.48}$$

Furthermore, this distribution is discrete and consists of finite number of mass points if some conditions on $(1 + \varsigma_2^2 x^2 + \varsigma_1^2 x)$ hold.

With the aforementioned theorem showing a finite number of mass points for finite A and P, we can use the search algorithm to find the optimal input distribution and the corresponding maximum mutual information for such constrained channel.

8.4.2 Capacity-achieving probability measurement with M-SDGN

The capacity-achieving probability distribution for the M-SDGN channel as in (8.41) subject to optical intensity constraints is discrete and nonuniform. It can be expressed

as

$$f_X(x) = \sum_{n=1}^{N} p_n \delta(x - x_n), \tag{8.49}$$

where N is the modulation order, p_n is the probability mass point with $\sum_{n=1}^{N} p_n = 1$, and $x_n \in [0, A]$, $(x_n < x_{n+1})$ is the intensity/amplitude mass point.

The capacity-achieving distribution for OCC channels under amplitude constraints is discrete and nonuniform. This distribution can be computed numerically by solving a complex non-linear optimization problem. In this problem, the mutual information is maximized over the input distributions with all constraints fulfilled. The amplitude mass point x_n, probability mass point p_n, and modulation order N are free parameters in the optimization problem. Efficient numerical optimization techniques can be applied to solve this problem and provide numerical solutions for the input distribution and channel capacity at different SNRs and APPR. In 2005, Chan *et al.* derived a necessary and sufficient condition for capacity-achieving probability measure [45]. Using this necessary and sufficient condition, they proposed an algorithm to find the capacity-achieving measure of a signal-dependent optical channel, which is traditionally difficult to analyze.

Theorem 2. (Necessity and Sufficiency) : *Let F_0 be an admissible input probability measure, i.e., $F_0 \in \mathcal{F}_X$. Let $\mathcal{H}(x) = H_{F_0}(Y|X) = \frac{1}{2}log(2\pi e\sigma^2) + \frac{1}{2}E\left[log(1 + 2\varsigma_2^2 x^2 + \varsigma_1^2 x)\right]$. Then, F_0 is capacity achieving if and only if there exists $\lambda \geq 0$ such that for all $x \in [0, A]$*

$$Q(x; F_0) - \mathcal{H}(x) - I_{F_0}(X;Y) - \lambda(x - P) \leq 0, \tag{8.50}$$

where $Q(x, F_0) = -\int_{\infty}^{\infty} f_{Y|X}(y|x) log f_Y(y; F_0) dy$.

Proposition 1. *Suppose F_0 is the capacity-achieving measure for the optical channel with M-SDGN with amplitude constraints as in (8.41). Then $x = 0$ is a point of increase of F_0.*

Corollary 1. *If F_0 is capacity achieving for channel (8.41) with amplitude constraints, and it satisfies (8.50), then*

$$\lambda = \left[I_{F_0}(X;Y) - Q(0; F_0) + \frac{1}{2}log(2\pi e\sigma^2)\right]/P,$$

where P is the average optical power. For each $n \in \{2, 3, \cdots\}$, let $\tau^{(n)}$ be an input probability measure in $\mathcal{F}_X$ that maximizes $I_{F_X}(X;Y)$ and has n or fewer points of increase.

Above theorem ensures a finite number of mass points of capacity-achieving input distribution. Using an approach similar to [45], the capacity-achieving measure of this channel can be found via the following search algorithm.

Search algorithm for capacity-achieving measures [45]

- Step 1: Set $n = 2$;
- Step 2: Solve for $\tau^{(n)}$;
- Step 3: Let $\lambda^{(n)} = \left[I(\tau^{(n)}) - Q(0;\tau^{(n)}) + \frac{1}{2}\log(2\pi e\sigma^2)\right]/P$. If $\lambda^{(n)} < 0$, increase n by 1 and go back to step 2;
- Step 4: Verify whether the inequality $Q(x;\tau^{(n)}) - I(\tau^{(n)}) - \mathcal{H}(x) - \lambda^{(n)}(x - P) \leq 0$ holds for all $x \in [0, A]$. If so, then $\tau^{(n)}$ is capacity achieving. Otherwise, increase n by 1 and go back to step 2.

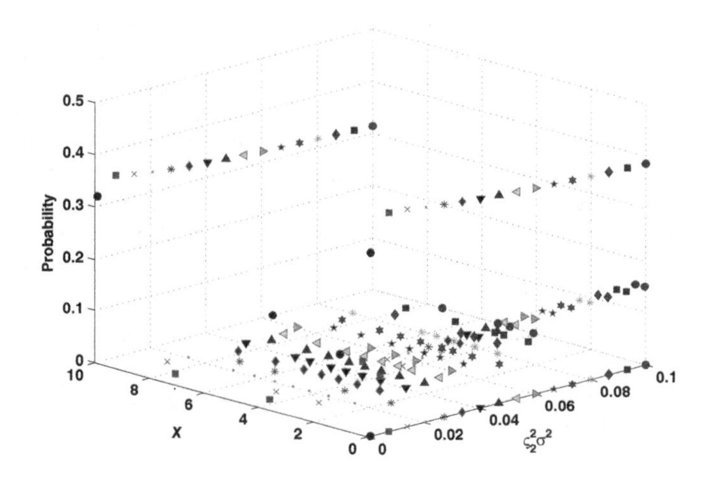

Figure 8.20 Capacity-achieving distribution for signal-dependent noise channel ($P = 4,\ A = 10,\ \varsigma_1^2\sigma^2 = 100\varsigma_2^2\sigma^2$).

Figure 8.20 shows the computed capacity of the capacity-achieving distribution for the M-SDGN channel using this algorithm, where x-axis denotes the values of $\varsigma_2^2\sigma^2$, y-axis denotes the input signal, and z-axis denotes the probability. If a point is indicated at the position $(\varsigma_2^2\sigma^2, x^*)$, then x^* is an amplitude mass point and the probability value is the probability mass point for optimal distribution. The simulation results demonstrate that $x = 0$ and $x = A$ are always the amplitude mass points in optimal input distribution. It has been proven in Proposition 1 that $x = 0$ is always an amplitude mass point. However, it is not clear whether $x = A$ is also an amplitude mass point. Moreover, the distance between two neighboring amplitude mass points varies significantly, which means that the optimal input distribution is nonuniform distribution.

Based on the redefined SNR and channel model, the capacity bounds for this M-SDGN channel can be derived using the aforementioned search algorithm to find the capacity-achieving input distribution and the corresponding channel capacity. Figure 8.21(a) shows the capacity results for single-pixel receiver-based OCC with peak and average-intensity constraints. It is observed that the capacity increases as the

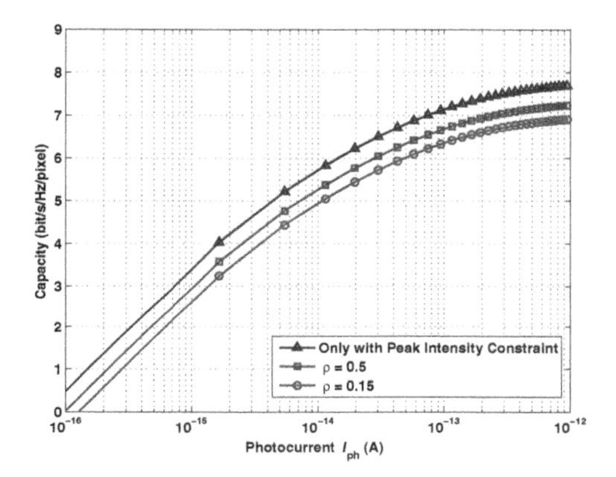

(a) Capacity bounds for OCC with optical intensity constraints under an ideal channel

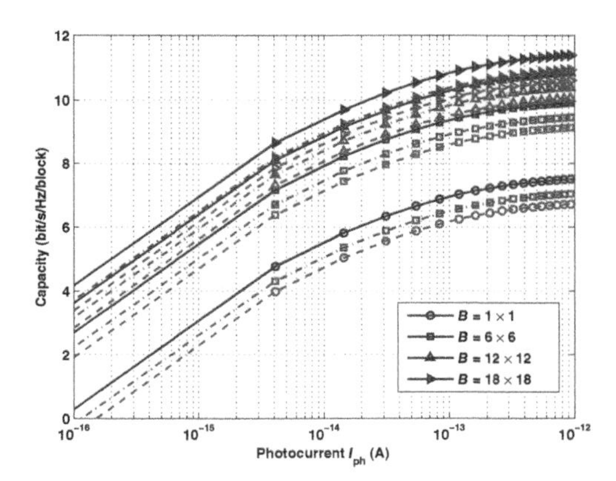

(b) The capacity bounds for OCC with different pixel block size under an ideal channel. Blue color-solid line (only peak intensity constraint), green color-dash dot line ($\rho = 0.5$), fuchsia color-dash line ($\rho = 0.15$)

Figure 8.21 Capacity bounds for OCC.

photocurrent gets larger. Moreover, the capacity bound increases as the APPR ρ becomes larger, especially when $\rho = 1$ which corresponds to the case with peak intensity constraint only. The channel can support 7.1 bit/s/Hz per pixel when photocurrent is $I_{\text{ph}} = 10^{-13}$ A, corresponding to SNR of 37 dB as in Fig. 8.14. Moreover, Fig. 8.21(b) shows the capacity bounds for a block-pixel receiver with different sizes under peak and average intensity constraints. It demonstrates that the channel capacity can be further improved by grouping multiple pixels into a block, and increases as the pixel block size increases. About 10.8 bit/s/Hz per block is possible when block size is $B = 18 \times 18$ and photocurrent is $I_{\text{ph}} = 10^{-13}$ A with general parameter settings. As the pixel block size increases, the capacity improvement slows down.

8.4.3
Capacity of imaging optical MIMO systems with bounded inputs

Generally, the indoor VLC systems are vulnerable to obstacles (shadowing). If directionality is offset, the system performance may dramatically degrade. Moreover, the high-data-rate parallel optical interconnects with low bit error rate (BER) typically require precise alignment. The optical MIMO technique allows the alignment to be achieved in electrical domain as it is not necessary for light from a source to strike a signal detector precisely. Therefore, the motivation for using optical MIMO in VLC is not only for capacity increase, but to alleviate the difficulty in achieving alignment physically by using electrical signal processing. However, it is difficult to apply the spatial-multiplexing MIMO scheme in an indoor VLC intensity channel with a dominant LOS link. Otherwise, a highly correlated channel matrix prevents from decoding the received signals in parallel at the receiver [48]. Optical MIMO can help to achieve high data rate by utilizing spatial diversity, but channel correlation is a major concern, and is sensitive to transmitter spacing and receiver position. Intuitively, ill-conditioned channel matrix may occur due to homogenous behaviors with different channel DC gains, which implies that the channel matrix is not full-rank or the condition number is very large.

As shown in Fig. 8.22, in order to reduce the correlation between the sub-channels in the channel matrix, various advanced receiver structures have been proposed. A simple way to decorrelate is to apply channel matrix pseudo-inversion at the receiver no matter whether the channel matrix is rank-deficient or not [8]. However, the problem is that these methods might result in noise amplification if some singular values of the channel matrix are very small. In [49], the non-imaging link-blocked receiver (LBR), spatially separated receiver (SSR) and the power imbalance between transmitters were proposed to reduce the channel correlation. However, the challenges are that the link-block is difficult to implement in practice by adaptively reflecting the change of blocking area, due to the user mobility and location changement, the larger receiver size, and limited performance improvement for SSR receiver. Recently, the mirror diversity receiver (MDR) was proposed to block the reception of light in one specific direction, and improve the channel gain in other directions by receiving the light reflected from a mirror between the PDs [48]. The problem is that the complex receiver structure limits its practical application. More recently, another advanced receiver, i.e., the angle diversity receiver (ADR) has been proposed to vary the orientation angles of PDs, so that the incident light from the specific direction can not reach the receiver plane or is directed out of the receiver FOV [50, 51]. However, these angle diversity receivers are bulky and impractical to incorporate hand-held devices. For imaging receivers, optical lens can help distinguish the light source images and reduce the channel correlation. This makes the imaging receiver to be one of the most efficient ways to implement optical MIMO. In [52, 53], the hemispherical lens and fisheye lens-based imaging receivers for VLC MIMO were proposed. They project the optical intensity signals onto the receiver PD array, yielding partial and complete separated light images. Consequently, a well-conditioned channel matrix and significant spatial diversity were achieved.

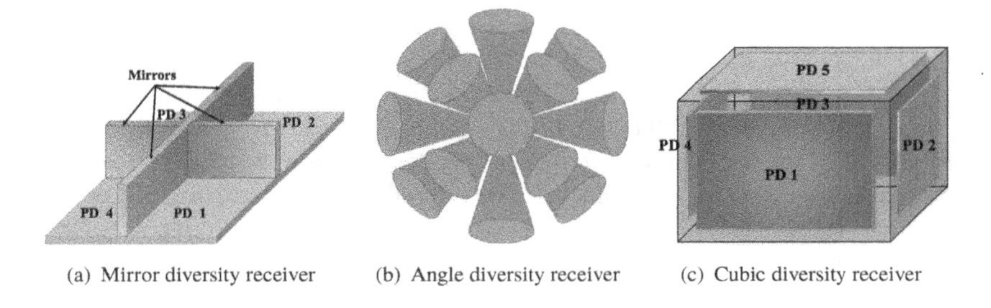

(a) Mirror diversity receiver (b) Angle diversity receiver (c) Cubic diversity receiver

Figure 8.22 Non-imaging MIMO receivers.

An important performance metric is the channel capacity, which specifies the highest rate of reliable information transmission over the imaging optical MIMO channel. In [11], Hranilovic *et al.* discussed the basic structure of channel model and capacity for the pixelated-MIMO. Exactly, the noise was modeled as SDGN and the channel was modeled as a 2-D Gaussian filter from prototype experimental measurement and data fitting. As for the channel capacity, they estimated the channel capacity by sphere-packing argument [54, 55] and water-pouring. However, the capacity contains only the on-axis case and does not take the optical intensity constraint into account. Ashok *et al.* predicted the OCC capacity under lens diffraction and perspective distortion [56]. However, the OCC channel was simply modeled as an additive input-independent Gaussian noise channel and interference was treated as noise. Thus the presented capacity is not very accurate. In [57], the upper and lower bounds on the capacity of constant parallel OWC channel with a total average intensity constraint were derived. Even though the parallel MIMO system was considered, the crosstalk between signals from different transmitters was neglected. In fact, electrical crosstalk and optical crosstalk are inevitable between pixels due to charge diffusion in APS [39], and diffraction limited optical subsystem in an image sensor. In [9], a modified SVD method was proposed. They applied coordinate system transformations on correlated channels to generate simultaneous independent links and maximize the capacity of the imaging optical MIMO channel while maintaining the target illumination. Moreover, the upper bound on capacity of the imaging SVD-VLC MIMO system was presented. These studies are still far more complete, and on-going research is necessary.

In the following, we consider the capacity of imaging optical MIMO channel with bounded inputs and total average intensity constraint. Unfortunately, extending the results of Smith in [43] to vector random variables is unattainable since the Identity Theorem cannot be directly applied. The theorem has shown strength for one-dimensional functions, but inefficiency in holomorphic function of several complex variables in a higher dimension space. In [58], the upper and lower bounds on the capacity of MIMO system with amplitude-limited inputs were derived by considering an equivalent channel via SVD, and by enlarging and reducing the corresponding feasible region of the channel input vector. Moreover, it demonstrates that the capacity-achieving distribution of an input-bounded vector Gaussian channel remains to have a finite number of discrete amplitudes [59, 61]. Following the arguments above, we first transform the coupled imaging optical MIMO channel into independent parallel channels, and then derive the capacity bound achieved using an exponential distribution or discrete input distribution. If CSI is available at the transmitter, the bounds have to be optimized with respect to intensity allocation over the parallel channels.

Note that, the availability of channel-state-information-at-the-transmitter (CSIT) is not a strong assumption in OCC, whose coherence time is typically much larger than the symbol duration. Thus, estimation and feedback of the CSI can be achieved in negligible time without considerably affecting performance. Furthermore, in a full-duplex system, CSI can be estimated directly at the transmitter if channel reciprocity applies [57].

8.4.3.1 Imaging optical MIMO system

The schematic diagram of the imaging optical MIMO system is shown in Fig. 8.23. In this system, M LED arrays provide indoor illumination and transmit signals using spatial multiplexing. An imaging receiver with well-designed lens is used to receive the optical signal. Light propagates from each transmitter array to the receiver as before, and each LED array is projected onto a detector array, where images may strike any pixel or group of pixels on the array, and be in arbitrary alignment with them. Each pixel on the detector array is a receiving element. We use matrix $\mathbf{H}$ to describe the optical connection between each pixel or pixel block and each transmitter LED array. With a well-designed lens, the imaging receiver can clearly separate the signals from different LEDs, achieve omnidirectional receiving and provide high spatial diversity for decoding of the MIMO signals. The received vector signal is given by

$$\boldsymbol{Y} = \boldsymbol{H}\boldsymbol{X} + \boldsymbol{Z}, \tag{8.51}$$

where $\boldsymbol{X} = [X_1 \ X_2 \ \cdots \ X_M]$ is the input signal vector whose element should satisfy the non-negative constraint and total average intensity constraint $||\boldsymbol{P}||_1 \leq P$, where $P_i = E(X_i), i \in \{1, 2, \cdots, M\}$. $\boldsymbol{H}$ is an $N \times M$ channel matrix whose entry $h_{n,m} \geq 0$ represents the channel gain from mth transmitter to nth receiver, $\boldsymbol{Y}$ is the received signal vector, and $\boldsymbol{Z}$ is the Gaussian noise vector of independent

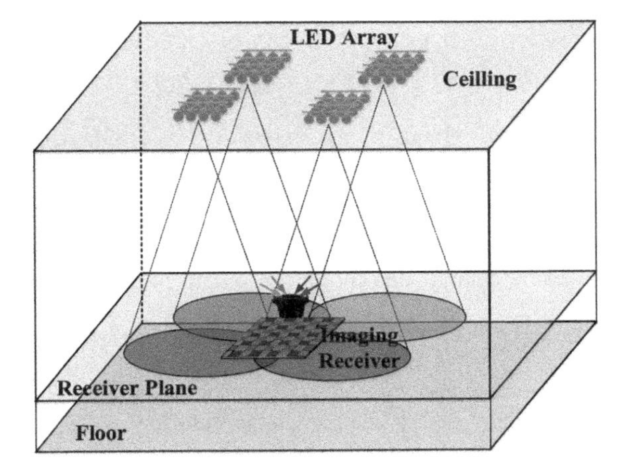

Figure 8.23 Diagram of the imaging optical MIMO system.

components for simplicity. The imaging optical MIMO channel $\boldsymbol{H}$ can be expressed as

$$\boldsymbol{H} = \begin{bmatrix} h_{1,1} & h_{1,2} & \cdots & h_{1,M} \\ h_{2,1} & h_{2,2} & \cdots & h_{2,M} \\ \vdots & \vdots & \ddots & \vdots \\ h_{N,1} & h_{N,2} & \cdots & h_{N,M} \end{bmatrix}. \tag{8.52}$$

8.4.3.2 Capacity of imaging optical MIMO systems

Consider bounded inputs and power constraints. Extending the results of Smith to vector random variables is unattainable since conditions required by the Identity Theorem are not satisfied [58, 59]. The SVD technique mentioned above can be applied under certain conditions, but does not impose any form of nonnegativity and power constraints. Thus a modified SVD-MIMO method is needed to maximize the data rate while maintaining the target power constraint. The imaging optical MIMO channel matrix $\boldsymbol{H}$ can be decomposed into rotation and scaling matrices using SVD as

$$\boldsymbol{H} = \boldsymbol{U}\boldsymbol{\Lambda}\boldsymbol{V}^*, \tag{8.53}$$

where $\boldsymbol{U}$ and $\boldsymbol{V}$ are unitary rotation matrices while $\boldsymbol{\Lambda}$ is a diagonal scaling matrix. Matrices $\boldsymbol{H}$ and $\boldsymbol{\Lambda}$ have the same rank $\Gamma \leq \min(M, N)$. The diagonal elements of $\boldsymbol{\Lambda}$, $(\lambda_1, \cdots, \lambda_k, \cdots, \lambda_\Gamma)$ are the singular values of matrix $\boldsymbol{H}$. The transmitted signal is constructed as

$$\boldsymbol{X} \triangleq \boldsymbol{V}\boldsymbol{S} + \boldsymbol{B}, \tag{8.54}$$

where $\boldsymbol{B} \in \mathbb{R}^M$ is a DC offset vector and $\boldsymbol{S} \in \mathbb{R}^M$ is the information-bearing symbols. The symbol of the codeword satisfies $S_i \in [-A_i, A_i]$ for any $A_i > 0$ and

$E[S_i] = 0$. Since $E[S_i] = 0$, then $E[X_i] = B_i = P_i$. To guarantee non-negativity of $\boldsymbol{X}$, it is required that $P_i = \sum_{i=1}^{M} |v_{i,j}|A_j$, where $v_{i,j}$ is the (i, j)th component of $\boldsymbol{V}$.

At the receiver, upon receiving $\boldsymbol{Y}$, the receiver is firstly subtracted by $\boldsymbol{HB}$ and then multiplies the signal by $\boldsymbol{U}^*$, and we have

$$\boldsymbol{U}^*\boldsymbol{Y} = \boldsymbol{\Lambda S} + \boldsymbol{U}^*\boldsymbol{Z}. \tag{8.55}$$

Define the new variable in rotated coordinate system as

$$\tilde{\boldsymbol{Y}} \triangleq \boldsymbol{U}^*\boldsymbol{Y}, \quad \tilde{\boldsymbol{Z}} \triangleq \boldsymbol{U}^*\boldsymbol{Z}, \tag{8.56}$$

where $\tilde{\boldsymbol{Y}}$ and $\tilde{\boldsymbol{Z}}$ are output and noise for the transformed system. The transformed simultaneous independent parallel link models are described by

$$\tilde{Y}_k = \lambda_k S_k + \tilde{Z}_k, \ 1 \leq k \leq \Gamma. \tag{8.57}$$

Then, the overall capacity lower bound is given by

$$\mathcal{C}(\boldsymbol{H}, P) = \max_{\boldsymbol{A} \in \mathcal{A}} \sum_{i=1}^{\Gamma} r(\lambda_i, A_i), \tag{8.58}$$

which is to be maximized with respect to A_i subject to

$$\sum_{i=1}^{N} \sum_{j=1}^{M} |v_{i,j}|A_j \leq P. \tag{8.59}$$

Note that, $r(\lambda_i, A_i)$ in (8.58) is the achievable rate or capacity bound over channel i using a DC-offset input signal with peak-constrained input (peak $2P$).

8.5 Outage capacity for OCC with misalignment

For practical OCC communication, pointing accuracy is an important issue in determining link performance and reliability. However, the relative motions between transceivers result in random optical beam sways, which in turn, cause pointing errors and signal fading at the receiver. In this case, availability of CSIT cannot be assumed. For such a system with no CSIT, and under a quasi-static channel, the capacity in the strict Shannon sense is zero. Performance, in this case, is captured by the outage probability [37, 57].

Since the typical time scale of the OCC fading process is much smaller than the bit transmission interval, it is realistic to model the OCC channel as a slow-fading channel. Then, the availability of CSIT is not a strong assumption. Channel can be estimated by transmitting a training sequence, and sent back from the receiver through a feedback link [37, 57]. Two scenarios are considered for transmission over slow fading channels, either utilizing a fixed or a variable rate at the transmitter. Here, the fixed rate scenario is considered. For a given fixed transmission rate, there is a finite probability that the transmitted rate exceeds the instantaneous mutual information of the channel, leading to an outage event. The outage event is mathematically described by the probability of outage. Since each data rate has a corresponding probability of outage, the pair of rate and outage probability are used together to describe the outage capacity.

In the SISO case, the instantaneous mutual information $\mathcal{R}(h)$ at given $\text{SNR}(h)$ associated with channel gain h and input distribution $f_X(x)$ is

$$\mathcal{R}(h) = I(X;Y|H = h). \tag{8.60}$$

The received $\text{SNR}(h)$ is random and given as

$$\text{SNR}(h) = \frac{(Ph)^2}{\sigma^2(1 + \varsigma_1^2 hP + \varsigma_2^2 h^2 P^2)}, \tag{8.61}$$

where P is the average signal intensity. The outage probability for a slow-fading channel with pointing error for intensity-modulation signaling can be computed using $f_h(h) = f_{h_p}(h_p)$. Then the outage probability at a given rate R_0 can be expressed as

$$P_{\text{out}}(R_0) = \text{Prob}(\mathcal{R}(h) < R_0). \tag{8.62}$$

Note that $\mathcal{R}(h)$ monotonically increases with h. The above expression is simplified as

$$P_{\text{out}}(R_0) = \text{Prob}(h < h_0), \tag{8.63}$$

where h_0 satisfies $\mathcal{R}(h_0) = R_0$.Therefore, the outage probability is the cumulative density function of h evaluated at h_0 and is expressed as

$$P_{\text{out}}(R_0) = \int_0^{h_0} f_h(h)dh. \tag{8.64}$$

The derivation of outage capacity for an imaging optical MIMO channel is similar to SISO case. We can study the capacity upper bounds with pointing errors and the probability distribution of the pointing error is incorporated into the fading channel.

8.6 Conclusion

As a new form of VLC, OCC employs the pervasive image sensors in consumer electronics as the receivers, and optical light sources (illumination LED, display or traffic light) as the transmitters. Image sensors are the natural multicolor, optical MIMO and anti-interference receivers, as well as a high-resolution object detector. Various challenges involved in an OCC system are addressed, including the limited frame rate, synchronization issue, shot noise effects, perspective distortions, pixel misalignment, and blur effect. Then, the channel characteristics and system performance, the pixel sensor structure, and the process during each phase of the sensor operation are discussed in detail. Besides, different noise sources are analyzed. Based on the noise model, the SNR, and a unified communication model for OCC are derived. Moreover, the OCC channel capacity under an ideal pixel-matched channel with bounded inputs and intensity constraints is presented. Preliminary study shows that capacity of 8–11 bit/s/Hz is possible under an ideal channel with a diversity structure. Combined with mobile computing, OCC has realized novel forms of sensing and communication applications, such as indoor localization, intelligent transportation, screen–camera communication, and privacy protection.

References

1 P. H. Pathak, X. Feng, P. Hu, and P. Mohapatra, "Visible light communication, networking, and sensing: A survey, potential and challenges," *IEEE Commun. Surv. Tuts.*, vol. 17, no. 4, pp. 2047–2077, Fourth Quarter 2015.

2 N. Saha, M. S. Ifthekhar, N. T. Le, and Y. M. Jang, "Survey on optical camera communications: Challenges and opportunities," *IET Optoelectron.*, vol. 9, no. 5, pp. 172–183, May 2015.

3 "The ieee 802.15.7r1 study group," [online], *http://www.ieee802.org/15/pub/IEEE%20802_15%20WPAN%2015_7%20Revision1%20Task%20Group.htm*.

4 W. Huang and Z. Xu, "Characteristics and performance of image sensor communication," *IEEE Photon. J.*, vol. 9, no. 2, pp. 1–19, Apr. 2017.

5 "Image sensors market analysis," [online], *http://www.grandviewresearch.com/industry-analysis/imagesensors-market*.

6 W. Hu, H. Gu, and Q. Pu, "Lightsync: Unsynchronized visual communication over screen-camera links," in *Proc. International Conference on Mobile Computing & Networking 2013* (Miami, FL), Sept. 30–Oct. 4, 2013, pp. 15–26.

7 W. Huang, P. Tian, and Z. Xu, "Design and implementation of a real-time CIM-MIMO optical camera communication system," *Opt. Exp.*, vol. 24, no. 21, pp. 24567–24579, Oct. 2016.

8 L. Zeng, D. C. O'Brien, H. Le Minh, G. E. Faulkner, K. Lee, D. Jung, Y. Oh, and E. T. Won, "High data rate multiple input multiple output (MIMO) optical wireless communications using white led lighting," *IEEE J. Sel. Areas Commun.*, vol. 27, no. 9, pp. 1654–1662, Dec. 2009.

9 P. M. Butala, H. Elgala, and T. D. Little, "SVD-VLC: A novel capacity maximizing VLC MIMO system architecture under illumination constraints," in *Proc. IEEE Global Communications Conference (GLOBECOM) Workshops 2013* (Atlanta, GA), Dec. 9–13, 2013, pp. 1087–1092.

10 T. Yamazato, M. Kinoshita, S. Arai, E. Souke, T. Yendo, T. Fujii, K. Kamakura, and H. Okada, "Vehicle motion and pixel illumination modeling for image sensor based visible light communication," *IEEE J. Sel. Areas Commun.*, vol. 33, no. 9, pp. 1793–1805, Sept. 2015.

11 S. Hranilovic and F. R. Kschischang, "A pixelated MIMO wireless optical communication system," *IEEE J. Sel. Topics Quantum Electron.*, vol. 12, no. 4, pp. 859–874, Jul./Aug. 2006.

12 S. D. Perli, N. Ahmed, and D. Katabi, "Pixnet: Interference-free wireless links using LCD-camera pairs," in *Proc. International Conference on Mobile Computing and Networking 2010* (Chicago, IL), Sept. 20–24, 2010, pp. 137–148.

13 M. R. H. Mondal and K. Panta, "Performance analysis of spatial OFDM for pixelated optical wireless systems," *Trans. Emerg. Telecommun. Technol.*, vol. 28, no. 2, pp. 1–13, May 2015.

14 M. S. Grewal, L. R. Weill, and A. P. Andrews, *Global Positioning Systems, Inertial Navigation, and Integration*, John Wiley & Sons, 2007.

15 N. U. Hassan, A. Naeem, M. A. Pasha, T. Jadoon, and C. Yuen, "Indoor positioning using visible LED lights: A survey," *ACM Comput. Surv.*, vol. 48, no. 2,

pp. 20:1–20:32, Nov. 2015.
16 S. Y. Jung, S. Hann, and C. S. Park, "TDOA-based optical wireless indoor localization using led ceiling lamps," *IEEE Trans. Consum. Electron.*, vol. 57, no. 4, pp. 1592–1597, Nov. 2011.
17 H. S. Liu and G. Pang, "Positioning beacon system using digital camera and LEDs," *IEEE Trans. Veh. Technol.*, vol. 52, no. 2, pp. 406–419, Mar. 2003.
18 Y. S. Kuo, P. Pannuto, K. J. Hsiao, and P. Dutta, "Luxapose: Indoor positioning with mobile phones and visible light," in *Proc. International Conference on Mobile Computing and Networking 2014* (Maui, HI), Sept. 7–11, 2014, pp. 447–458.
19 Z. Yang, Z. Wang, J. Zhang, C. Huang, and Q. Zhang, "Wearables can afford: Light-weight indoor positioning with visible light," in *Proc. International Conference on Mobile Systems, Applications, and Services 2015* (Florence, Italy), May 18–22, 2015, pp. 317–330.
20 T. Yamazato, I. Takai, H. Okada, T. Fujii, T. Yendo, S. Arai, M. Andoh, T. Harada, K. Yasutomi, K. Kagawa, and S. Kawahito, "Image-sensor-based visible light communication for automotive applications," *IEEE Commun. Mag.*, vol. 52, no. 7, pp. 88–97, Jul. 2014.
21 I. Takai, S. Ito, K. Yasutomi, K. Kagawa, M. Andoh, and S. Kawahito, "LED and CMOS image sensor based optical wireless communication system for automotive applications," *IEEE Photon. J.*, vol. 5, no. 5, pp. 6801418–6801418, Oct. 2013.
22 A. Wang, Z. Li, C. Peng, G. Shen, G. Fang, and B. Zeng, "Inframe++: Achieve simultaneous screen-human viewing and hidden screen-camera communication," in *Proc. International Conference on Mobile Systems, Applications, and Services 2015*, (Florence, Italy), May 18–22, 2015, pp. 181–195.
23 T. Li, C. An, X. Xiao, A. T. Campbell, and X. Zhou, "Real-time screen-camera communication behind any scene," in *Proc. International Conference on Mobile Systems, Applications, and Services 2015* (Florence, Italy), May 18–22, 2015, pp. 197–211.
24 J. Chen, K. Venkataraman, D. Bakin, B. Rodricks, R. Gravelle, P. Rao, and Y. Ni, "Digital camera imaging system simulation," *IEEE Trans. Electron Devices*, vol. 56, no. 11, pp. 2496–2505, Nov. 2009.
25 J. E. Farrell and B. A. Wandell, "I2. 2: Invited paper: Image systems simulation," in *SID Symposium Digest of Technical Papers*, vol. 46, no. 1, pp. 180–183, Wiley Online Library, 2015.
26 A. El Gamal and H. Eltoukhy, "CMOS image sensors," *IEEE Circuits Devices Mag.*, vol. 21, no. 3, pp. 6–20, May/Jun. 2005.
27 P. B. Denyer, D. S. Renshaw, G. Wang, M. Y. Lu, and S. Anderson, "On-chip CMOS sensors for VLSI imaging systems." in *VLSI*, vol. 91, pp. 157–166, 1991.
28 E. R. Fossum, "Active pixel sensors: Are CCDS dinosaurs?" in *Proc. IS&T/SPIE's Symposium on Electronic Imaging: Science and Technology 1993* (San Jose, CA), Jan. 31, 1993, pp. 2–14.
29 B. Fowler, A. El Gamal, and D. X. Yang, "A CMOS area image sensor with pixel-level a/d conversion," in *Proc. IEEE International Solid-State Circuits Conference 1994* (Ulm, Germany), Sept. 20–22, 1994, pp. 226–227.
30 O. Yadid-Pecht, "Geometrical modulation transfer function for different pixel active area shapes," *Opt. Eng.*, vol. 39, no. 4, pp. 859–865, Apr. 2000.
31 Y. Reibel, M. Jung, M. Bouhifd, B. Cunin, and C. Draman, "CCD or CMOS camera noise characterisation," *Eur. Phys. J. AP*, vol. 21, no. 1, pp. 75–80, Nov. 2002.
32 M. Konnik and J. Welsh, "High-level numerical simulations of noise in CCD and CMOS photosensors: Review and tutorial," *arXiv preprint arXiv:1412.4031*, Dec. 2014.
33 R. M. Gagliardi and S. Karp, *Optical Communications*, New York: Wiley-Interscience, 1976.
34 J. C. Chau and T. D. Little, "Analysis of cmos active pixel sensors as linear shift-invariant receivers," in *Proc. IEEE International Conference on Communications Workshops (ICCW) 2015* (London, UK), Jun. 8–12, 2015, pp. 1398–1403.
35 S. M. Moser, "Capacity results of an optical intensity channel with input-dependent Gaussian noise," *IEEE Trans. Inf. Theory*, vol. 58, no. 1, pp. 207–223, Jan. 2012.
36 A. A. Farid and S. Hranilovic, "Channel capacity and non-uniform signalling for free-space optical intensity channels," *IEEE*

J. Sel. Areas Commun., vol. 27, no. 9, pp. 1553–1563, Dec. 2009.

37 A. A. Farid and S. Hranilovic, "Diversity gain and outage probability for MIMO free-space optical links with misalignment," *IEEE Trans. Commun.*, vol. 60, no. 2, pp. 479–487, Feb. 2012.

38 I. Moreno and C.-C. Sun, "Modeling the radiation pattern of leds," *Opt. Exp.*, vol. 16, no. 3, pp. 1808–1819, 2008.

39 I. Djite, M. Estribeau, P. Magnan, G. Rolland, S. Petit, and O. Saint-Pe, "Theoretical models of modulation transfer function, quantum efficiency, and crosstalk for CCD and CMOS image sensors," *IEEE Trans. Electron Devices*, vol. 59, no. 3, pp. 729–737, Mar. 2012.

40 A. A. Farid and S. Hranilovic, "Outage capacity optimization for free-space optical links with pointing errors," *J. Lightw. Technol.*, vol. 25, no. 7, pp. 1702–1710, 2007.

41 F. Yang, J. Cheng, and T. A. Tsiftsis, "Free-space optical communication with nonzero boresight pointing errors," *IEEE Trans. Commun.*, vol. 62, no. 2, pp. 713–725, 2014.

42 J. G. Smith, *On the information capacity of peak and average power constrained Gaussian channels*, University of California, 1969.

43 J. G. Smith, "The information capacity of amplitude-and variance-constrained sclar Gaussian channels," *Inf. Control*, vol. 18, no. 3, pp. 203–219, Apr. 1971.

44 A. Tchamkerten, "On the discreteness of capacity-achieving distributions," *IEEE Trans. Inf. Theory*, vol. 50, no. 11, pp. 2773–2778, Nov. 2004.

45 T. H. Chan, S. Hranilovic, and F. R. Kschischang, "Capacity-achieving probability measure for conditionally Gaussian channels with bounded inputs," *IEEE Trans. Inf. Theory*, vol. 51, no. 6, pp. 2073–2088, Jun. 2005.

46 B. Mamandipoor, K. Moshksar, and A. K. Khandani, "On the sum-capacity of Gaussian MAC with peak constraint," in *Proc. IEEE International Symposium on Information Theory (ISIT) 2012* (Cambredge, MA), Jul. 1–6, pp. 26–30.

47 A. ElMoslimany, "A new communication scheme implying amplitude limited inputs and signal dependent noise: System design, information theoretic analysis and channel coding," Ph.D. dissertation, Arizona State University, 2015.

48 K. H. Park, W. G. Alheadary, and M. S. Alouini, "A novel mirror diversity receiver for indoor MIMO visible light communication systems," in *Proc. IEEE International Symposium on Personal, Indoor, and Mobile Radio Communications (PIMRC) 2016* (Valencia, Spain), Sept. 4–7, 2016, pp. 1–6.

49 T. Fath and H. Haas, "Performance comparison of MIMO techniques for optical wireless communications in indoor environments," *IEEE Trans. Commun.*, vol. 61, no. 2, pp. 733–742, Feb. 2013.

50 A. Nuwanpriya, S. W. Ho, and C. S. Chen, "Indoor MIMO visible light communications: Novel angle diversity receivers for mobile users," *IEEE J. Sel. Areas Commun.*, vol. 33, no. 9, pp. 1780–1792, Sept. 2015.

51 P. F. Mmbaga, J. Thompson, and H. Haas, "Performance analysis of indoor diffuse VLC MIMO channels using angular diversity detectors," *J. Lightw. Technol.*, vol. 34, no. 4, pp. 1254–1266, Feb. 2016.

52 T. Q. Wang, Y. A. Sekercioglu, and J. Armstrong, "Analysis of an optical wireless receiver using a hemispherical lens with application in MIMO visible light communications," *J. Lightw. Technol.*, vol. 31, no. 11, pp. 1744–1754, Jun. 2013.

53 T. Chen, L. Liu, B. Tu, Z. Zheng, and W. Hu, "High-spatial-diversity imaging receiver using fisheye lens for indoor MIMO VLCs," *IEEE Photon. Technol. Lett.*, vol. 26, no. 22, pp. 2260–2263, Nov. 2014.

54 S. Hranilovic and F. R. Kschischang, "Capacity bounds for power- and band-limited optical intensity channels corrupted by Gaussian noise," *IEEE Trans. Inf. Theory*, vol. 50, no. 5, pp. 784–795, May 2004.

55 A. Farid and S. Hranilovic, "Capacity bounds for wireless optical intensity channels with Gaussian noise," *IEEE Trans. Inf. Theory*, vol. 56, no. 12, pp. 6066–6077, Dec. 2010.

56 A. Ashok, S. Jain, M. Gruteser, N. Mandayam, W. Yuan, and K. Dana, "Capacity of pervasive camera based communication under perspective distortions," in *Proc. IEEE International Conference on Pervasive Computing and Communications 2014* (Budapest, Hungary), Mar. 24–28, 2014, pp. 112–120.

57 A. Chaaban, Z. Rezki, and M. S. Alouini, "Fundamental limits of parallel optical wireless channels: Capacity results and outage formulation," *IEEE Trans. Commun.*, vol. 65, no. 1, pp. 296–311, Jan. 2017.

58 A. ElMoslimany and T. Duman, "On the capacity of multiple-antenna systems and parallel Gaussian channels with amplitude-limited inputs," *IEEE Trans. Commun.*, vol. 64, no. 7, pp. 2888–2899, Jul. 2016.

59 B. Rassouli and B. Clerckx, "On the capacity of vector Gaussian channels with bounded inputs," in *Proc. IEEE International Conference on Communications (ICC) 2015* (London, UK), Jun. 8–12, 2015, pp. 4030–4035.

60 B. Mamandipoor, K. Moshksar, and A. K. Khandani, "Capacity-achieving distributions in Gaussian multiple access channel with peak power constraints," *IEEE Trans. Inf. Theory*, vol. 60, no. 10, pp. 6080–6092, Oct. 2014.

61 T. H. Chan, S. Hranilovic, and F. R. Kschischang, "Capacity achieving probability measure of an input-bounded vector Gaussian channel," in *Proc. IEEE International Symposium on Information Theory (ISIT) 2003* (Yokohama, Japan), Jun. 29–Jul. 4, 2003, pp. 371–371.

62 T. M. Cover and J. A. Thomas, *Elements of Information Theory*, John Wiley & Sons, 2012.

9
Optical Camera Communication: Modulation and System Design

In this chapter, we review modulation schemes and discuss system design issues in optical camera communication (OCC). We also point out impairment factors in each modulation scheme and the corresponding mitigation methods. Then we address synchronization challenges, and in particular introduce two synchronization schemes, namely the per-line tracking and inter-frame coding, and rateless coding. We then turn our attention from theory to practice by investigating practical modulation schemes and multiplexing techniques in the system design. The spatial, color and intensity dimensions are fully explored to create high-dimensional signal constellations and parallel communication channels. Moreover, the color-intensity modulation (CIM) and multiple-input multiple-output (MIMO) configurations are built. Finally, we present designs and implementations of a real-time CIM-MIMO OCC system.

In Section 9.1, the problem about how to design the capacity-achieving nonuniform discrete signaling for signal-dependent noise channels under the optical intensity constraints in practice is discussed. To implement a capacity-achieving system, a nonuniform source distribution is required for an OCC system. It can be accomplished by employing multilevel coding (MLC) and multi-stage decoding (MSD) with a deterministic mapper applied to multiple binary linear codes. Furthermore, a nonuniform mapper coupled with a binary low density parity check (LDPC) code can be used to generate the desired input distribution. This scheme has a lower complexity compared with non-binary LDPC code constructions and requires a single encoder and decoder. Thus it is free of error propagation and has less latency in decoding.

However, some challenges in the aforementioned design still exist. Link feedback is necessary and the computational complexity of the sum-product-based joint demapper/decoder algorithm is very high. Thus, other modulation schemes are developed in OCC. In Section 9.2, modulation schemes in different domains are introduced, including the undersampling-based modulation schemes and the rolling-shutter-effect-based modulation schemes in the time-frequency domain, color-intensity modulation in the color space, and the spatial orthogonal frequency division multiplexing (OFDM) and spatial wavelet packet division multiplexing (WPDM) in the spatial-frequency domain. Then, the nonideal factors in each mod-

Visible Light Communications: Modulation and Signal Processing. First edition. Zhaocheng Wang, Qi Wang, Wei Huang, and Zhengyuan Xu. Published 2017 by John Wiley & Sons, Inc.

ulation scheme are considered, and the effects of impairment factors on various modulation schemes are analyzed in Section 9.3. Specifically, the linear misalignment, geometry perspective distortion, blur effect, and vignetting in spatial OFDM are considered. And the corresponding techniques to mitigate the impairment factors are discussed, including equalization, perspective correction, adaptive coding, and modulation. Synchronization is another important aspect for a practical OCC system. The difficulty in frame synchronization mainly arises from frame diversity and variability. In Section 9.4, two synchronization schemes, the per-line tracking and inter-frame coding, and rateless coding are discussed. These methods tackle the synchronization issues by decoding imperfect frames and recovering every lost frame at the receiver.

Based on those designs and considerations, an experimental OCC platform has been built and its components are described in Section 9.5. In particular, spatial, color, and intensity dimensions are fully utilized to create high-dimensional signal constellations and parallel communication channels for a real-time CIM-MIMO OCC system. In this way, the data rate is significantly increased and bit error rate (BER) performance improves. Particularly, some solutions to tackle several challenges in the system design are introduced, including unstable frame rate, joint nonlinearity and crosstalk, flicker noise, and rolling shutter.

9.1 Coding and decoding

Typically, an OCC system adopts intensity-modulation direct-detection (IM/DD), where the desired information is modulated onto the optical intensity and transmitted to the receiver over a free space link. The pixel, being basically a power detection unit and a fundamental element in an image sensor, responds to the instantaneous field count rate process. Its output appears as a shot noise process, whose count rate is proportional to the instantaneous received power. Pulse amplitude modulation (PAM) is one of the most popular intensity modulation schemes developed for an optical communication system, and on-off keying (OOK), as a binary-level version of PAM, is also widely used.

Almost all existing PAM-based schemes belong to uniform signaling with equal-probability symbols. It performs well for a channel with additive white Gaussian noise (AWGN), independent of the signal. However, it has been proven that the capacity-achieving input probability for signal-dependent Gaussian noise channels subject to optical intensity constraints follows a nonuniform discrete distribution. Meanwhile, binary linear codes can be applied directly for channels with uniform input distribution, but channel coding with nonuniform input distribution is more complex. The method to implement nonuniform distribution signal was first proposed by Gallager [1], where a deterministic mapper was employed at the output of a binary encoder to generate symbols following a nonuniform distribution. Another approach to induce the nonuniform distribution is to design LDPC codes over GF(q) with $q > 2$ [2]. Furthermore, an inverse Huffman code type mapper was proposed to generate the nonuniform distribution [3]. Although these approaches yield substantial performance improvement, the higher complexity in both code design and decoding limits their applications. Moreover, the complexity of soft decoding is also prohibitive.

In this section, we introduce the widely used methods to realize a capacity-achieving system. The MLC and MSD with a deterministic mapper [4] are applied to induce the nonuniform source distribution. However, MLC/MSD requires multiple encoders and decoders, which result in error propagation and decoding latency. To address this problem, Cao *et al.* [5] proposed a nonuniform mapper coupled with a binary LDPC code to generate the desired input distribution. They also provided a joint demapper/decoder design based on a sum-product algorithm. This scheme only requires a single encoder and decoder with lower complexity compared with non-binary LDPC code constructions. It also offers less decoding latency and no propagation error.

9.1.1 Multilevel coding and multi-stage decoding

The MLC/MSD structure consists of two parts: mapping/demapping and encoding/decoding modules [4, 6]. Figure 9.1 shows a diagram for an optical channel including a mapper. Every N-bits from the independent bit stream with uniform distribution form a group $\boldsymbol{W} = [W_1 \ \cdots \ W_N]$. And then, the bits are mapped to a symbol X through the mapping function $\mathcal{M}$, to form nonuniform distribution symbols. The output symbols have probability $m/2^N$, where $m = 1, \cdots, 2^N - 1$ is an integer. Thus, there exist a set of nonuniform distributions that can be induced for a given N. Note that the mapping $\boldsymbol{W} \xrightarrow{\mathcal{M}} X$ is not necessarily one-to-one. However, the mutual information $I(X;Y)$ between the channel input and output is unaffected, since $\boldsymbol{W} \to X \to Y$ is a Markov chain and $\mathcal{M}(\boldsymbol{W}) = X$ is a deterministic mapper. Thus, for a given deterministic mapping function $\mathcal{M}$, the information rate can be realized and achieved even when the mapping is not reversible.

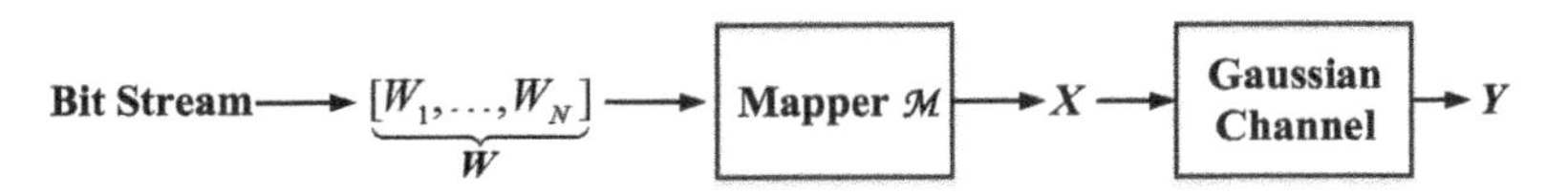

Figure 9.1 Illustrative diagram for an optical channel with a mapper [4].

Figure 9.2 shows a block diagram for the MLC system with a deterministic mapper. Assuming a bit stream with independent and equal-probability bits, k-bits from the stream are divided into N sub-streams each with k_i bits, where $i = 1, \cdots, N$, such that $k = \sum_{i=1}^{N} k_i$. The ith sub-stream is encoded by a linear binary code of rate R_i. The codeword length of each encoder output is fixed to $n = k_i/R_i$. The outputs of the encoders are arranged in a vector $\boldsymbol{W} = [W_1 \ \cdots \ W_N]$, where W_i denotes the ith encoder output bit. The vector $\boldsymbol{W}$ is mapped to a symbol using a deterministic mapper function $\mathcal{M}$, which can output the desired nonuniform probability signal.

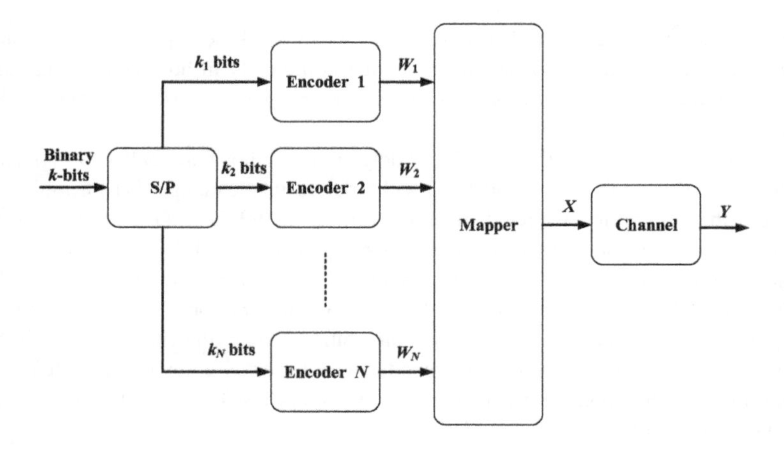

Figure 9.2 Schematic block diagram for MLC and mapper [4].

Based on the chain rule, the mutual information can be expressed in terms of the sub-channel rates as

$$I(\boldsymbol{W}; Y) = \sum_{i=1}^{N} R_i, \tag{9.1}$$

where the sub-channel rates are given by

$$R_i = I(W_i; Y | W_1, \cdots, W_{i-1}). \tag{9.2}$$

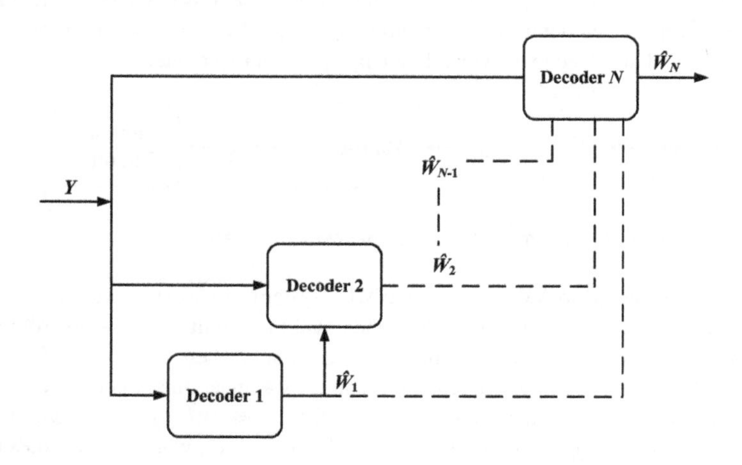

Figure 9.3 Schematic block diagram for multi-stage decoders (MSD) [4].

At the receiver, the received codeword is decoded sequentially to recover the transmitted bits. Figure 9.3 illustrates the sequential decoding strategy of the MSD. The first decoder utilizes the received signal Y to decode W_1 as $\hat{W}_1$. Given that the codeword is decoded correctly, the second decoder utilizes both Y and estimated $\hat{W}_1$ to obtain the estimated $\hat{W}_2$ since the second encoder operates at rate $R_2 = I(W_2; Y|W_1)$. Repeat this process till $\hat{W}_N$ is estimated. Note that if a codeword is decoded incorrectly, an error occurs and propagates, resulting in decoding error in the estimated transmitted data. Due to the sequential decoding processing in MSD, time latency is inevitable.

Example: To illustrate how the nonuniform signal is generated, we assume that the nonuniform distribution has two mass points $\{0, A\}$ with probabilities of $p(0) = 7/8$ and $p(A) = 1/8$. This system can be constructed using $N = 3$ encoders with a deterministic mapping function $\mathcal{M}$, given by

$$\boldsymbol{W} = [W_1\ W_2\ W_3] \xrightarrow{\mathcal{M}} X : X = \begin{cases} A, & \text{if } W_1 = W_2 = W_3 = 1, \\ 0, & \text{otherwise.} \end{cases} \tag{9.3}$$

The detailed diagram of the mapping function is presented in Fig. 9.4. To achieve the rate of $R_1 = I(W_1; Y)$, the bit $W_1 = 0$, which is the most significant bit, is mapped to $X = 0$ in all cases, while $W_1 = 1$ is mapped to $X = 0$ with the probability of 3/4.

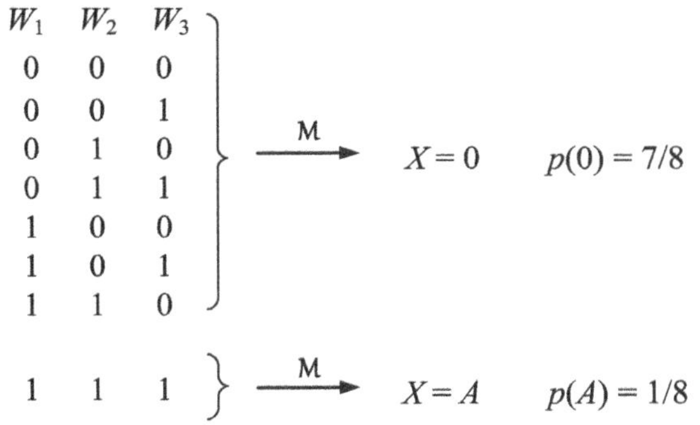

W_1	W_2	W_3			
0	0	0			
0	0	1			
0	1	0	$\xrightarrow{M}$	$X=0$	$p(0)=7/8$
0	1	1			
1	0	0			
1	0	1			
1	1	0			
1	1	1	$\xrightarrow{M}$	$X=A$	$p(A)=1/8$

Figure 9.4 Mapping function over $\mathcal{X} = \{0, A\}$ to induce $p(0) = 7/8$ and $p(A) = 1/8$.

9.1.2 Single-level coding and joint decoding

Although a mapper coupled with MLC/MSD can induce nonuniform signaling, this method suffers from error propagation and decoding latency, and requires multiple encoders and decoders. Alternatively, a single code can be used to encode all bits and the mapper is employed to induce the correct distribution.

Figure 9.5 shows the block diagram of the single encoder with a mapper. When a stream of independent message U composed of k-bits with uniformly distributed

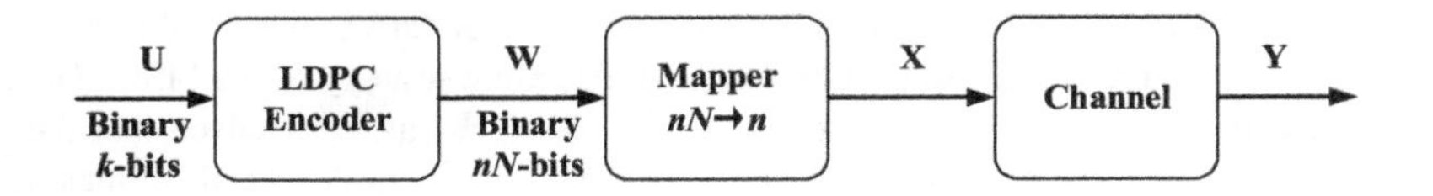

Figure 9.5 System model for the single encoding and mapping scheme.

inputs to the LDPC encoder and the length of the LDPC code is nN, the modulation order becomes 2^N. The parameter n takes an integer value to ensure the capacity $C > k/n$. The output bits of the encoder are $(W_1^{(i)}, \cdots, W_N^{(i)})$, for $i = 1, \cdots, n$. Since the LDPC code is a linear code, the output distribution of the symbols in $\boldsymbol{W}$ can be assumed to be uniform. After that, the output of the encoder $\boldsymbol{W}$ is mapped to a symbol using a deterministic mapper function $\mathcal{M}$ to induce the desired distribution. This mapper is easy to implement since all probable masses are constrained to be the form of $k/2^N$. Thus, each block of N coded bits is sent to the mapper to yield a single channel input X with the desired probability distribution. At the receiver, demapping and decoding are conducted jointly via the sum-product algorithm [5].

Example : Consider $N = 2$. The mapper function $\mathcal{M}$ induces the following distribution

$$(W_1^i, W_2^i) \xrightarrow{\mathcal{M}} X : X = \begin{cases} A, & \text{if } W_1^i = W_2^i = 1, \\ 0, & \text{otherwise.} \end{cases} \tag{9.4}$$

The equivalent channel seen by bit W_1 (and W_2 due to the symmetry of the mapper) can be obtained by marginalizing the conditional probability density function (PDF)

$$\begin{aligned} f_{Y|W}(y|w_1 = 1) &= \sum_{w_2} f_{Y|X}(y, w_2|w_1 = 1) \\ &= \frac{1}{2} f_{Y|X}(y|A) + \frac{1}{2} f_{Y|X}(y|0), \end{aligned} \tag{9.5}$$

$$f_{Y|W}(y|w_1 = 0) = f_{Y|X}(y|0), \tag{9.6}$$

where $f_{Y|X}(\cdot|\cdot)$ is the channel conditional PDF.

Joint LDPC decoding and demapping can be represented in the factor graph, as shown in Fig. 9.6. Then, the message passing on this graph using the sum-product algorithm can demap and decode the bits jointly [5]. The lower part of the graph represents a traditional LDPC code and the mapping function $\mathcal{M}$ is represented by the triangular nodes. Both $w^{(i)}$ and x_i are binary in this example. Following the standard sum-product algorithm, the message from the mapper to the message bit $w_1^{(i)}$ is

$$\begin{aligned} \mu_{\mathcal{M}\to w_1^{(i)}}(w_1^{(i)} = 1) =& \mu_{x_i\to\mathcal{M}}(x_i = A)\mu_{w_2^{(i)}\to\mathcal{M}}(w_2^{(i)} = 1) \\ &+ \mu_{x_i\to\mathcal{M}}(x_i = 0)\mu_{w_2^{(i)}\to\mathcal{M}}(w_2^{(i)} = 0). \end{aligned} \tag{9.7}$$

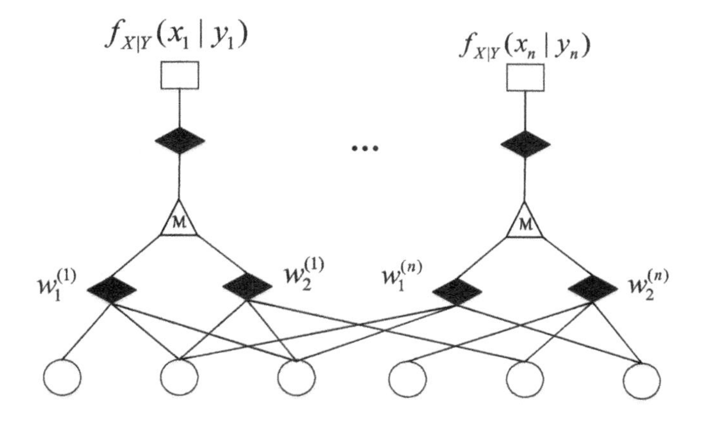

Figure 9.6 The factor graph for joint demapping and decoding.

$$\begin{aligned}\mu_{\mathcal{M}\to w_1^{(i)}}(w_1^{(i)}=0) =& \mu_{x_i\to\mathcal{M}}(x_i=0)\mu_{w_2^{(i)}\to\mathcal{M}}(w_2^{(i)}=1)\\ &+\mu_{x_i\to\mathcal{M}}(x_i=0)\mu_{w_2^{(i)}\to\mathcal{M}}(w_2^{(i)}=0).\end{aligned} \tag{9.8}$$

An analogous message from $\mathcal{M}$ to $w_2^{(i)}$ can also be derived similarly. Then, the message from x_i to $\mathcal{M}$ can be written in a log-likelihood ratio (LLR) form as

$$m_{x_i\to\mathcal{M}} = \ln\frac{\mu_{x_i\to\mathcal{M}}(x_i=0)}{\mu_{x_i\to\mathcal{M}}(x_i=A)} = \ln\frac{f(x_i=0|y_i)}{f(x_i=A|y_i)} = \ln\frac{3f(y_i|x_i=0)}{f(y_i|x_i=A)}. \tag{9.9}$$

All other message passing for the LDPC code takes place in the standard manner. Due to the symmetry of the mappers in $w_1^{(i)}$ and $w_2^{(i)}$, the update rules for both are the same. After several rounds of message passing, a hard decision is performed for each $w^{(i)}$.

9.2 Modulation schemes

In the previous section, we introduce how to design the capacity-achieving nonuniform signaling in practice. However, some challenges in the aforementioned design still exist. Link feedback is necessary and the computational complexity of the joint demapper/decoder is very high. Thus, other modulation schemes are employed in OCC.

9.2.1 Undersampling-based modulation

An OCC system design requires consideration of human perception of light source flicker. Usually, the commercial camera's frame rate is lower than the transmission frequency. To achieve a flicker-free communication, undersampling-based modulation methods are proposed to ensure no perceptual intensity fluctuation. Figure 9.7 shows the cutoff frequencies of human eye and camera. The human eye has a cutoff frequency in the vicinity of 100 Hz, whereas the camera's cutoff frequency can significantly exceed 100 Hz depending on the exposure speed setting. In regard to the observability of a "blinking light", the signal waveform based on undersampling modulation can be captured by a camera with the appropriate exposure setting, but not perceived by the human eye because the camera's exposure speed is much faster than the eye can perceive.

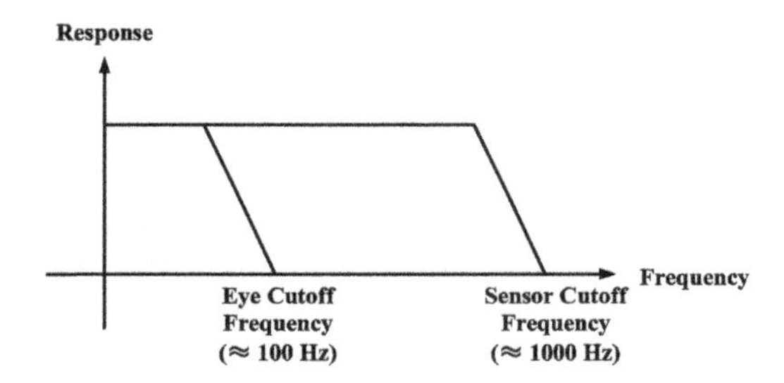

Figure 9.7 The cutoff frequencies of the human eye and the camera.

Undersampling frequency shift on-off keying (UFSOOK) is a form of direct current (DC) balanced differential coding [7]. Similar to frequency shift keying (FSK), mark and space ON-OFF keying frequencies indicate the coding bits. The mark (logic 1) and space (logic 0) frequencies are selected such that when undersampled by a low frame rate camera, the mark/space frequencies distorted by low pass frequencies can be further processed to decode the bit values. The mark frequency is defined as an integer multiple of the camera frame rate f_{camera} plus/minus one half, i.e., $(n \pm \frac{1}{2}) \times f_{\text{camera}}$, and the space frequency is defined as an integer multiple of the camera frame rate, i.e., $n \times f_{\text{camera}}$.

In UFSOOK, two transmitted frame samples represent one bit, which does not effectively utilize the sampled values. Figure 9.8(a) depicts an example of the UFSOOK pattern, where y-axis indicates whether the light is turned ON or OFF. In this particular example, a logic 1 (mark frequency) is selected to be transmitted for seven cycles of 105 Hz OOK, which is 3.5 times the camera frame rate (the camera frame rate is 30 Hz), followed by a logic 0 (space frequency) that is chosen to be transmitted at a rate 4 times of the camera frame rate. The frame head is transmitted with a frequency f_{FH}, which is much higher than the cutoff frequency $f_{\text{max-camera}}$ of a camera. Therefore, the light appears as half ON (average) in the received frame. The camera captures continuous frames at the position of the dash sampling strobes, with each UFSOOK symbol sampled twice at the frequency of 30 Hz. Thus, the camera subsampling of the mark frequency results in the light to appear blinking OFF then ON, and the camera subsampling of the space frequency results in the light to be OFF for both samples. In principle, subsampling of the mark frequency will result in aliasing that causes the light to appear blinking (OFF-ON or ON-OFF), and subsampling of the space frequency will result in aliasing that causes the light to be in a steady state (either ON or OFF). Adhering to the stated rules, it always results in an even number of OOK cycles per bit for a space frequency and an odd number of cycles for a mark frequency. Hence, the "code" is always balanced.

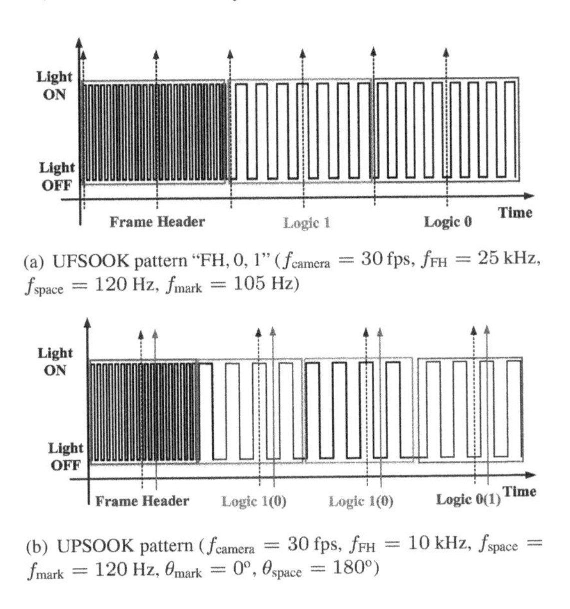

(a) UFSOOK pattern "FH, 0, 1" ($f_{\text{camera}} = 30$ fps, $f_{\text{FH}} = 25$ kHz, $f_{\text{space}} = 120$ Hz, $f_{\text{mark}} = 105$ Hz)

(b) UPSOOK pattern ($f_{\text{camera}} = 30$ fps, $f_{\text{FH}} = 10$ kHz, $f_{\text{space}} = f_{\text{mark}} = 120$ Hz, $\theta_{\text{mark}} = 0°$, $\theta_{\text{space}} = 180°$)

Figure 9.8 The example of UFSOOK pattern and UPSOOK pattern [8].

Undersampling phase shift on-off keying (UPSOOK) modulation is similar to the phase shift keying (PSK) [8], where the mark (logic 1) and space (logic 0) have the same frequency and amplitude, but the phases of the corresponding carrier signal are opposite. In UPSOOK, frame head is also transmitted at the frequency f_{FH}, which is much higher than the camera's cutoff frequency $f_{\text{max-camera}}$. The mark and space are represented by square waves of the same frequency $f_{\text{mark}} = f_{\text{space}} = n \times f_{\text{camera}}$ but different phase (e.g., $\theta_{\text{mark}} = 0°$, $\theta_{\text{space}} = 180°$), where n is an even integer ($f_{\text{max-eye}} < f_{\text{mark}} < f_{\text{max-camera}}$). Figure 9.8(b) depicts an example of the UPSOOK pattern, where f_{camera} = 30 Hz, and $f_{\text{mark}} = f_{\text{space}}$ = 120 Hz. Under this setting, one bit per four frames can be transmitted. Since it is likely that the sample phase of the camera is out of control, there might be a random phase difference between the transmitter and the camera. Therefore, at the receiver side, it is uncertain to determine whether the received “1” or “0” represents mark or space, and a framing strategy is needed to eliminate this uncertainty. The phase uncertainty problem will only cause an error when receiving a mark or a space signal, while it has no effect on the frame header signal. Data is sent according to the frame structure as shown in Fig. 9.9(b), where each q-PAM (including OOK) symbol is packed into the payload of a data frame with a header named as a start frame delimiter (SFD) for asynchronous communication and nonlinear compensation at the receiver. Specifically, SFD is composed of three parts labeled as A, B, and C. Part A denotes the start of a data frame. When the camera captures the UPSOOK modulated signal in Part A, the recorded transmitter will be in the half-ON (HO) state. Part B is the mark symbol, which can be used to indicate the phase uncertainty. Part C has the length of L symbols for L-level undersampling phase shift pulse amplitude modulation (UPSPAM) signal, and is designed to obtain the transceiver nonlinear curve. In UPSOOK, Part C is unnecessary in SFD. In such a way, the error caused by phase uncertainty can be detected by examining the second received symbol of a frame. There will be two possible states of the second symbol in the received data frame. If the second received symbol in the received frame is fully ON, it means that phase uncertainty does not introduce error. If the second symbol in received frame is OFF, it means all the fully ON symbols in the frame should be OFF, and all the OFF symbols should be fully ON. This procedure can correct the error introduced by phase uncertainty, which can also be considered as a special forward error correction (FEC) scheme.

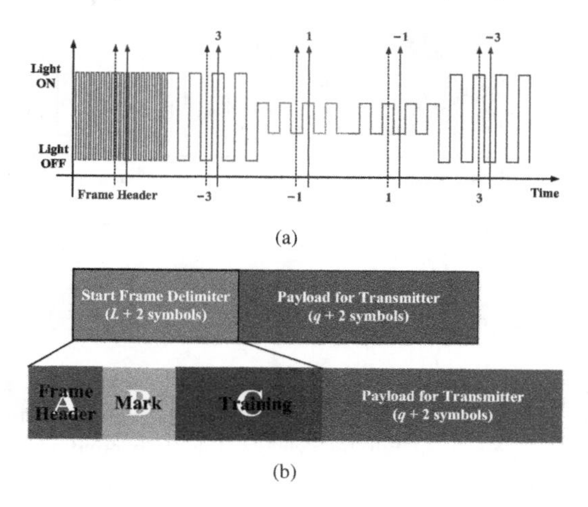

Figure 9.9 An example of the 4-UPSPAM signal and the data frame structure.

As introduced above, both the UFSOOK and UPSOOK modulation schemes employ the undersampling technique to sample high-frequency signals and obtain the transmitter’s states for data recovery. However, the spectral efficiencies are 0.5 bit/s/transmitter and 1 bit/s/transmitter for UFSOOK and UPSOOK, respectively. In order to improve the spectral efficiency without increasing the number of transmitters, UPSPAM is proposed for high efficiency and non-flicking OCC [9]. Instead of directly transmitting baseband PAM signal, in UPSPAM, a square wave intensity-modulated PAM at the carrier frequency of $f_s = n \times f_{\text{camera}}$ is transmitted. Figure 9.9(a) illustrates the waveform of 4-UPSPAM symbols. If we assume the original binary data stream is [1 1 1 0 0 1 0 0] and $n = 4$, then the corresponding 4-PAM signal using Gray coding is [3 1 -1 -3]. Correspondingly, the modulated 4-UPSPAM signal is [3 -3 3 -3 3 -3 3 -3 1 -1 1 -1 1 -1 1 -1 -1 1 -1 1 -1 1 -1 1 -3 3 -3 3 -3 3 -3 3]. Since it is likely that the sampling phase of camera at the receiver side is out of control, there may exist a random phase difference between the transmitter and the camera, which can be shown in Fig. 9.9(a). Note that the subsampled result (shown on the top of Fig. 9.9(a)) using blue solid strobes is the same as the original data, but the obtained result sampled with red dot strobes has an inverse sign from the original data. Thus, at the receiver, the uncertainty can also be eliminated by a framing strategy as shown in Fig. 9.9(b). Part C in SFD, which has the length of L symbols for the L-UPSPAM signal, is designed to obtain the transceiver nonlinear curve. After obtaining the transceiver nonlinearity, curve pre-compensation or post-compensation techniques can be used to mitigate the nonlinear effect [10].

9.2.2 Rolling shutter effect-based modulation

As mentioned above, the complementary metal oxide semiconductor (CMOS) image sensors, most commonly used in today's consumer electronics, exhibit a phenomenon referred to as rolling shutter [11–15]. The image sensor consists of a matrix of photodiodes, where each photodiode converts the incident photons into voltage. In order to reduce the design complexity and power consumption, or to accelerate the read rate in readout circuit, the rolling shutter image sensors expose only one scanline of photodiodes at a time and read the output. This scanning of photodiodes, one scanline after another in sequence, is referred to as the rolling shutter. Generally speaking, the rolling shutter effect is undesirable, because it is challenging for symbol demodulation, and inevitably introduces intensity fluctuation, leading to performance loss. However, the property of the rolling shutter can actually be utilized for data transmission from a light source to a rolling-shutter-based image sensor.

We first introduce the principle of the rolling shutter [16]. As shown in Fig. 9.10, the exposure in CMOS image sensor is controlled by the row-reset and row-select signals generated by the row address decoder, where each row becomes photosensitive after row reset, and stops collecting photons and starts reading out data once a row-select signal is detected. Since there is only one row of readout circuits, the readout timings for different rows cannot overlap. Thus, the level of the signal generated by image sensor depends on the amount of incident light on the photodetectors, in terms of both intensity and duration. During the scanning process, each scanline of the sensor array is exposed, sampled, and stored sequentially. When this procedure is completed, the scanlines are merged together in order to form a single image.

Various effects can be observed due to rolling shutter operation, such as the skew in

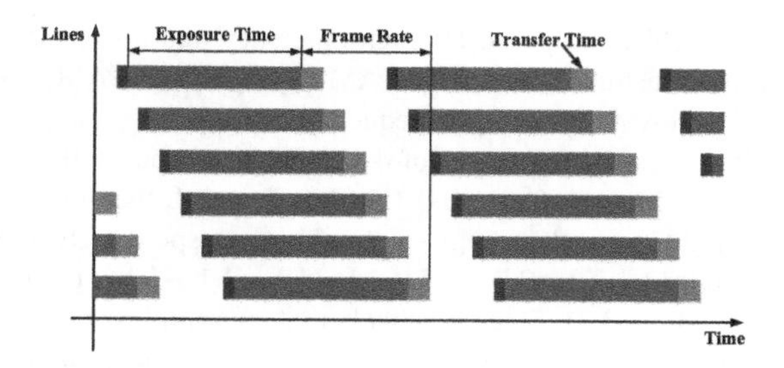

Figure 9.10 Timing chart for the rolling shutter image sensor.

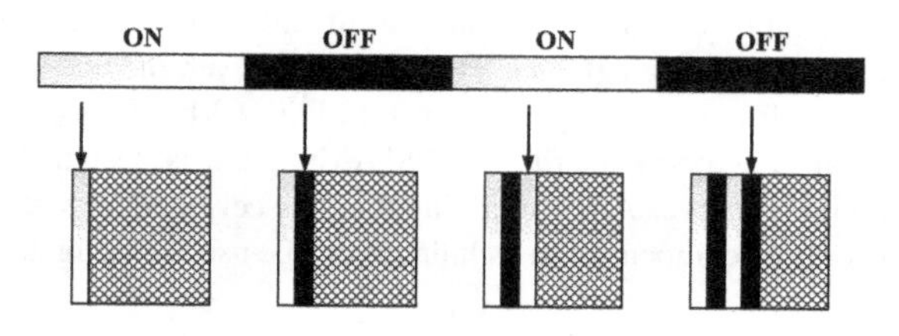

Figure 9.11 Data decoding using the rolling shutter scheme.

images of a moving object [12]. While this may seem undesirable, it can also be utilized for data communication. When the transmission frequency of the light source is lower than the rolling shutter's scanning frequency but higher than the frame rate, bands of different light intensity appear in the image sensor as shown in Fig. 9.11. When the light source is "ON", the image sensor captures a bright frame and the CMOS image sensor exposes one array of this image shown as the first white line in the image. The light source then changes to the "OFF" state and the second scanline is enabled, which results in the first black line in the image. The aforementioned operations continue until all the scanlines are exposed and the image is completed. Thus, it demonstrates that as the state of light source alternates between "ON" and "OFF", the corresponding image sensor produces an image frame by alternating bands of pixels with bright dark shades due to the rolling shutter effect. If the light source's "ON" and "OFF" states are used to represent bit "1" and "0" respectively, then the OOK modulation is realized, which is shown in Fig. 9.12. In practice, since OOK only utilizes light source's white light, it is less robust to ambient light noise. Moreover, OOK can also induce human perceivable light source flickering for long runs of 0s or 1s in the transmission data.

In practice, the width of bright or dark bands is proportional to the symbol rate of the transmitter and the rate of the image sensor to capture preview images [11, 15]. By adjusting these values, an array of images with bands of different widths and intensity can be obtained. Using simple image processing techniques, these bands can be converted into a binary array, from which useful information can be extracted. Thus, if different symbols are conveyed by ON-OFF bands at different frequencies,

Figure 9.12 OOK, FSK, and CSK modulation using the rolling shutter effect.

then FSK modulation is realized. Figure 9.12 shows a frame with two FSK symbols. In particular, FSK modulation can reduce the demodulation error due to long symbol duration and multiple ON-OFF bands in each symbol.

As introduced in the previous chapter, a CMOS image sensor is a natural multicolor receiver enabled by a color filter array, and color shift keying (CSK) modulation can be performed in OCC. With the ability of commodity cameras to detect a wide range of colors, it is possible to use high-order constellations with CSK [14]. In the case of one transmitter, different duty cycles or transmission frequencies with a steady voltage between 0 and full, allow us to control the brightness of the transmitter. If we use three pulse width modulation (PWM) signals to control the multicolor transmitter elements, such as one tri-LED containing the inner red, green, and blue LEDs with different duty cycles respectively, the tri-LED will generate an accumulated color and produce a desired CSK symbol. At the receiver, the CMOS image sensor receives the color symbols in the form of different color bands in a recorded frame as shown in Fig. 9.12. The receiver then compares the received color to reference symbol's color for demodulation. The higher order CSK modulation along with shorter symbol duration provides higher data rates compared to previously studied modulation approaches like FSK. Thus, it is possible that the joint effect of CSK modulation does not impact the human perceivable color for consistent white illumination of the LED, and the color flicker problem in conventional CSK can be eliminated.

For rolling-shutter-effect-based modulation schemes, there are still many open challenges. The first one is the flicker effect. When an LED or screen is used for data communication, it concurrently serves for illumination or display. The OOK and FSK utilize white light during the ON period. If the transmission frequency is high enough, human will perceive no illumination fluctuation. However, if the data symbols are transmitted in the form of different color light, the color changes can be perceived by human eye. Hence, even when the color symbols are transmitted, the human perceivable color of illumination should remain white. The second one is the inter-frame data loss problem. The commercial cameras available in the consumer electronics market cannot capture image frames in a successive manner. They require a certain amount of time to process the captured frame. Thus, the symbols transmitted during this inter-frame interval are not received by the camera. Techniques to recover the symbols are needed to ensure reliable communication. When adopting commercial image sensors as receivers, it is necessary to take into account the diversity of these cameras, such as color filters, type and arrangement, and frame rate. Due to the variety, the same transmitted CSK symbol can be perceived differently by different cameras. It is essential to design an adaptive mechanism to reduce the demodulation errors.

9.2.3 Spatial OFDM

It is known that optical MIMO systems have the potential to provide higher data rate than their single-input single-output (SISO) counterparts. However, it is also demonstrated that a non-imaging optical MIMO system provides little diversity gain due to the ill-conditioned channel matrix. For an imaging receiver, optical lens can help distinguish the source light images and reduce the channel correlation, which makes the imaging receiver one of the most efficient ways to apply optical MIMO. The pixel-based system is a typical imaging MIMO systems, where a series of image frames are transmitted, and a lens along with an array of pixelized photodetectors is used to capture images at the receiver. The pixelated imaging optical MIMO system is shown in Fig. 9.13, where a 2-D array of optical elements sends information encoded as a sequence of images. Some intensity modulators (IM) such as the liquid crystal displays (LCDs) or LED arrays can be used as transmitters. A charge-coupled device (CCD) or CMOS camera can be used as a direct detection (DD)-based receiver. The imaging optical MIMO systems are attractive candidates for many applications in highly dense contention scenarios, or near-field communication (NFC) applications such as mobile advertisement, data exchange and secure communication in military applications. A possible outdoor application of an imaging optical MIMO system is in intelligent transportation system (ITS), where LED traffic lights and LED automobile headlights can be used to transmit driver assistance information, and a camera mounted in a vehicle is used to detect the signal.

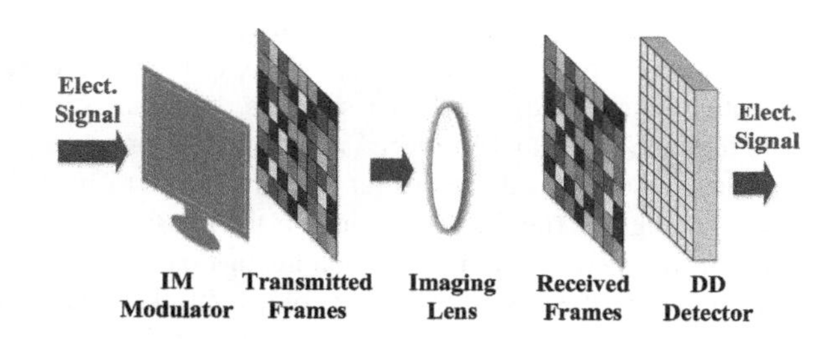

Figure 9.13 Illustration of an imaging optical MIMO communication system.

Recently, spatial orthogonal frequency division multiplexing (spatial OFDM) was proposed, which encodes information in the spatial-frequency domain [6, 17–20]. Such a design is inspired by the popular OFDM transmission scheme that encodes data in time frequencies. The spatial OFDM modulation forms images by transmitting data in spatial-frequency bins subject to a loading algorithm. By inserting training sequences before the data transmission, the receiver can determine the signal-to-noise ratio (SNR) in each spatial-frequency bin. With this information, the transmitter can perform power allocation and load the spatial-frequency bins appropriately. This technique allows for efficient transmission over the spatial-frequency selective optical channel. More importantly, a spatial OFDM system is capable of tackling the perspective distortion by detecting and correcting such distortions from the communication perspective. By designing and adaptive coding at the transmitter, the receiver can decode the images after applying simple correction algorithms with a low computational complexity. This is different from perspective transformation, which is widely studied in computer vision.

The transmitted signals in spatial OFDM are constrained to be non-negative for IM/DD-based imaging optical MIMO system. DC-biased optical OFDM (DCO-OFDM) and asymmetrically clipped optical OFDM (ACO-OFDM) are two forms of optical OFDM that have been developed for 1-D VLC. Accordingly, the 2-D spatial DCO-OFDM (SDCO-OFDM) [6, 17, 18, 21] and spatial ACO-OFDM (SACO-OFDM) [19] can be adopted in an imaging optical MIMO system. Similar to 1-D optical OFDM, SDCO-OFDM uses a DC bias to convert the bipolar spatial OFDM signal into unipolar, and thus suffers poor power efficiency. The SACO-OFDM has been shown to be more efficient in terms of electrical as well as optical power, where only the odd spatial frequencies are used to carry data and the resultant bipolar signals are clipped at zero to generate a unipolar signal.

The concept of a generalized spatial OFDM system is shown in Fig. 9.14. Consider SACO-OFDM first. For each of the transmitted SACO-OFDM frames, the input data is mapped onto the $N_1 \times N_2$ matrix $\mathbf{X}$ of constellation symbols given by

$$\mathbf{X} = \begin{pmatrix} 0 & X_{0,1} & 0 & \cdots & X_{0,N_2-1} \\ \vdots & \vdots & \vdots & \ddots & \vdots \\ 0 & X_{N_1-1,1} & 0 & \cdots & X_{N_1-1,N_2-1} \end{pmatrix}. \tag{9.10}$$

Each element of $\mathbf{X}$ denotes the symbol encoded in the corresponding spatial-frequency subcarrier. Therefore, X_{k_1,k_2} represents the signal on the (k_1, k_2)th subcarrier, where $0 \leq k_1 < N_1$ and $0 \leq k_2 < N_2$. In order to ensure a real-valued matrix $\mathbf{x}$ as the 2-D inverse fast Fourier transform (IFFT) output of $\mathbf{X}$, Hermitian symmetry is maintained for $\mathbf{X}$. Assuming that both N_1 and N_2 are even integers, the Hermitian symmetry for $\mathbf{X}$ can be defined as

$$X_{k_1,k_2} = X^*_{N_1-k_1,N_2-k_2}, \tag{9.11}$$

where k_1 and k_2 are the row and column indices respectively, and '$*$' is the complex conjugate operator. In a SACO-OFDM system, data are mapped onto only odd index columns (or odd subcarriers) of $\mathbf{X}$, while the even index columns (or even subcarriers) are set to zero. The use of only odd subcarriers ensures that the desired signal is not affected by the clipping noise due to asymmetrical clipping. Next, the bipolar output $\mathbf{x}$ from the 2-D IFFT processing of $\mathbf{X}$ is generated, and the elements of $\mathbf{x}$ are given by

$$x_{l_1,l_2} = \frac{1}{N_1 N_2} \sum_{k_1=0}^{N_1-1} \sum_{k_2=0}^{N_2-1} X_{k_1,k_2} \exp\left(\frac{j2\pi k_1 l_1}{N_1} + \frac{j2\pi k_2 l_2}{N_2}\right), \tag{9.12}$$

where (l_1, l_2) is the 2-D spatial index, with $0 \le l_1 < N_1$ and $0 \le l_2 < N_2$. The bipolar signal x_{l_1,l_2} is then converted to an unipolar signal s_{l_1,l_2} by asymmetrical clipping at the zero amplitude level, as

$$s_{l_1,l_2} = \begin{cases} 0, & x_{l_1,l_2} < 0, \\ x_{l_1,l_2}, & x_{l_1,l_2} \ge 0. \end{cases} \tag{9.13}$$

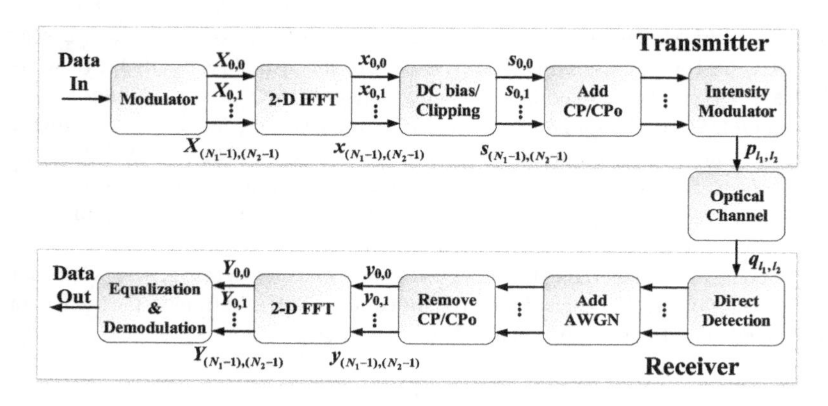

Figure 9.14 Block diagram of a spatial OFDM system [27].

For SDCO-OFDM, $\mathbf{X}$ is also constrained by Hermitian symmetry. However, in SDCO-OFDM, data is encoded onto both odd and even subcarriers of $\mathbf{X}$ given by

$$\mathbf{X} = \begin{pmatrix} X_{0,0} & X_{0,1} & X_{0,2} & \cdots & X_{0,N_2-1} \\ \vdots & \vdots & \vdots & \ddots & \vdots \\ X_{N_1-1,0} & X_{N_1-1,1} & X_{N_1-1,2} & \cdots & X_{N_1-1,N_2-1} \end{pmatrix}, \tag{9.14}$$

where $X_{0,0}$ is not used for data transmission. To generate an unipolar signal x_{l_1,l_2} as required in an IM/DD system, a DC bias b_{DC} is added to x_{l_1,l_2} so that the resultant signal does not take a negative value. The DC biased signal can be described as

$$x^b_{l_1,l_2} = x_{l_1,l_2} + b_{\text{DC}}, \tag{9.15}$$

where $b_{\text{DC}} = \mu\sigma_x$ is the DC bias, μ is a proportionality constant and $\sigma_x = \sqrt{E\{x^2_{l_1,l_2}\}}$ is the power of x_{l_1,l_2}. Since the bipolar signal x_{l_1,l_2} before clipping has a Gaussian distribution, there may exist some negative time domain samples even with a large DC bias. To remove this negative part, the resulting signal is clipped at zero which generates clipping noise. The clipping noise can be reduced by using a high level of DC bias at the cost of increased average optical power. So it is required to select an optimum DC bias $b_{\rm DC}$ to have low clipping noise with comparatively small power wastage. The clipped signal is given by

$$s_{l_1,l_2} = Kx_{l_1,l_2} + b_{\rm DC} + d_{l_1,l_2}, \tag{9.16}$$

where K is clipping attenuation and is approximately equal to $1 - Q(b_{\rm DC}/\sigma_x)$ with $Q(\cdot)$ denoting the Q-function, and d_{l_1,l_2} is the clipping noise. Since K can easily be compensated at the receiver, we set $K = 1$ for simplicity.

For both SACO-OFDM and SDCO-OFDM, cyclic extension in the form of a cyclic prefix (CP) and a cyclic postfix (CPo) is appended around the edges of clipped signal matrix. The last part of the transmitted frame represents the modulated signals $\mathbf{s}$. Assuming the electrical-to-optical conversion efficiency is ς, the intensity of the transmitted signal p_{l_1,l_2} is given by

$$p_{l_1,l_2} = \varsigma s_{l_1,l_2}. \tag{9.17}$$

The transmitted optical signal is affected by spatial distortion before being collected by the pixels. The intensity of the received signal q_{l_1,l_2}, as a function of p_{l_1,l_2}, is determined by the joint effects of various spatial distortions. The received intensity value is then converted back to an electrical signal by the pixel-based detector. Moreover, the electrical signal suffers from the channel noise z_{l_1,l_2} composed of shot noise and thermal noise modeled as AWGN. Next, the CP and CPo are removed from the noisy signal, yielding the following received electrical signal y_{l_1,l_2} as

$$y_{l_1,l_2} = R_p \cdot p_{l_1,l_2} \otimes h_{l_1,l_2} + z_{l_1,l_2} = R_p q_{l_1,l_2} + z_{l_1,l_2}, \tag{9.18}$$

where R_p is the responsivity of the photo-detecting elements, h_{l_1,l_2} is the system point spread function (PSF), $\otimes$ represents the 2-D linear convolution, and z_{l_1,l_2} is channel noise. The PSF is linear, spatially invariant, and determined by the joint effects of impairment factors. After frame synchronization, channel estimation, and equalization, 2-D FFT is then performed to obtain a spatial-frequency domain signal Y_{k_1,k_2}, which is further demodulated to recover the data.

9.2.4 Spatial WPDM

To generate real-valued and positive signals, various optical OFDM schemes for the IM/DD system have been proposed. Note that, generating the real-valued signals imposes extra loss of at least 50% reduction in spectral efficiency due to Hermitian symmetry redundancy. Moreover, the cyclic prefix as time-domain guard interval, further reduces the transmission spectral efficiency. The other disadvantages of OFDM-based transmission are the high peak-to-average power ratio (PAPR) and high side lobe. On the other hand, WPDM has been proposed as an alternative for OFDM [22, 23]. As one type of filter bank multicarrier (FBMC), WPDM employs orthogonal wavelet packet functions for symbol modulation. Similarly to OFDM, it provides orthogonality between subcarriers, whilst the basis functions are wavelet packet functions with finite length in time domain. The advantages of WPDM over OFDM lie in higher spectral and power efficiency, lower PAPR, and stronger resistance to inter-symbol-interference (ISI). For these merits, WPDM has been proposed for ultra-wideband (UWB) communication, power line communication (PLC), and optical fiber communication. It is also a candidate waveform for 5G transmission system. In 2015, WPDM was introduced in 1-D VLC with optimal waveform design in the presence of LED dispersion [24], showing the capability of enhancing the performance over OFDM, in terms of superior out-of-band power leakage suppression, lower PAPR, stronger resistance to the LED nonlinearity, and channel dispersion. Moreover, the optimized waveform for the LED dispersion achieves performance gain over the conventional wavelet basis function.

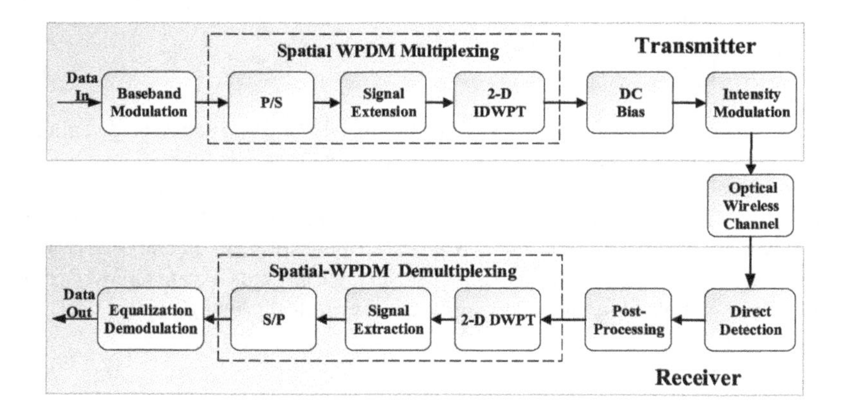

Figure 9.15 Block diagram of a spatial WPDM system.

Considering an imaging optical MIMO communication system where data is mapped onto the spatial-frequency domain, the 1-D optical WPDM scheme can also be extended to spatial WPDM. Figure 9.15 shows the block diagram of a generalized spatial WPDM system. In a similar way to the 2-D IFFT and 2-D FFT processing in spatial OFDM system, the 2-D inverse discrete wavelet packet transform (IDWPT) and the 2-D DWPT are performed at the transmitter and receiver, respectively. Although the WPDM modulation can be implemented by fast iterative Mallat algorithm for infinite-length signals, the boundary effect occurs because of finite length when the transmitted signal is convolved with the wavelet filters. This will lead to aliasing in signal reconstruction. To alleviate this effect, the input signal has to be extended via adding zeros before the WPDM modulation by symmetric boundary extension [24]. The signal extension and extraction are needed in WPDM implementation. Otherwise, the error floor will significantly degrade the system performance. In addition, another significant difference between spatial WPDM and spatial OFDM is that the waveform bases for OFDM schemes are defined in the complex domain, but WPDM bases can be defined either in the real domain or complex domain, which is determined by the scaling and dilatation filters. For IM/DD system, the spatial WPDM signal must be real and positive. It is thus more preferable to define the WPDM bases in the real domain and adopt M-PAM modulation on each subcarrier. Moreover, WPDM belongs to overlapped transformation, yielding better spectral concentration. Meanwhile, WPDM waveforms overlap in the time domain. There is no need for CP/CPo which could improve the bandwidth efficiency.

In Japan, the concept of spatial WPDM was adopted in road-to-vehicle VLC from the LED array to the high-speed camera [25, 26]. The measurement results from field trials at a speed of 30 km/h show that the spatial WPDM is more robust to the spatial nonideal conditions such as perspective distortion and blur effect, and can support long distance communication. In particular, one can adaptively choose the transceiver structure attributed to the flexibility and adaptability of the orthogonal bases. The number of subcarriers no longer needs to be power of 2 for each dimension, and the optimal orthogonal basis waveforms can be designed to meet the requirement. It is expected that the optimal spatial WPDM system is more robust to spatial distortion and blur effect, and has superior out-of-band power leakage suppression performance, lower PAPR, and high spectral efficiency.

9.3 System impairment factors

The system performances are impaired by a number of factors inherent to OCC channels, including linear misalignment between the transmitted and received images, perspective distortion, which is the geometry distortion of the original object, illumination fall-off at the received images termed as blur effect, vignetting, which is the blurring effect because of the limited diffraction optics subsystem. Some other problems such as temporal and spatial synchronization, and rotational misalignment are also inevitable. Theoretical and simulation results for their effects in the presence of received noise are analyzed subsequently in this section. Moreover, the associated mitigation techniques such as temporal and spatial frame synchronization, blur-adaptive coding, and perspective correction are introduced.

9.3.1 Impairment factors in spatial OFDM

In this subsection, we analyze some spatial impairment factors in the spatial OFDM-based imaging optical MIMO system, including linear misalignment [19, 27], perspective distortion due to geometry distortion [18, 28], blur effect [28, 29], and vignetting [30] in the presence of channel noise. Moreover, the associated mitigation techniques such as equalization, blur-adaptive coding, and perspective correction are introduced.

9.3.1.1 Linear misalignment

The performance of imaging optical MIMO system can be impaired by linear misalignment, since it is practically impossible to perfectly align the pixel in image sensor with the transmitted image in spatial domain [19, 27]. The misalignment occurs as an integer multiple of the side-length of the pixels or a fractional number of pixels. We theoretically analyze and numerically simulate the effect of integer and fractional linear misalignment on spatial OFDM.

For an imaging optical MIMO system, the receiver samples the incoming image in the spatial domain. The spatial sampling rate depends on the spacing of the receiver pixels, and each pixel is a spatial sample. To simplify the analysis of linear misalignment effect, the blur effect or magnification is not taken into account. Besides, the number of data-carrying transmitted pixels is equal to the number of received pixels, and all the pixels are square-shaped. When the misalignment is in one or two dimensions, a receiving pixel will be affected by two or four transmitted pixels, respectively. Figure 9.16 illustrates the fractional misalignment in both dimensions. Figure 9.16(a) shows a small frame of 4×4 pixels without CP and CPo, whilst Fig. 9.16(b) indicates that a transmitted pixel is divided conceptually into four segments by the borders of the received pixels. The area of each square-shaped transmitted pixel is $A_r = m^2$, where m is the length of a pixel. Then the four segments have the normalized areas of $A_1 = m_r m_c/m^2$, $A_2 = m_r(m - m_c)/m^2$, $A_3 = (m - m_r)m_c/m^2$, and $A_4 = (m - m_r)(m - m_c)/m^2$, where m_r and m_c are the sampling offsets in two dimensions. The intensity of a received pixel q_{l_1,l_2} is contributed by up to four transmitted pixels based on their normalized areas

$$\begin{aligned} q_{l_1,l_2} = {} & A_4 p_{(l_1+\Delta l_1),(l_2+\Delta l_2)} + A_3 p_{(l_1+\Delta l_1),(l_2+\Delta l_2+1)} \\ & + A_2 p_{(l_1+\Delta l_1+1),(l_2+\Delta l_2)} + A_1 p_{(l_1+\Delta l_1+1),(l_2+\Delta l_2+1)}, \end{aligned} \tag{9.19}$$

where Δl_1 and Δl_2 are the number of offset pixels along the two dimensions. Therefore, the total length of misalignment Δld_1 and Δld_2 in each dimension can be expressed as

$$\Delta ld_1 = (\Delta l_1 + \frac{m_r}{m})m, \text{and } \Delta ld_2 = (\Delta l_2 + \frac{m_c}{m})m. \tag{9.20}$$

Now we analyze the effect of fractional misalignment in the spatial-frequency domain. The received electrical signal after the deduction of CP and CPo can be obtained by modifying as (9.18)

$$\begin{aligned} y_{l_1,l_2} = {} & A_4 s_{(l_1+\Delta l_1),(l_2+\Delta l_2)} + A_3 s_{(l_1+\Delta l_1),(l_2+\Delta l_2+1)} \\ & + A_2 s_{(l_1+\Delta l_1+1),(l_2+\Delta l_2)} + A_1 s_{(l_1+\Delta l_1+1),(l_2+\Delta l_2+1)} + z_{l_1,l_2}. \end{aligned} \tag{9.21}$$

To simplify analysis, we assume that $\varsigma = 1$ in (9.17) and $R_p = 1$ in (9.18). s_{l_1,l_2} is the transmitted signal before the addition of CP and CPo. For SACO-OFDM, the generated unipolar signal s_{l_1,l_2} can be expressed as X_{k_1,k_2} after using the formula

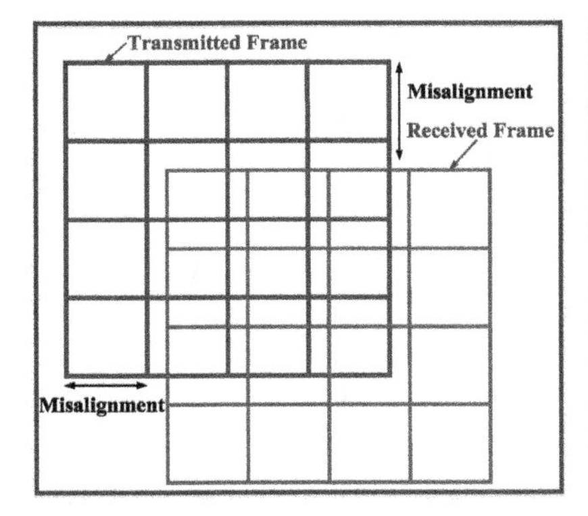

(a) The transmitted and received frames

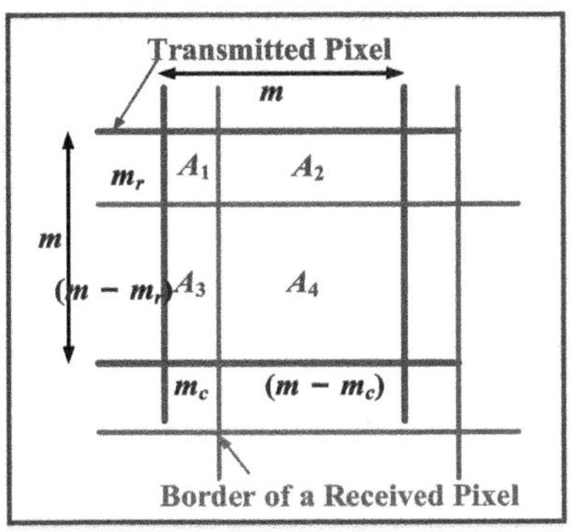

(b) A transmitted pixel and a received pixel

Figure 9.16 Illustration of the linear misalignment between transmitted and received frames [27].

of 2-D IFFT, as

$$s_{l_1,l_2} = \frac{1}{N_1 N_2} \sum_{k_1=0}^{N_1-1} \sum_{k_2=0}^{N_2-1} X_{k_1,k_2} \exp\left(\frac{j2\pi k_1 l_1}{N_1} + \frac{j2\pi k_2 l_2}{N_2}\right), \tag{9.22}$$

where (k_1, k_2) represents the 2-D spatial frequency. By applying (9.22) into (9.21), y_{l_1,l_2} is formulated as

$$y_{l_1,l_2} = \frac{1}{N_1 N_2} \sum_{k_1=0}^{N_1-1} \sum_{k_2=0}^{N_2-1} W''_{k_1,k_2} \exp\left(\frac{j2\pi k_1 l_1}{N_1} + \frac{j2\pi k_2 l_2}{N_2}\right) + z_{l_1,l_2}, \tag{9.23}$$

where

$$\begin{aligned} W''_{k_1,k_2} = X_{k_1,k_2} \exp\left(\frac{j2\pi k_1 \Delta l_1}{N_1} + \frac{j2\pi k_2 \Delta l_2}{N_2}\right)\bigg\{ A_4 + A_3 \exp\left(\frac{j2\pi k_2}{N_2}\right) \\ + A_2 \exp\left(\frac{j2\pi k_1}{N_1}\right) + A_1 \exp\left(\frac{j2\pi k_1}{N_1} + \frac{j2\pi k_2}{N_2}\right)\bigg\}. \end{aligned} \tag{9.24}$$

To obtain the received constellations Y_{k_1,k_2}, 2-D FFT is performed on (9.23), After some rearrangement, we obtain

$$Y_{k_1,k_2} = X_{k_1,k_2} H'_{k_1,k_2} H''_{k_1,k_2} + Z_{k_1,k_2}, \tag{9.25}$$

where Z_{k_1,k_2} is the AWGN noise in the spatial-frequency domain. Moreover,

$$H'_{k_1,k_2} = \exp\left(\frac{j2\pi k_1 \Delta l_1}{N_1} + \frac{j2\pi k_2 \Delta l_2}{N_2}\right). \tag{9.26}$$

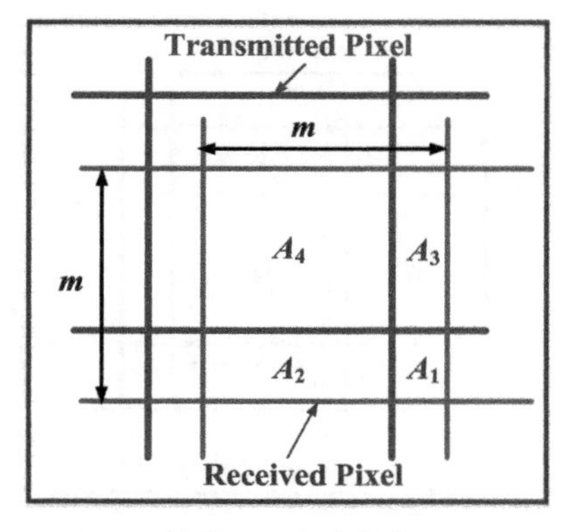

(a) One received pixel

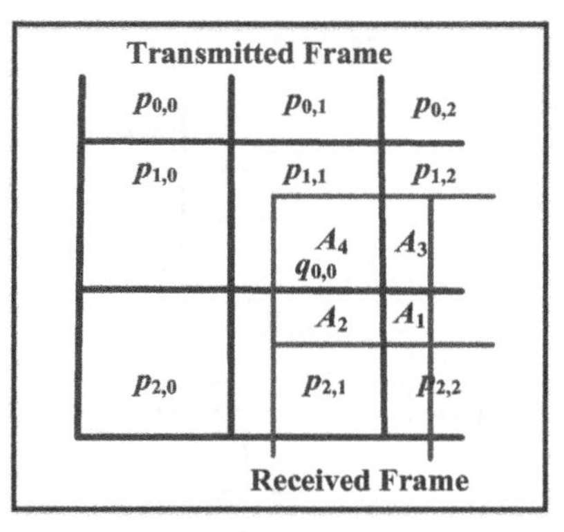

(b) A transmitted and a received frame

Figure 9.17 Illustration of a received pixel depending on four transmitted pixels.

and

$$\begin{aligned} H''_{k_1,k_2} &= A_4 + A_3 \exp\left(\frac{j2\pi k_2}{N_2}\right) + A_2 \exp\left(\frac{j2\pi k_1}{N_1}\right) \\ &+ A_1 \exp\left(\frac{j2\pi k_1}{N_1} + \frac{j2\pi k_2}{N_2}\right). \end{aligned} \tag{9.27}$$

The exponential terms in H'_{k_1,k_2} and H''_{k_1,k_2} will both cause phase rotation. The rotation angle depends linearly on both the subcarrier indices (k_1, k_2) and the extent of the misalignment $(\Delta l_1, \Delta l_2)$. In addition to the phase shift, H''_{k_1,k_2} also induces attenuation, which is a function of A_1, A_2, A_3, and A_4 that depend on the magnitude of the spatial sampling offset. Besides, the attenuation also depends on the spatial frequency. It means that, higher spatial frequency subcarriers may experience greater attenuation, since they change more rapidly with spatial distance than lower frequency subcarriers.

Figure 9.18 shows the constellation of four-phase quadrature amplitude modulation (4-QAM) SACO-OFDM signals in the presence of integer and fractional misalignment. It is observed from Fig. 9.18(a) that, when only integer misalignment is considered, the received constellation points experience only phase rotation, which depends only on the spatial frequency. However, in the case of fractional misalignment, the constellation points experience not only phase rotation but also amplitude attenuation as shown in Fig. 9.18(b). Similar to the rotation, the amplitude attenuation is also dependent on the spatial subcarrier index.

Figure 9.19(a) illustrates the amplitude attenuation of the received constellation points versus spatial subcarrier index of a SACO-OFDM system in the case of fractional misalignment with a sampling offset of 70%. From Fig. 9.19(a), the Nyquist rows thus, the $\frac{N_1}{2}$th and $(\frac{N_1}{2} + 1)$th subcarriers, suffer greater attenuation than the first row subcarrier, which indicates that higher spatial subcarriers suffer greater at-

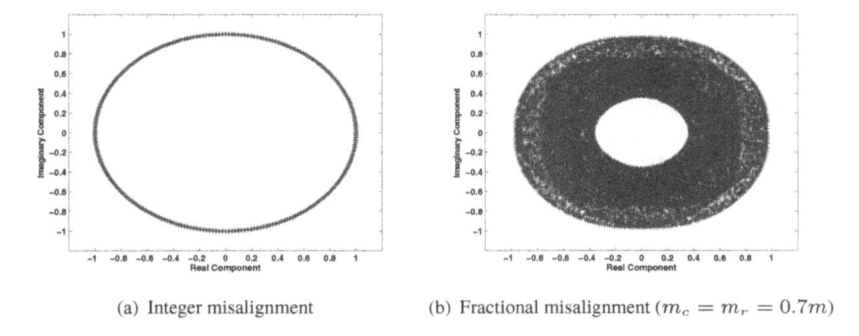

(a) Integer misalignment (b) Fractional misalignment ($m_c = m_r = 0.7m$)

Figure 9.18 Constellation of SACO-OFDM impaired by linear misalignment.

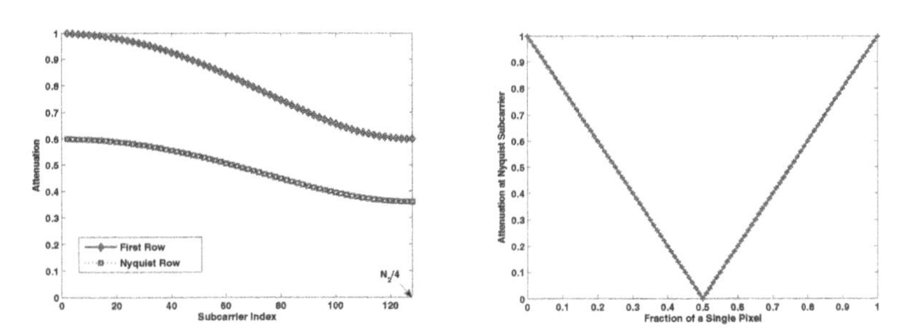

(a) Attenuation versus subcarrier index in one row of 4-QAM SACO-OFDM frame (b) Attenuation at the highest subcarrier with different offset in SACO-OFDM [27]

Figure 9.19 Attenuation versus subcarrier index and fraction of misalignment.

tenuation than lower spatial subcarriers. Moreover, the attenuation of the subcarrier depends on the amount of the spatial sampling offset as shown in Fig. 9.19(b). It illustrates that the attenuation at the subcarrier is the greatest and equal to 50% of a pixel. Theoretically, the value of $|H''_{k_1,k_2}|$ is minimum in (9.27) for the case of $m_r = m_c = 0.5m$. In such a case, the intensity of a received pixel depends equally on the four transmitted pixels, leading to maximum spatial averaging effect.

9.3.1.2 **Perspective distortion**

The perspective distortions are caused by the nature of the imaging system mechanism and depend on the perspective of the image sensor. They manifest as deformation in size and shape of the captured object on the image, resulting in visual compression or magnification of the transmitter's projection on the image [18, 28]. Moreover, when the transmitted element is at an out-of-focus distance from the imaging lens, these distortions become prominent and lead to interference between adjacent transmitted pixels on the image, namely the inter-pixel interference (IPI). The perspective distortions cause the transmitted elements to deform in size when the transmitter is not at the focus of the image sensor, and in shape when it is not frontally aligned (viewed at an angle) with the image sensor. If the transmitted element is at the focus, and assuming the transmitter and image sensor have the same resolution, its image on the image sensor should occupy the same area as one pixel. Actually, the light rays from the transmitted element may not end exactly on image sensor pixel boundaries, and there is some surrounding area that accumulates interference. This area of misalignment and geometry of the imaged transmitted element will be perspective-dependent and accounts for distortion due to the perspective scaling of the pixel area.

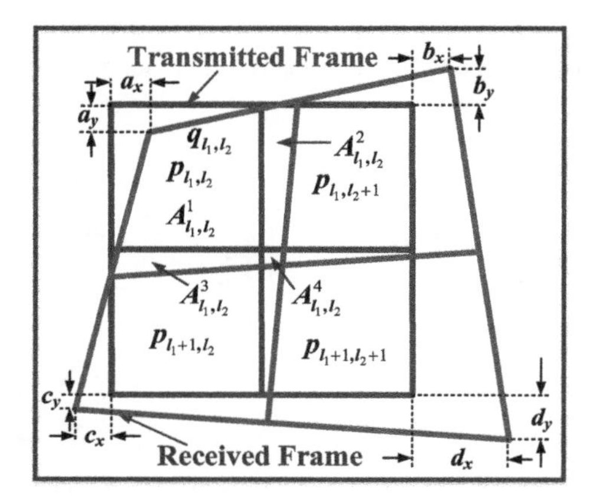

Figure 9.20 Illustration of a received frame with geometry perspective distortion.

In this section, we only focus on the geometry distortion but ignore the blur distortion. The illustration of received geometry perspective distortion is provided in Fig. 9.20. Similar to the linear misalignment, the received pixel depends on up to four transmitted pixels due to blur effect, and the intensity of a received pixel q_{l_1,l_2} can be expressed as a function of the intensities of the transmitted pixels. Noteworthily, for the received pixel q_{l_1,l_2}, the four corresponding transmitted pixels have the normalized areas of $A_1^{l_1,l_2}$, $A_2^{l_1,l_2}$, $A_3^{l_1,l_2}$, and $A_4^{l_1,l_2}$, which are determined by the shift offsets of (a_x, a_y), (b_x, b_y), (c_x, c_y), and (d_x, d_y). Moreover, the normalized areas of interference pixels are spatial dependent, different from the linear misalignment. If $a_x = b_x = c_x = d_x$ and $a_y = b_y = c_y = d_y$, then all the pixels experience the constant shift, and the normalized areas of interference pixels are spatial independent. Specifically, linear misalignment is a special case of geometry perspective distortion with a constant shift offset. The received q_{l_1,l_2} can be obtained from the transmitted pixel p_{l_1,l_2} as

$$q_{l_1,l_2} = A_1^{l_1,l_2} p_{(l_1+\Delta l_1^{(1)}),(l_2+\Delta l_2^{(1)})} + A_2^{l_1,l_2} p_{(l_1+\Delta l_1^{(2)}),(l_2+\Delta l_2^{(2)}+1)}$$
$$+ A_3^{l_1,l_2} p_{(l_1+\Delta l_1^{(3)}+1),(l_2+\Delta l_2^{(3)})} + A_4^{l_1,l_2} p_{(l_1+\Delta l_1^{(4)}+1),(l_2+\Delta l_2^{(4)}+1)},$$

(9.28)

where $(\Delta l_1^{(1)}, \Delta l_1^{(2)}, \Delta l_1^{(3)}, \Delta l_1^{(4)})$ and $(\Delta l_2^{(1)}, \Delta l_2^{(2)}, \Delta l_2^{(3)}, \Delta l_2^{(4)})$ are the number of full pixels along the two dimensions, which are determined by the sampling/shift offsets. So, the expression of the received electrical signal after the deduction of CP and CPo, y_{l_1,l_2} can be obtained by modifying (9.21) as

$$\begin{aligned} y_{l_1,l_2} &= A_1^{l_1,l_2} s_{(l_1+\Delta l_1^{(1)}),(l_2+\Delta l_2^{(1)})} + A_2^{l_1,l_2} s_{(l_1+\Delta l_1^{(2)}),(l_2+\Delta l_2^{(2)}+1)} \\ &+ A_3^{l_1,l_2} s_{(l_1+\Delta l_1^{(3)}+1),(l_2+\Delta l_2^{(3)})} \\ &+ A_4^{l_1,l_2} s_{(l_1+\Delta l_1^{(4)}+1),(l_2+\Delta l_2^{(4)}+1)} + z_{l_1,l_2}. \end{aligned} \tag{9.29}$$

For SACO-OFDM, after some mathematical derivation, y_{l_1,l_2} can be simplified as

$$y_{l_1,l_2} = \frac{1}{N_1 N_2} \sum_{k_1=0}^{N_1-1} \sum_{k_2=0}^{N_2-1} U''_{k_1,k_2} \exp\left(\frac{j2\pi k_1 l_1}{N_1} + \frac{j2\pi k_2 l_2}{N_2}\right) + z_{l_1,l_2}, \tag{9.30}$$

where

$$\begin{aligned} U''_{k_1,k_2} &= X_{k_1,k_2} \left\{ A_1^{l_1,l_2} \exp\left(\frac{j2\pi k_1 \Delta l_1^{(1)}}{N_1} + \frac{j2\pi k_2 \Delta l_2^{(1)}}{N_2}\right) \right. \\ &+ A_2^{l_1,l_2} \exp\left(\frac{j2\pi k_1 \Delta l_1^{(2)}}{N_1} + \frac{j2\pi k_2 (\Delta l_2^{(2)}+1)}{N_2}\right) \\ &+ A_3^{l_1,l_2} \exp\left(\frac{j2\pi k_1 (\Delta l_1^{(3)}+1)}{N_1} + \frac{j2\pi k_2 \Delta l_2^{(3)}}{N_2}\right) \\ &\left. + A_4^{l_1,l_2} \exp\left(\frac{j2\pi k_1 (\Delta l_1^{(4)}+1)}{N_1} + \frac{j2\pi k_2 (\Delta l_2^{(4)}+1)}{N_2}\right) \right\}. \end{aligned} \tag{9.31}$$

After 2-D FFT processing, the received constellations Y_{k_1,k_2} can be equivalently expressed as

$$Y_{k_1,k_2} = X_{k_1,k_2} H^{\star\star}_{k_1,k_2} + Z_{k_1,k_2}, \tag{9.32}$$

where Z_{k_1,k_2} is the AWGN in the spatial-frequency domain, and

$$\begin{aligned} H^{\star\star}_{k_1,k_2} &= A_1^{l_1,l_2} \exp\left(\frac{j2\pi k_1 \Delta l_1^{(1)}}{N_1} + \frac{j2\pi k_2 \Delta l_2^{(1)}}{N_2}\right) \\ &+ A_2^{l_1,l_2} \exp\left(\frac{j2\pi k_1 \Delta l_1^{(2)}}{N_1} + \frac{j2\pi k_2 (\Delta l_2^{(2)}+1)}{N_2}\right) \\ &+ A_3^{l_1,l_2} \exp\left(\frac{j2\pi k_1 (\Delta l_1^{(3)}+1)}{N_1} + \frac{j2\pi k_2 \Delta l_2^{(3)}}{N_2}\right) \\ &+ A_4^{l_1,l_2} \exp\left(\frac{j2\pi k_1 (\Delta l_1^{(4)}+1)}{N_1} + \frac{j2\pi k_2 (\Delta l_2^{(4)}+1)}{N_2}\right). \end{aligned} \tag{9.33}$$

The exponential components in $H^{\star\star}_{k_1,k_2}$ will cause phase rotation and attenuation. The rotation angle depends linearly on the subcarrier indices (k_1, k_2), the extent of the geometry distortion $(\Delta l_1, \Delta l_2)$, and the spatial position (l_1, l_2). For a particular spatial-frequency, the amount of attenuation is determined by $A_1^{l_1,l_2}$, $A_2^{l_1,l_2}$, $A_3^{l_1,l_2}$, and $A_4^{l_1,l_2}$ which depend on the magnitude of the spatial shift/sampling offsets (a_x, a_y), (b_x, b_y), (c_x, c_y), and (d_x, d_y), as shown in Fig. 9.20.

9.3.1.3 Blur effect

Here, we consider only the blur effect at the receiver, while the rest of impairment factors is not taken into account except for the AWGN. To clearly illustrate the blur effect, the number of transmitted pixels is assumed to be equal to the number of received pixels. In a practical image-sensor-based receiver, imperfect focus or limited diffraction of the optics subsystem results in blurring which can be modeled by a 2-D Gaussian distribution in the spatial domain. Moreover, for the receiver made up of an array of pixel-based photodetecting elements, the PSF is a discrete rather than a continuous function. In a practical system, the presence of blurring leads to fluctuations in the intensity pattern, and the intensity is distributed over a larger space. The blurring-degraded PSF h_{l_1,l_2} can be described as a 2-D discrete symmetrical Gaussian distribution with a spread of standard deviation σ and a dimension of $(2P+1) \times (2P+1)$, where P is the length of both directions with respect to the center of the distribution [21, 29]. It is assumed that $P <$ CP and $P <$ CPo. The discrete blurring-degraded PSF h_{l_1,l_2} is given by

$$h_{l_1,l_2} = h_{0,0} \exp\left(\frac{-l_1^2 - l_2^2}{2\sigma^2}\right), \tag{9.34}$$

where $h_{0,0}$ is the maximum amplitude of the distribution. Considering this blurring effect in terms of h_{l_1,l_2}, the intensity of each received pixel becomes the summation of the intensity of $(2P+1) \times (2P+1)$ transmitted pixels weighted by the corresponding PSF elements. Hence we have

$$q_{l_1,l_2} = p_{l_1,l_2} \otimes h_{l_1,l_2} = \sum_{u=-P}^{P} \sum_{v=-P}^{P} h_{u,v} p_{l_1+u,l_2+v}. \tag{9.35}$$

Substituting (9.35) into (9.18) and using (9.17), we obtain

$$y_{l_1,l_2} = \sum_{u=-P}^{P} \sum_{v=-P}^{P} h_{u,v} s_{l_1+u,l_2+v} + z_{l_1,l_2}, \tag{9.36}$$

where we assume $\varsigma = 1$ in (9.17) and $R_p = 1$ in (9.18) to simplify the analysis.

Applying 2-D IFFT to S_{k_1,k_2}, the transmitted signal s_{l_1+u,l_2+v} can be expressed as

$$s_{l_1+u,l_2+v} = \frac{1}{N_1 N_2} \sum_{k_1=0}^{N_1-1} \sum_{k_2=0}^{N_2-1} S_{k_1,k_2} \exp\left(\frac{j2\pi k_1(l_1+u)}{N_1} + \frac{j2\pi k_2(l_2+v)}{N_2}\right).$$

$$(9.37)$$

Substituting (9.37) into (9.36) and with some rearrangement, we obtain

$$\begin{aligned} y_{l_1,l_2} &= \frac{1}{N_1 N_2} \sum_{k_1=0}^{N_1-1} \sum_{k_2=0}^{N_2-1} S_{k_1,k_2} \exp\left(\frac{j2\pi k_1 l_1}{N_1} + \frac{j2\pi k_2 l_2}{N_2}\right) \\ &\times \left\{ \sum_{u=-P}^{P} \sum_{v=-P}^{P} h_{u,v} \exp\left(\frac{j2\pi k_1 u}{N_1} + \frac{j2\pi k_2 v}{N_2}\right) \right\} + z_{l_1,l_2} \\ &= \frac{1}{N_1 N_2} \sum_{k_1=0}^{N_1-1} \sum_{k_2=0}^{N_2-1} S_{k_1,k_2} H^d_{k_1,k_2} \exp\left(\frac{j2\pi k_1 l_1}{N_1} + \frac{j2\pi k_2 l_2}{N_2}\right) \\ &+ z_{l_1,l_2}, \end{aligned} \tag{9.38}$$

where the term $H^d_{k_1,k_2}$ represents the blur effect, given by

$$\begin{aligned} H^d_{k_1,k_2} &= \sum_{u=-P}^{P} \sum_{v=-P}^{P} h_{u,v} \exp\left(\frac{j2\pi k_1 u}{N_1} + \frac{j2\pi k_2 v}{N_2}\right) \\ &\overset{(a)}{=} h_{0,0} + 2\sum_{u=1}^{P} h_{u,0} \cos\left(\frac{2\pi k_1 u}{N_1}\right) + 2\sum_{v=1}^{P} h_{0,v} \cos\left(\frac{2\pi k_2 v}{N_2}\right) \\ &+ 4\sum_{u=1}^{P}\sum_{v=1}^{P} h_{u,v} \cos\left(\frac{2\pi k_1 u}{N_1}\right) \cos\left(\frac{2\pi k_2 v}{N_2}\right), \end{aligned} \tag{9.39}$$

where (a) is due to the symmetric of PSF in 2-D [21]. The term y_{l_1,l_2} in (9.38) is the 2-D IFFT of $S_{k_1,k_2} H^d_{k_1,k_2}$ with z_{l_1,l_2}. After using 2-D FFT, the term Y_{k_1,k_2} is given by

$$Y_{k_1,k_2} = S_{k_1,k_2} H^d_{k_1,k_2} + Z_{k_1,k_2}, \tag{9.40}$$

where Z_{k_1,k_2} is the AWGN in the spatial-frequency domain.

For SDCO-OFDM, it is assumed that the clipping noise in (9.16) is $d_{l_1,l_2} = 0$. Applying 2-D FFT to the SDCO-OFDM signal given in (9.16) results in

$$S_{k_1,k_2} = X_{k_1,k_2} + B_{\text{DC}}\, \delta_{k_1,k_2}, \tag{9.41}$$

where $B_{\text{DC}} = N_1 N_2 b_{\text{DC}}$ and δ_{k_1,k_2} is the 2-D Kronecker delta function defined as

$$\delta_{k_1,k_2} = \begin{cases} 1, & (k_1,k_2) = (0,0), \\ 0, & (k_1,k_2) \neq (0,0). \end{cases} \tag{9.42}$$

Recall that, in spatial OFDM, the zeroth subcarrier, $(k_1, k_2) = (0, 0)$, is not used to carry data. Hence (9.41) results in $S_{k_1,k_2} = X_{k_1,k_2}$ for all the received subcarriers.

Similarly, $S_{k_1,k_2} = X_{k_1,k_2}$ for the data-carrying subcarriers of SACO-OFDM. So, for both SDCO-OFDM and SACO-OFDM, one can conclude that

$$S_{k_1,k_2} = X_{k_1,k_2}. \tag{9.43}$$

Substituting (9.43) into (9.40) gives

$$Y_{k_1,k_2} = X_{k_1,k_2} H^d_{k_1,k_2} + Z_{k_1,k_2}. \tag{9.44}$$

From (9.44), it can be observed that the blur effect only attenuates the original subcarrier X_{k_1,k_2} by a factor of $H^d_{k_1,k_2}$. Moreover, the attenuation is subcarrier dependent as shown in (9.39). Thus, the higher spatial frequency subcarriers will experience greater attenuation.

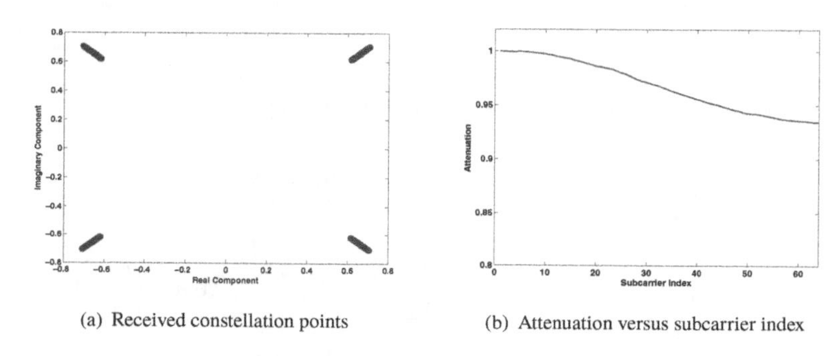

(a) Received constellation points

(b) Attenuation versus subcarrier index

Figure 9.21 Received constellation for SACO-OFDM and Gaussian PSF due to blur effect with $\sigma = 0.35$.

Figure 9.21(a) illustrates the received constellation for the case of 4-QAM SACO-OFDM, where $N_1 = N_2 = 256$, and the PSF has a standard deviation $\sigma = 0.35$. It shows that the blur effect will only attenuate the received constellation points, rather than the phase rotation. Moreover, the attenuation is subcarrier dependent as shown in Fig. 9.21(b). It shows that the amplitude attenuation of the constellation points reduces gradually from unity at DC spatial frequency to 0.934 at the highest spatial frequency. This observation verifies that the constellations are attenuated depending on the spatial frequencies and the variance of the PSF distribution.

9.3.1.4 **Vignetting effect**

For imaging optical MIMO communication, the system is also impaired by vignetting effect which is the gradual illumination fall-off from the center to the corners of the received images [30]. The level of vignetting depends on the geometry of the lens optics, aperture setting, and other optical properties of the receiver. Especially, the cosine-fourth radiometric effect is the most prominent factor that contributes to vignetting.

Here, we consider only the vignetting effect at the receiver, while the rest of the impairment factors is not taken into account except for the addition of AWGN. Figure 9.22 shows the setup of the imaging optical MIMO system, where $a_1b_1c_1d_1$ is

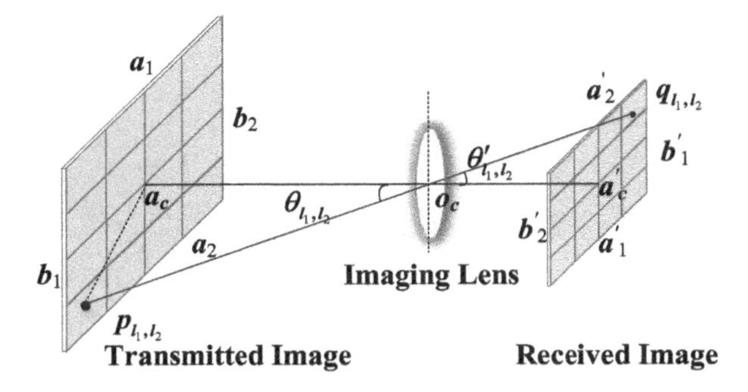

Figure 9.22 Illustration of imaging optical MIMO system using an optical lens.

the transmitted frame and $a_1'b_1'c_1'd_1'$ is the corresponding received frame. The centre of the receiver lens is o_c. The transmitted intensity signal from any pixel-based transmitted element p_{l_1,l_2} creates an angle θ_{l_1,l_2} to the optical axis $a_c o_c$. The corresponding received pixel q_{l_1,l_2} is also off-axis by θ_{l_1,l_2}. For simplicity, it is assumed that the lens is lossless and gives a constant magnification across the whole transmitting image. The aperture of the lens is small. With these assumptions, the intensity of the received pixel q_{l_1,l_2} scaled by the vignetting effect is as follows

$$q_{l_1,l_2} = \cos^4\theta_{l_1,l_2} \cdot p_{l_1,l_2}, \tag{9.45}$$

where the vignetting function $\cos^4\theta_{l_1,l_2}$ for any pixel p_{l_1,l_2} can be expressed as

$$\cos\theta_{l_1,l_2} = \frac{a_c o_c}{\sqrt{(a_c o_c)^2 + (a_c a_{l_1,l_2})^2}}, \tag{9.46}$$

where $a_c a_{l_1,l_2}$ is the distance between the centre of the transmitted image a_c and the pixel p_{l_1,l_2}, and $a_c o_c$ is the distance between the lens centre and a_c. From (9.45) and (9.46), it is clear that the vignetting function tends to approach unity when $a_c o_c \gg a_c a_{l_1,l_2}$ or $a_c o_c \approx o_c a_{l_1,l_2}$. Therefore, the size of the transmitted image and its distance from the lens centre have an impact on the maximum level of vignetting. Consequently, the amount of vignetting for any particular transmitter and receiver setup depends on two ratios: $\gamma_1 = \frac{a_1 a_2}{a_c o_c}$ and $\gamma_2 = \frac{b_1 b_2}{a_c o_c}$, where $a_1 a_2$ and $b_1 b_2$ are the height and width of the transmitted image. The imaging optical MIMO system has potential usage in short range communication where the ratios γ_1 and γ_2 are not negligible. Thus the level of vignetting must be considered.

To simplify the analysis, we assume that $\varsigma = 1$ in (9.17) and $R_p = 1$ in (9.18). Substituting (9.45) into (9.18) gives

$$y_{l_1,l_2} = \cos^4\theta_{l_1,l_2} \cdot s_{l_1,l_2} + z_{l_1,l_2}. \tag{9.47}$$

Applying a 2-D FFT to (9.47), then rearranging and separating out the component at

(k_1', k_2'), it gives

$$Y_{k_1',k_2'} = \frac{1}{N_1 N_2}\left\{S_{k_1',k_2'}\left(\sum_{l_1=0}^{N_1-1}\sum_{l_2=0}^{N_2-1}\cos^4\theta_{l_1,l_2}\right) + \sum_{k_1=0}^{N_1-1}\right.$$
$$\times \sum_{k_2=0}^{N_2-1} S_{k_1,k_2} \cdot \sum_{l_1=0}^{N_1-1}\sum_{l_2=0}^{N_2-1}\cos^4\theta_{l_1,l_2}\exp\left(\frac{j2\pi(k_1-k_1')l_1}{N_1}\right. \tag{9.48}$$
$$\left.\left.+\frac{j2\pi(k_2-k_2')l_2}{N_2}\right)\right\} + Z_{k_1',k_2'}, \quad (k_1,k_2)\neq(k_1',k_2').$$

For SACO-OFDM, only the odd subcarriers are used to carry data. When an SACO-OFDM signal is asymmetrically clipped, all the clipping noise falls on the even subcarriers. Although clipping halves the amplitude of the odd subcarriers, it can be corrected by multiplying by a factor of two before transmission. This leads to the following frequency symbol

$$S_{k_1,k_2} = \begin{cases} X_{k_1,k_2}, & \text{when } k_2 \text{ is odd,} \\ D_{k_1,k_2}, & \text{when } k_2 \text{ is even,} \end{cases} \tag{9.49}$$

where D_{k_1,k_2} is the clipping noise in the spatial-frequency domain. For SDCO-OFDM, S_{k_1,k_2} can be obtained by applying 2-D FFT to (9.16), where we assume $K = 1$ and the clipping noise $d_{l_1,l_2} = 0$ to simplify analysis, as

$$S_{k_1,k_2} = X_{k_1,k_2} + B_{\mathrm{DC}}\delta_{k_1,k_2}. \tag{9.50}$$

Substituting (9.49) and (9.50) into (9.48), respectively, both SACO-OFDM and SDCO-OFDM experience attenuation $H_{k_1',k_2'}$, and ICI term $I_{k_1',k_2'}$, as

$$Y_{k_1',k_2'} = X_{k_1',k_2'}H_{k_1',k_2'} + I_{k_1',k_2'} + Z_{k_1',k_2'}, \tag{9.51}$$

where

$$H_{k_1',k_2'} = \frac{1}{N_1 N_2}\sum_{l_1=0}^{N_1-1}\sum_{l_2=0}^{N_2-1}\cos^4\theta_{l_1,l_2}. \tag{9.52}$$

For SACO-OFDM, the ICI term $I_{k_1',k_2'}$ is given by

$$I_{k_1',k_2'} = \frac{1}{N_1 N_2}\sum_{k_1=0}^{N_1-1}\sum_{k_2=0}^{N_2-1}(X_{k_1,k_2}+D_{k_1,k_2})\sum_{l_1=0}^{N_1-1}$$
$$\sum_{l_2=0}^{N_2-1}\cos^4\theta_{l_1,l_2}\exp\left(\frac{j2\pi(k_1-k_1')l_1}{N_1}\right. \tag{9.53}$$
$$\left.+\frac{j2\pi(k_2-k_2')l_2}{N_2}\right), \quad (k_1,k_2)\neq(k_1',k_2').$$

For SDCO-OFDM, the ICI term $I_{k_1',k_2'}$ is given by

$$I_{k_1',k_2'} = \frac{1}{N_1N_2}\sum_{k_1=0}^{N_1-1}\sum_{k_2=0}^{N_2-1} X_{k_1,k_2}\sum_{l_1=0}^{N_1-1}\sum_{l_2=0}^{N_2-1}\cos^4\theta_{l_1,l_2}\cdot \exp\left(\frac{j2\pi(k_1-k_1')l_1}{N_1}+\frac{j2\pi(k_2-k_2')l_2}{N_2}\right),\quad (k_1,k_2)\neq(k_1',k_2'). \tag{9.54}$$

In Eqs. (9.52)–(9.54), the attenuation term, $H_{k_1',k_2'}$ is independent of (k_1',k_2'), while the inter-carrier interference (ICI) term, $I_{k_1',k_2'}$ is a function of X_{k_1,k_2} and $\cos^4\theta_{l_1,l_2}$.

The analysis of ICI term can be simplified by defining an ICI complex gain $F_{\tilde{k}_1,\tilde{k}_2}$, given by

$$F_{\tilde{k}_1,\tilde{k}_2} = \frac{1}{N_1N_2}\sum_{l_1=0}^{N_1-1}\sum_{l_2=0}^{N_2-1}\cos^4\theta_{l_1,l_2}\exp\left(\frac{j2\pi\tilde{k}_1l_1}{N_1}+\frac{j2\pi\tilde{k}_2l_2}{N_2}\right),\quad (\tilde{k}_1,\tilde{k}_2)\neq(0,0), \tag{9.55}$$

where $(\tilde{k}_1 = k_1 - k_1')$ and $(\tilde{k}_2 = k_2 - k_2')$ represent the spacing between the subcarriers S_{k_1,k_2} and $S_{k_1',k_2'}$ in the two dimensions, and the term $S_{k_1,k_2}F_{\tilde{k}_1,\tilde{k}_2}$ denotes the ICI contribution of S_{k_1,k_2} to $S_{k_1',k_2'}$. Figure 9.23(a) shows the absolute value of $F_{\tilde{k}_1,\tilde{k}_2}$ versus $\tilde{k}_2$ for $N_1 = N_2 = 256$, $\tilde{k}_1 = 0$, and for three levels of vignetting. It is apparent that the value of $|F_{\tilde{k}_1,\tilde{k}_2}|$ is significant only for small $\tilde{k}_2$, and the peak is at $\tilde{k}_2 = \pm 1$. For other values of $\tilde{k}_1$, they have a similar form but the absolute values are smaller.

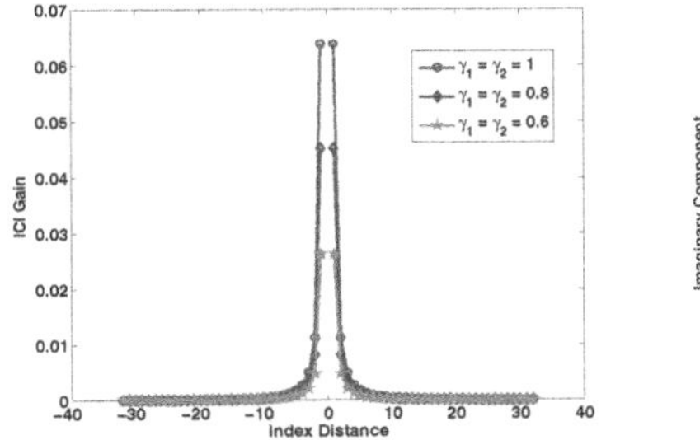

(a) The ICI gain $|F_{\tilde{k}_1,\tilde{k}_1}|$ versus index distance $\tilde{k}_2$

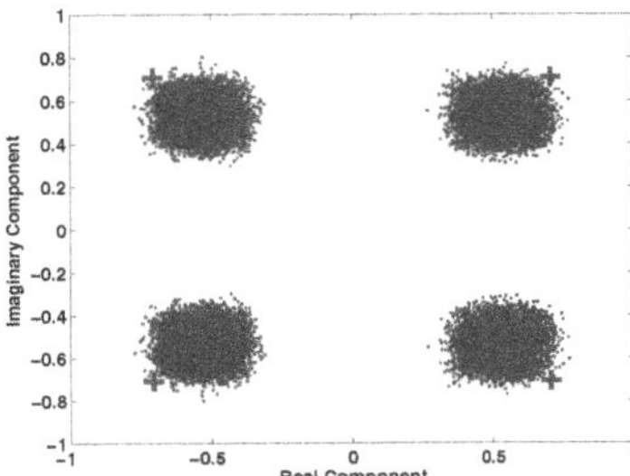

(b) Constellation for a 4-QAM SACO-OFDM under vignetting with $\gamma = 1$ [30]

Figure 9.23 The ICI gain and received constellation for spatial OFDM in the presence of vignetting.

Consequently, for both SACO-OFDM and SDCO-OFDM, the zeroth subcarrier has the largest amplitude and therefore makes the greatest ICI contribution than any other subcarriers. Moreover, the subcarriers close to the zeroth one experience severe ICI. This is more pronounced in SDCO-OFDM, where the zeroth subcarrier has a very large amplitude resulting from the DC bias. Figure 9.23(b) shows the received constellations for the case of 4-QAM SACO-OFDM, using 256×256 subcarriers, CP = CPo = 10%, and $\gamma_1 = \gamma_2 = 1$ as shown in Fig. 9.22. Noteworthily, the large amplitudes for even subcarriers are not displayed in Fig. 9.23(b), since SACO-OFDM uses only odd subcarriers for data carrying. Moreover, the effect of noise is ignored to illustrate the effect of vignetting. It can be seen that, the constellation points will experience both phase rotation and amplitude attenuation with vignetting effect.

9.3.2 Impairment mitigation techniques

As discussed above, the general case of the linear misalignment and geometry perspective distortion produces pixel-dependent sampling offset and spatial dependent normalized areas of interference pixels. It thus induces both amplitude attenuation and phase rotation. The rotation angle is determined by the subcarrier index, the extent of the geometry distortion, and the spatial position. The amplitude attenuation is subcarrier dependent, and the attenuation level also depends on the magnitude of the spatial shift/sampling offsets. Under significant perspective distortion, the equalizer cannot completely compensate the attenuation and angle rotation effect if the sampling offsets are large. Therefore, perspective correction is necessary before equalization. Another impairment, the blur effect causes attenuation on the higher spatial frequency subcarriers. Similar to the fractional misalignment, the attenuation can be corrected by an equalizer at the cost of noise enhancement. Furthermore, the vignetting also causes attenuation in the spatial-frequency domain. However, different from the linear misalignment and blur effect, the attenuation is the same for all subcarriers, and independent of spatial frequency. Moreover, the vignetting also causes ICI which brings an extra noise term in the received subcarriers. This ICI noise is enhanced by the equalizer, and interacts with the attenuations due to all the impairments.

Note that, the four components of attenuation and phase rotation are independent because of the different underlying mechanisms. The overall attenuation and phase rotation for a given subcarrier can be calculated by multiplying the individual attenuation and phase rotation factors jointly.

9.3.2.1 Equalization

Based on the property of spatial impairment factors, linear misalignment, geometry perspective distortion, blur effect, and vignetting all affect the system performance in a similar manner. In order to overcome the linear misalignment and blur effect, one can equalize the received signal with the unused higher subcarriers. To minimize the vignetting effect, a small number of lower subcarriers is kept null, and a vignetting estimation and equalization technique is applied.

In conventional 1-D OFDM, OFDM can dramatically simplify the equalization task by turning the frequency-selective channel into a flat channel. Similarly, in the spatial-frequency domain, a simple one-tap equalizer is needed to estimate the channel and recover the data [6, 17]. In the following, utilizing a decision feedback equalizer (DFE) for mitigating the nonideal spatial effect at the receiver side will be addressed.

For short-range imaging optical MIMO links where the receiver plane is assumed to be parallel to the transmitter plane, the channel noise is dominated by the thermal noise due to readout circuit and shot noise due to ambient light. The transmitter is a spatial light modulator, LCDs or LEDs of size $N_1 \times N_2$ pixels with pixel space equal to D_T in both dimensions. The image sensor has a resolution of $M_1 \times M_2$ and a sampling rate D_R in both dimensions. The full Nyquist region of size is $\frac{1}{D_T} \times \frac{1}{D_T}$ and $\frac{1}{D_R} \times \frac{1}{D_R}$ for transmitter and receiver, respectively. It imposes a limit on the maximum spatial frequency that can be supported by each side. Its spatial frequency response is termed as optical transfer function (OTF) to characterize the spatial property, similar to PSF in the temporal impulse response of the channel. The imaging optical MIMO channel can be modeled in general as

$$y[l_1, l_2] = x(\mathcal{T}\{x, y\}) \otimes h(x, y) + z(x, y)|_{x=l_1 D_R, y=l_2 D_R}, \tag{9.56}$$

where $h(\cdot)$ denotes the spatial PSF, including magnification (or blur) effect, rotation (or geometry perspective distortion), and D_R is the spatial sampling rate (pixel spacing) at the receiver in both spatial dimensions. The transformation $\mathcal{T}(\cdot)$ denotes the remapping of the coordinates caused by spatial misalignment, magnification, and rotation perspective. In the following, the magnification and rotation perspective are addressed, while the spatial misalignment is excluded since the resulting linear phase shift in the spatial frequency domain can be easily equalized.

The blur effect or spatial magnification will expand the image in space, equivalent to compression in spatial frequency. This joint spatial distortion in the continuous spatial frequency domain can be described by

$$\begin{pmatrix} \tilde{f}_1 \\ \tilde{f}_2 \end{pmatrix} = \frac{1}{\kappa} \begin{pmatrix} \cos\theta & \sin\theta \\ -\sin\theta & \cos\theta \end{pmatrix} \begin{pmatrix} f_1 \\ f_2 \end{pmatrix}, \tag{9.57}$$

where $(\tilde{f}_1, \tilde{f}_2)$ are the transformed spatial frequency coordinates, (f_1, f_2) are the spatial frequency coordinates of the transmitter. The spatial magnification is denoted by κ, and θ is the rotation angle. After spatial discrete multitone (SDMT) modulation based on continuous tones, the Nyquist region of the transmitted frames is divided into discrete spatial frequency bins. The continuous spectrum of the transmitted frames in the continuous spatial frequency domain $X(f_1, f_2)$ is given by

$$\begin{aligned} X(f_1, f_2) = & \frac{1}{N_1 N_2 D_T^2} \sum_{k_1=0}^{N_1-1} \sum_{k_2=0}^{N_2-1} \Bigg[X[k_1, k_2] \\ & \cdot \delta\left(f_1 - \frac{k_1}{N_1 D_T}, f_2 - \frac{k_2}{N_2 D_T}\right)\Bigg], \end{aligned} \tag{9.58}$$

where $X[k_1, k_2]$ denotes the SDMT frame in the discrete spatial frequency. Actually, the effect of the pixel shape is coupled with the OTF measurement. Thus the transmitted image can be treated as a continuous waveform in 2-D space.

During reception by the image sensor, the transmitted image is magnified and rotated. The spatial frequency spectrum (or signal) of the transformed image is given by

$$\tilde{X}(f_1, f_2) = \frac{1}{N_1 N_2 D_T^2 \kappa^2} \sum_{k_1=0}^{N_1-1} \sum_{k_2=0}^{N_2-1} \Big[X[k_1, k_2]\delta(f_1 - \nu_1, f_2 - \nu_2)\Big], \tag{9.59}$$

where

$$\nu_1 = \frac{1}{\kappa}\left(\frac{k_1 \cos\theta}{N_1 D_T} + \frac{k_2 \sin\theta}{N_2 D_T}\right), \tag{9.60}$$

and

$$\nu_2 = \frac{1}{\kappa}\left(-\frac{k_1 \sin\theta}{N_1 D_T} + \frac{k_2 \cos\theta}{N_2 D_T}\right). \tag{9.61}$$

At the receiver, the captured image is multiplied by a windowing function of size $M_1 D_R \times M_2 D_R$, where the windowing function signifies the finite extent of the image sensor. The resulting signal is given by

$$\begin{aligned} Y(f_1, f_2) =& \frac{H(f_1, f_2)}{N_1 N_2 D_T^2 \kappa^2} \sum_{k_1=0}^{N_1-1} \sum_{k_2=0}^{N_2-1} \Big[X[k_1, k_2] \\ & \cdot W(f_1 - \nu_1, f_2 - \nu_2)\Big] + \tilde{Z}(f_1, f_2), \end{aligned} \tag{9.62}$$

where $W(\cdot)$ is the 2-D Fourier transform of the finite extent window, $H(\cdot)$ is the channel OTF, and $\tilde{Z}(f_1, f_2) = Z(f_1, f_2) \otimes W(f_1, f_2)$ is the 2-D Fourier transform of the windowed channel noise. This signal is sampled in the spatial domain with a sampling period equal to D_R in both dimensions. The received sampled signal can be expressed as

$$\begin{aligned} Y[\tilde{k}_1, \tilde{k}_2] =& H[\tilde{k}_1, \tilde{k}_2] \sum_{k_1=0}^{N_1-1} \sum_{k_2=0}^{N_2-1} \Big(X[k_1, k_2] \\ & U[\tilde{k}_1, \tilde{k}_2; k_1, k_2]\Big) + \tilde{Z}[\tilde{k}_1, \tilde{k}_2], \end{aligned} \tag{9.63}$$

where

$$U[\tilde{k}_1, \tilde{k}_2; k_1, k_2] = \frac{W\left(\frac{\tilde{k}_1}{M_1 D_R} - \nu_1, \frac{\tilde{k}_2}{M_2 D_R} - \nu_2\right)}{N_1 N_2 D_T^2 \kappa^2 M_1 M_2 D_R^2}. \tag{9.64}$$

In (9.63), the received data in the spatial frequency bin $[\tilde{k}_1, \tilde{k}_2; k_1, k_2]$ is a combination of the transmitted data in all the discrete spatial frequency bins. And the joint effect is referred to as spatial frequency inter-channel interference (SF-ICI). The joint function $U(\cdot)$ is a spatial frequency-dependent function, determined by the window, magnification, and rotation perspective. It is desirable to have a narrow main-lobe and high side-lobe attenuation. However, there is a tradeoff between the width of the main-lobe and the attenuation of the side-lobes. Since the SF-ICI exists in a small band of neighboring frequencies and can easily be equalized, a wide main-lobe and high side-lobe attenuation window such as Blackman window is preferred.

Bin-by-bin detection for the SF-ICI channel

For bin-by-bin detection, the received signal in space is multiplied by a rectangular window function, and the detected signal is given by

$$\hat{X}[\tilde{k}_1, \tilde{k}_2] = \mathcal{D}^{-1}\left\{ \frac{Y[\tilde{k}_1, \tilde{k}_2;]}{H[\tilde{k}_1, \tilde{k}_2]U[\tilde{k}_1, \tilde{k}_2; k_1, k_2]} \right\}, \tag{9.65}$$

where $\mathcal{D}^{-1}(\cdot)$ represents a decision device. Note that, by using a rectangular windowing function, the SF-ICI effect is assumed to be negligible to simplify system design, since most of the signal energy is concentrated in one frequency bin. That means, when SF-ICI is zero, bin-by-bin detection is the optimal detection strategy.

DFE for the SF-ICI channel

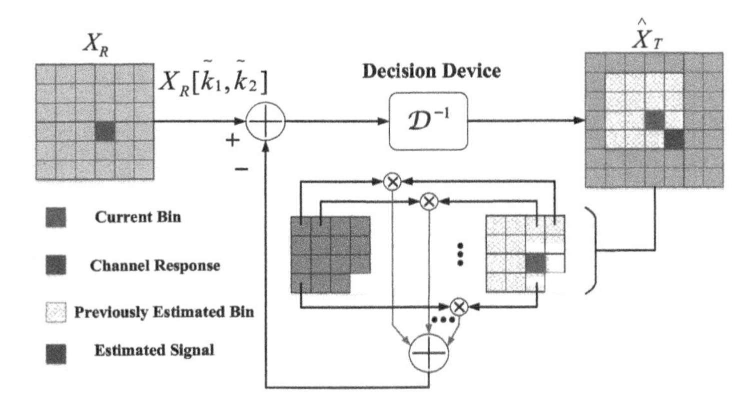

Figure 9.24 Diagram of decision feedback equalization in spatial frequency domain.

Since magnification and rotation perspective affect the system in a similar manner, it is assumed that the system undergoes magnification only to simplify the receiver design. By designing a proper window, the SF-ICI can be limited to a small band of frequencies, and (9.63) can be approximated by

$$\hat{Y}[\tilde{k}_1,\tilde{k}_2] \approx H[\tilde{k}_1,\tilde{k}_2] \sum_{k_1=\tilde{k}_1-\tau_{g_1}}^{\tilde{k}_1+\tau_{h_1}} \sum_{k_2=\tilde{k}_2-\tau_{g_2}}^{\tilde{k}_2+\tau_{h_2}} \Big(X[k_1,k_2] \, U[\tilde{k}_1,\tilde{k}_2;k_1,k_2] \Big) + \tilde{Z}[\tilde{k}_1,\tilde{k}_2]. \tag{9.66}$$

The above equation implies that, by choosing a proper window, the interference in each frequency bin $[\tilde{k}_1,\tilde{k}_2]$ can be limited in its neighboring bins $[-\tau_{g_1},\tau_{h_1}]$, and $[-\tau_{g_2},\tau_{h_2}]$. Note that, since the inference in $Y[\tilde{k}_1,\tilde{k}_2]$ is non-causal, it cannot be eliminated completely from $Y[\tilde{k}_1,\tilde{k}_2]$. To overcome non-causality, the DFE is introduced to the system as shown in Fig. 9.24. By introducing a spatial delay, the signal $\hat{X}[\tilde{k}_1+\tau_{h_1},\tilde{k}_2+\tau_{h_2}]$ is estimated by observing its SF-ICI contribution to the received signal $Y[\tilde{k}_1,\tilde{k}_2]$. Note that, for DFE equalization, $\hat{X}$ must be known for all frequencies preceding the spatial frequency bin $[\tilde{k}_1+\tau_{h_1},\tilde{k}_2+\tau_{h_2}]$ relative to a decoding path.

To improve the performance of the DFE, windowing function design is needed. In [17], two window design criteria are introduced. The first one is to design a window with high SNR, and the second one is to make most of the window energy to be used in the feedback loop. Thus, for SF-ICI limited channels, the complex windowing with DFE can be used to equalize the SF-ICI in the spatial frequency domain with excellent BER performance and low complexity.

9.3.2.2 **Perspective correction**

Perspective transformation is widely studied in computer vision and usually requires higher computation complexity. However, for the imaging optical MIMO communication system, the off-line algorithms are not applicable and the system cannot tolerate significant time delay. The system can only tolerate minor perspective distortions typically acceptable to the human eyes. On the other hand, it offers more design flexibility if we can encode the imaged object in a manner that simplifies correction for perspective distortion. Due to the constraints mentioned above, the traditional perspective correction algorithms cannot be used in imaging optical MIMO systems directly. Then, a generalized spatial OFDM sampling correction algorithm is introduced [18].

The intuition underlying these approaches is as follows. We can approach the perspective distortion problem as a sampling problem from the communication perspective [18]. Specifically, the intensities of transmitted elements refer to the signal samples at the transmitter. The received intensities in a pixelized receiver refer to the signal samples at the receiver. When the transmitter, such as LCD, and camera-based receiver, are located at an angle, some parts of the LCD are closer to the camera, and hence occupy a relatively bigger space in the image. This means a higher sampling rate for this part. Parts of the transmitted elements further away from the camera occupy a relatively smaller space in the image, which means they are sampled at a lower rate. To tackle the perspective distortion problem, the spatial OFDM-based imaging optical MIMO system needs to find the relationship between the sampling points on the image sensor and those on the transmitted elements, and resample the received image at the location that best reflects the originally transmitted samples. To decode the signal properly, the receiver needs to resample the signal as closely as possible to the transmitter's samples. Based on this intuition, a generalized OFDM sampling correction algorithm can be used to tackle the perspective distortion.

A basic property of the Fourier transform is that a time shift corresponds to the phase offset in the frequency domain. Given this property, it is relatively simple to figure out how the sampling shift offsets affect the encoded data as in Eqs. (9.31)–(9.32). Consider a general case where we sample a spatial OFDM symbol with a perspective distortion as shown in Fig. 9.20. The corner shifts would result in phase offsets at the receiver. Specifically, spatial OFDM symbol x_{l_1,l_2} is generated by taking a 2-D IFFT on the complex signal X_{k_1,k_2}, and the receiver samples the spatial symbol with a shift. To decode the spatial OFDM symbol from these samples, the receiver takes a 2-D FFT. However, because the samples are shifted, the 2-D FFT does not reproduce exactly the original complex signal, rather than a phase-shifted version, denoted as $X_{k_1,k_2}H^{\star\star}_{k_1,k_2} + Z_{k_1,k_2}$, of the original complex signal. Thus, each complex signal experiences attenuation and phase shift $H^{\star\star}_{k_1,k_2}$ as in (9.33), where $H^{\star\star}_{k_1,k_2}$ is determined by the shift offsets of (a_x, a_y), (b_x, b_y), (c_x, c_y), and (d_x, d_y). Thus, if the unknown attenuation and phase shift $H^{\star\star}_{k_1,k_2}$ need to be estimated, there are totally eight unknowns to estimate (a_x, a_y), (b_x, b_y), (c_x, c_y), and (d_x, d_y). It means that we need at least eight equations to estimate these eight offsets. We solve this problem by considering a super-symbol consisting of four symbols as shown in Fig. 9.20.

A 2×2 spatial OFDM super symbol can be used by taking the 2-D IFFT on the complex $X^r_{k_1,k_2}$, where (k_1, k_2) are the subcarrier indices with $0 \le k_1 < N_1$ and $0 \le k_2 < N_2$, $r \in \{(l_1, l_2), (l_1, l_2+1), (l_1+1, l_2), (l_1+1, l_2+1)\}$ denotes the symbol index or the spatial position. This symbol is sampled with relatively small x and y corner offsets of (a_x, a_y), (b_x, b_y), (c_x, c_y), and (d_x, d_y) at its four corners. Let $\Delta\theta^r_x$ be the difference between the phase shifts experienced by $X^r_{k_1,k_2}$ and $X^r_{k_1',k_2}$, and $\Delta\theta^r_y$ be the difference between the phase shifts experienced by $X^r_{k_1,k_2}$ and $X^r_{k_1,k_2'}$, then we have [18]

$$\begin{pmatrix} \Delta\theta_x^{(l_1,l_2)} \\ \Delta\theta_x^{(l_1,l_2+1)} \\ \Delta\theta_x^{(l_1+1,l_2)} \\ \Delta\theta_x^{(l_1+1,l_2+1)} \end{pmatrix} = \frac{2\pi(k_1 - k_1')}{16N_1} \begin{pmatrix} 9 & 3 & 3 & 1 \\ 3 & 9 & 1 & 3 \\ 3 & 1 & 9 & 3 \\ 1 & 3 & 3 & 9 \end{pmatrix} \begin{pmatrix} a_x \\ b_x \\ c_x \\ d_x \end{pmatrix}, \tag{9.67}$$

and

$$\begin{pmatrix} \Delta\theta_y^{(l_1,l_2)} \\ \Delta\theta_y^{(l_1,l_2+1)} \\ \Delta\theta_y^{(l_1+1,l_2)} \\ \Delta\theta_y^{(l_1+1,l_2+1)} \end{pmatrix} = \frac{2\pi(k_2 - k_2')}{16N_2} \begin{pmatrix} 9 & 3 & 3 & 1 \\ 3 & 9 & 1 & 3 \\ 3 & 1 & 9 & 3 \\ 1 & 3 & 3 & 9 \end{pmatrix} \begin{pmatrix} a_y \\ b_y \\ c_y \\ d_y \end{pmatrix}. \tag{9.68}$$

Using the pilot bins, the receiver computes the attenuation and phase shift experienced by each known pilot. Then, the offsets (a_x, a_y), (b_x, b_y), (c_x, c_y), and (d_x, d_y) are estimated with the help of Eqs. (9.67) and (9.68). Afterwards, the receiver can resample each spatial OFDM symbol at the correct sampling points and compute the 2-D FFT again. Then the received signals are demodulated to obtain the transmitted bits.

9.3.2.3 Adaptive coding and modulation

The joint blur effect and perspective distortions eliminate sharp transitions in an image and cause nearby pixels to blend together. This process is similar to low pass filtering which attenuates the high-frequency subcarriers in spatial OFDM. Generally, the signal amplitudes of the transmitted signal and its received version are functions of the frequency index, as shown in Fig. 9.25. The transmitted signal is chosen to have the same energy at all frequencies. At the receiver, the high frequencies are heavily attenuated and can only support low-order modulation signals. In some extreme cases, they cannot even be used for transmitting information. Conversely, the low frequencies have only slight attenuation. They can deliver information with almost no error and can support high-order modulation transmission. All other frequencies experience significant attenuation but can still be used to transmit some information [18].

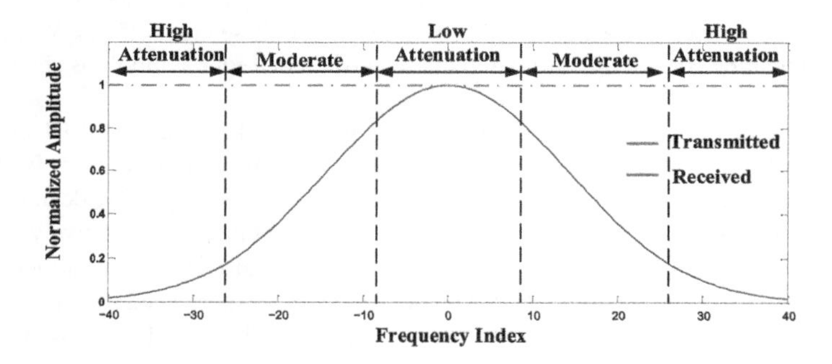

Figure 9.25 The transmitted and received signal amplitudes across different frequencies due to the blur effect.

It is clear that spatial OFDM belongs to spatial-frequency domain modulation. It can naturally deal with different frequencies experiencing different attenuations. Spatial OFDM completely suppresses very high frequencies, which are not used to transmit information or only transmit low-order modulation signal. Frequencies that experience low or moderate attenuation are used to transmit information and protected with error correction code such as Reed–Solomon (RS) code. An error correction code is chosen for each frequency with different redundancy commensurate with the attenuation each frequency experiences. Moreover, due to geometry distortion, the extent of attenuation varies not only with frequency, but also with the spatial position of the pixels/symbols in a received frame. Generally, symbols away from the center of the frame are not in the plane of focus and hence experience increased attenuation due to blur effect. Thus, a spatial OFDM system can exploit this property to optimize the redundancy of the error correction code and the modulation order [18]. Specifically, the symbols at the center of the frame will have low redundancy and can support high-order modulation, while the symbols away from the center of the frame will have high redundancy.

9.4 Synchronization in OCC

9.4.1 Synchronization challenges

To achieve a high throughput, the imperfect frame synchronization issues must be addressed in a practical OCC system. Frame synchronization is a prerequisite for effective communication. However, perfect synchronization is difficult in OCC. Synchronization failure is mainly caused by frame rate mismatch and a phase offset between the transmitted element and the image sensor. Figure 9.26 shows some examples. Typically, only a frame within one transmission period is a clean frame, and thus can be decoded. If a captured frame spans across different transmission frames, it becomes a mixed frame. Moreover, since most CMOS-based cameras employ rolling shutter which scans one line at a time, it makes frame mixing vary on a per-line basis. Such a frame has to be dropped usually. The frames may also get lost due to discrete shutter sampling, if the transmitted element's refresh rate is higher than the camera's capture rate. Only after receiving all the transmitted frames cleanly can a receiver successfully decode the original message (cases 1 and 3).

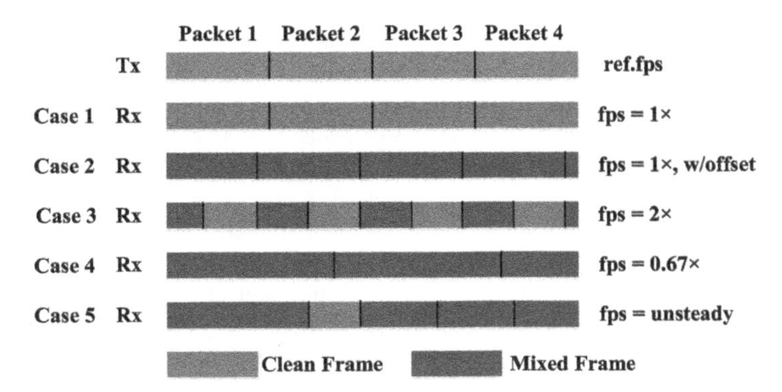

Figure 9.26 Mismatched frame rates between transmitter–receiver pairs (inter-frame intervals are omitted).

The difficulty in frame synchronization arises from frame rate diversity and variability [31]. From Nyquist sampling theorem, a receiver should capture frames at a rate at least twice the transmitted rate, to guarantee that it could receive all frames cleanly (case 3). Unfortunately, in terms of rate capability and visual experience, Nyquist sampling rate is unavailable for a commercial image sensor. Some modern transmitted elements such as LCD screens usually have a 60 fps or even higher. The other transmitted elements such as LED array or traffic light can support higher rate. However, most of the commercial consumer cameras have lower frame rates of up to 30 fps. Moreover, many cameras embedded in smartphones exhibit unsteady frame rate due to hardware design, firmware setting, and application program interface (API) constraints.

Variability and unsteady inter-frame intervals: Compared to the transmitted elements, the received frame rate exhibits more variability. The camera needs sufficient light exposure to capture legible frames. The time it takes to harvest sufficient light depends on the image sensor's sensitivity and the lighting conditions, in particular, the contrast and intensity levels. Therefore, it is necessary to adopt a different frame rate adaptive to specific contrast and intensity levels, and the camera normally adjusts this rate automatically to ensure the visual quality of the captured frames. Since the readout rate is limited by the image sensor readout circuit hardware, perfect acquisition of the matched frame rate is impossible. Even if we control the frame rate by the application software and disregard the imaging quality in the OCC mode, the unsteady frame rate is inevitable due to the limited readout rate in the readout circuit and the post-processing unit in an image sensor.

Operation range and diversity across devices: The maximum frame rate is determined by the readout rate in the image sensor and the hardware quality, such as the sensor quality and heat dissipation capability of the device. Different cameras can support different maximum frame rates. Moreover, the camera will automatically compensate for the changed average intensity of each transmitted frame by adjusting the frame rate. Thus, the frame rate varies across different transmitted frames for a given camera, and for the same transmitted frame across different cameras.

Usually, if each frame for mixing is homogeneous, the mixed frame is also homogeneous. We often obtain mixtures of different frames at the same captured frame rate. Moreover, since most image sensors embedded in cameras are rolling-shutter-based CMOS image sensors, the proportion of mixture varies by line even for the same frames. Generally, only completely clean frames can be decoded. Therefore, the goal of frame synchronization is to adjust the frame rates on both sides to capture enough clean frames to decode all original information.

Typical frame synchronization approaches let the receiver capture frames at a rate twice as fast as the transmitter to ensure a good frame every other frame. However, there are several fundamental drawbacks for these approaches. First, they are inefficient. The transmitter-side capacity is under-utilized, while the receiver drops a large number of mixed frames that actually contain much useful information. Second, they assume an initial sender setup based on the receiver capability, which, in practice, requires feedback and needs to tailor to individual receivers. Third, they are still unreliable in the presence of receiver rate fluctuation, which makes it difficult to work with a general phone camera. Since each frame is already in the form of discrete samples, we could have equal frame rates on both sides in theory. The key to achieve this is to decode imperfect frames and recover lost frames. The former is complicated due to inherent image quality issues. The latter can be easily implemented with erasure coding across the original frames, which can also help to reduce decoding errors. Except the oversampling approach, there exist some other schemes to tackle the synchronization issue, such as per-line tracking and inter-frame coding [31], and rateless coding [33].

9.4.2 Per-line tracking and inter-frame coding

To achieve faster and more flexible frame synchronization, and allow smooth communication between the transmitter and cameras with different frame rate, the per-line tracking and inter-frame coding synchronization scheme was proposed in [31], which features two main components. First, this approach adopts intra-frame per-line color tracking to decode the mixed frames. As a result, any received frame is decodable and useful. Second, it employs inter-frame erasure coding and frame-based tracking to recover lost frames and to prevent incorrect frames. Based on this scheme, all captured frames containing useful information can contribute to the overall frame recovery, and result in reliable information transmission.

9.4.2.1 Line-based overlap tracking

To tackle frame mixing due to rolling shutter effect, the pre-designed line-based overlap tracking bars are inserted into each frame in the shutter rolling direction. With these tracking bars, the percentage of overlap in each line can be tracked by comparing with the reference colors. Figure 9.27 illustrates the layout of the color pattern, where four bars are placed in each transmitted frame, two white and two black. Each vertical line represents a tracking bar, while each horizontal line represents the color pattern of a line in the frame. The color pattern shifts by one bar in each successive transmitted frame, and the overall patterns repeat every four frames. As a result, all four combinations are present in any received frame, and the four blocks in each line serve as reference colors for per-frame decoding. The idea for these schemes is analogous to training sequence placement in traditional 1-D radio frequency (RF) communications.

For any block in a mixed frame, the decoding algorithm proceeds in two stages [31]. First, the exact transmitted frames are determined by using the tracking bar encoding pattern. By identifying the component frames per line, we are able to handle practically any phase offsets between the transmitted and the captured frames. And then, the code block color was compared with the reference colors to determine the code block sequence in the original frames concerned. Taking the color distortion into account, the inference-based approach is preferred, which allows us to decode multiple transmitted frames from a single mixed frame. Otherwise, we look up the block at the same position in the previous captured frame to help resolve the ambiguity. If that still fails, we mark this block unknown, and leave it to the cross-frame correction stage of inter-frame erasure decoding.

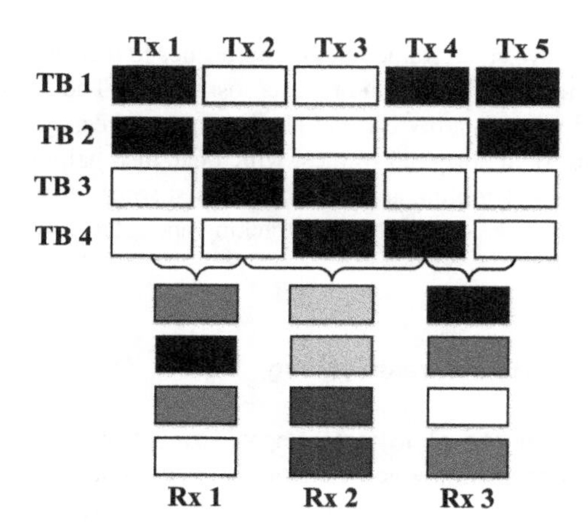

Figure 9.27 Per-frame tracking bar pattern.

9.4.2.2 **Inter-frame erasure coding**

One of the key ideas of inter-frame coding is to apply erasure coding across the original frames, so that every transmitted frame may contain information from multiple original frames [31]. It means that any partial or mixed frame is just a version of the coded frame and can be decoded as long as we keep track of the coding coefficients. Therefore, once the image sensor captures enough linearly independent frames, all original frames can be recovered.

Ideally, each encoded frame should be a linear combination of all original frames, but linearly independent of any other encoded frames. This would allow us to recover all N original frames from any N encoded frames. Generally, a sparse and deterministic coding matrix is necessary for low complexity coding. To guarantee no more than one frame loss per three frames and speed up decoding, we can divide all original frames to groups of three consecutive frames. Then, generate the transmitted frame sequence as $[F_1, F_2, F_3, F_1 \otimes F_2, F_2 \otimes F_3, F_1 \otimes F_3]$, where the first half of the frame sequence F_1, F_2, and F_3 contains the original frames, and the second half includes the encoded frames. The chosen coding matrix is suboptimal to make a linearly independent transmitted frame sequence. However, even seemingly linearly dependent encoded frames will be helpful when per-frame decoding error happens.

An erasure coder treats each frame as a stream of bits and encodes it to a coded bit stream with the same length. Therefore, the inter-frame coding is independent of the base code per frame, and can be viewed as an outer code over a 2-D barcode. The coded bits within a frame can then be mapped to colored blocks and the blocks are arranged in some pattern following the per-frame base code. After frame tracking, the original frames are determined. Once the receiver captures enough coded frames and the corresponding rows of the coding matrix are known, decoding is straightforward by using a few XOR to recover the original frames. Since only part of the frame is lost, decoding can be performed only for the lines with missing information.

9.4.2.3 Unsynchronized system design

At the transmitter, an erasure-coded bit stream is obtained by erasure coder and the transmitted frames are generated following bit-to-block mapping according to the code block layout. To keep track of the transmitted frames with low complexity, a code block as in COBRA system [32] can be optimized and designed. The tracking bars as shown in Fig. 9.27 can be inserted into each frame which is perpendicular to the scanning direction in order to track the mixed frames. Finally, the encoded frames are transmitted at a given transmission frequency.

At the receiver, after identifying the beginning coded sequence, we invoke per-frame decoding and cross-frame correction. And then, we perform per-frame decoding process to decode the coded blocks. Once the decoder detects a frame in the second half of the transmitted frame sequence in the received frame, we start cross-frame correction after erasure decoding each new frame and filling in any missing blocks at the relevant positions if possible. Both frame capture and decoding terminate if all original frames have been decoded. The blocks are then mapped to an output bit stream.

From the signal processing perspective, an erasure coding approach is essentially a form of undersampling, where the sampling rate is lower than the Nyquist rate. And it is an effective approach to tackle the synchronization issue. In the last subsection, another special erasure coding approach, named as rateless coding approach, will be introduced for an asynchronous system.

9.4.3 Rateless coding

Similar to synchronization scheme using the per-line tracking and inter-frame coding, which could decode mixed frames, recover lost frames and guard against incorrectly decoded frames, there exists another synchronization scheme to recover the original data by rateless coding approach [33]. In a practical OCC system, the link quality varies according to many factors, including ambient light, unsteady frame rate, perspective distortion, and trembling of user's hands. Therefore, the systems need to adapt to the link diversity. For rateless coding, it fits well with the channels in which interruptions occur frequently since the encoded bits of the same original information are generated continuously till the successful transmission is completed [34]. By using rateless codes, protocol complexity and packet delay can be reduced due to the fact that only the correctly received information needs the feedback information. If rateless coding is adopted to convert the original data into a stream of encoded bits, the receiver can extract information from the captured frames, even with error bits. Since every encoded bit contains useful information of the original data, each correctly decoded bit is useful for the link throughput. Once the receiver accumulates sufficient amount of clean bits, it can recover the original data by rateless decoding. Thus, the transmitter is able to automatically adapt the data rate to the dynamic channels with different qualities, including the asynchronous link.

The conventional rateless erasure codes, such as Luby Transform (LT) code [35] and Raptor code [36], are both based on blocks of bits. The transmitted block (or packet) is assumed to be either correctly received or totally lost. A critical factor that affects the system performance is the block size. If the block size is too large, more blocks may be discarded even though they contain a large proportion of clean bits. Otherwise, more overhead is required. And the block size is adaptively adjusted according to the channel coherence as in conventional RF communication. However, for OCC links, it is difficult to set an appropriate block size due to the lack of feedback about channel coherence from cameras. Thus, it is difficult to build an efficient erasure channel. The other rateless codes for AWGN channels, like Spinal codes and soft decoding, avoid the block size setting by bit-level coding. However, the computational complexity is too high in the decoding process, due to the intensive floating-point iteration operations. To tackle the challenges, the light-weight rateless coding on bit-level erasure channel is proposed [33], which extracts soft hint from every received symbol to assess how likely each bit is correctly decoded during demodulation. An erasure channel can be established by discarding the symbols with a soft hint lower than a threshold, where the soft hint can be interpreted as distance in signal space between the received symbol's constellation and the decoded symbol's constellation points [33]. And then, light-weight rateless erasure coding can be used over the bit-level erasure channels.

By incorporating a set of proper modulation techniques, it enables an accurate soft hint estimation and establishes a bit-level erasure channel with minimized false positive (the ratio between the number of wrongly reserved bits and the total number of received bits) and false negative (the ratio between the number of wrongly erased bits and the total number of received bits). Furthermore, the rateless coding scheme adopts a new light-weight bit-level rateless coding scheme that tolerates the false positive of the bits provided by the erasure channel. The transmitter encodes the data frames based on a systematic rateless code and the receiver decodes the received frames by XOR operations at the frame level. The uncertain bit positions are taken into account, which enables bit-level rateless decoding. A majority vote decoding algorithm can be used to eliminate the impact of false bits and guarantee the decoding efficiency with slight computational complexity increase.

At the transmitter, the encoder encodes data by a systematic rateless code at the frame level. The transmitter divides the original bit stream into a sequence of bit frames, then the bit frames are encoded by a FEC code, like RS code, LDPC code, or extended irregular repeat accumulate (eIRA) code as shown in Fig 9.28, to generate some parity-check frames for error correction. The intermediate frames (including the original bit frames and the parity check frames) are further encoded to produce a stream of rateless frames, each of which is calculated by XORing a certain number of randomly chosen intermediate frames. The encoded frames including both the intermediate frames and the XORed frames are transmitted in series. The systematic design allows the system to approach the channel capacity when the link quality is high. In every transmitted frame, the coding information is inserted, including the number of original frames and the seed of random number generator, which allows the receiver to reproduce the generation equations (coefficient matrix) of encoded frames.

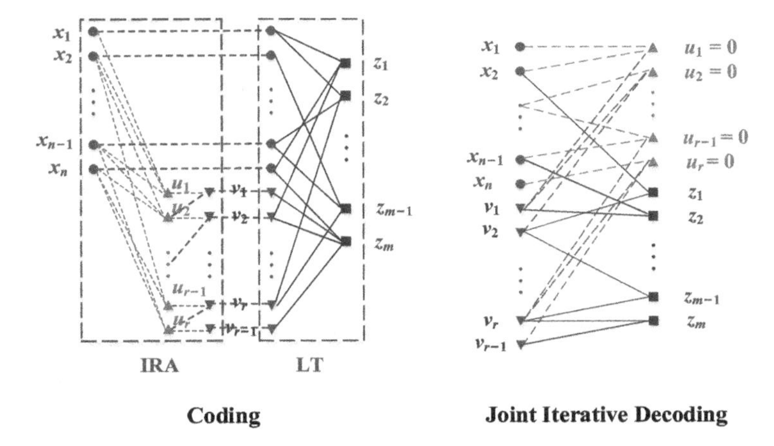

Figure 9.28 The architecture of eIRA-Raptor coding and joint iterative decoding.

Upon the reception of its first frame, the receiver reproduces the generation equations. By solving the linear generation equations (the number of equations is n times larger than the number of variables), every intermediate frame has multiple instances, each of which is expressed by XORing several encoded frames. When an encoded frame is received, it is plugged into the expressions of the related instances. As more encoded frames are received, more instances are calculated. For one bit position, all calculated instances of different intermediate frames may have different results. A majority vote decoding algorithm is performed to find the correct value of each bit. We record the occurrence frequency of "1" and "0" at every bit position in all calculated instances. The bits are set to the value with the highest occurrence frequency. Therefore, for every bit position, the error or "x" bits in one or few instances can be overweighed by the other correct instances.

As mentioned above, the CMOS-based cameras in the current smart devices capture an image by the rolling shutter scheme. If the frame rate of the transmitter is high, some received frames may contain the contents of multiple transmitted frames and the lines captured during the transition of the transmitted frames experience severe blur effects. With the rateless coding scheme, the system can automatically recover from the bit loss of blurred symbols by accumulating more rateless encoded bits. Moreover, the rateless coding scheme is able to automatically adapt to different communication environments and significantly improve the throughput of OCC systems. Furthermore, we can optimally design the code-rate, code-length, or degree distribution for OCC according to the link quality variations. In addition, the intra-frame diversity causes link quality variation, and thus considerable decoding errors. In order to recover the erroneous symbols, interleaving operation can be incorporated before transmission. By interleaving, the bits at different positions experience a similar bit reception rate on average. All bit positions succeed in decoding almost at the same time. Even if a few bits of the intermediate frames are incorrect, the original bit frames can be recovered by the parity-check frames.

9.5 OCC system experimental platform

9.5.1 Design and implementation of a real-time OCC system

In OCC, most efforts are focused on increasing data rate and communication distance by maximally utilizing the spatial, frequency, intensity, and color dimensions, such as undersampling-based modulation schemes, rolling shutter effect-based modulation schemes, spatial OFDM/DMT, and CSK, as introduced in Section 9.2. The modulation methods in the literature can perform very well under a stable frame rate. However, the system performance will dramatically degrade due to the frame rate fluctuation of a commercial image sensor.

In this part, a real-time OCC system design by using an LED display and an image sensor will be introduced. To solve the frame synchronization problem encountered in a typical commercial sensor, the LED display is refreshed at a much lower rate than the sensor frame rate, and upsampled signal processing techniques are developed to recover data at the receiver. Each LED has independently controlled RGB chips, and each Bayer-pattern sensor unit yields RGB parallel outputs. Thus, CSK [39, 40] and multilevel PAM [41] are combined in the design of the proposed color-intensity modulation (CIM). Meanwhile, the pixels in the sensor are partitioned into detection sub-blocks, and each block is fit with the image of the LED at the display and separates three colors as well. Thus, an MIMO channel configuration is constructed.

In the following, by using an RGB LED array and millions of pixels, we can design a real-time CIM-MIMO OCC system [37]. To tackle the critical issues in a practical OCC system, including the unstable frame rate, the joint transceiver nonlinearity and color crosstalk effect, flicker noise, and rolling shutter, a redundant transmission and upsampled reception technique is applied to make a smooth communication, even with the inevitable frame instability. Moreover, signal constellation and iterative training sequences are designed to tackle the joint nonlinearity and color crosstalk effect, and rolling shutter effect, respectively. Therefore, spatial, color, and intensity dimensions are fully utilized to create a suboptimal high-dimensional signal constellation and parallel communication channels. Applying a commercial 16×16 LED array with 192 data-carrying LEDs using 256-CIM at a refresh rate of 82.5 Hz and a mobile phone camera at a frame rate of 330 fps, the proposed experimental system achieves a real-time data rate of 126.72 kbps over communication distance up to 1.4 m without any external optical assistance.

9.5.1.1 CIM-MIMO modulation and signal detection framework

In the CIM-MIMO framework, signals in terms of triplet LED driving currents are mapped into the intensity and color of emitted light of each RGB LED unit in a K-element array. Considering (L_r, L_g, L_b) intensity levels for each color, totally $M = L_r \times L_g \times L_b$ CIM symbols, each of which contains $B = \log_2 M$ data bits, are generated and described as the collection of triplets

$$\mathcal{S} = \{s_1, s_2, \cdots, s_M\}, \tag{9.69}$$

where the mth CIM symbol is given by

$$s_m = [i_r^m \; i_g^m \; i_b^m], \tag{9.70}$$

with i_r^m, i_g^m, and i_b^m representing the driving currents of red, green, and blue LEDs, respectively.

Following electrical–optical–electrical conversion, the blended-color CIM signal is detected by the active pixel sensor (APS) with Bayer-pattern color filter, to produce an output photocurrent. To make a full-rank matrix, the sensor area is partitioned into N detection blocks, where $N \geq K$. Each detection block is a collection of pixels and the output photocurrent is the average output photocurrent for these pixels. In an extreme case, a detection block only contains a single pixel. The $3N \times 3K$ channel matrix for CIM-MIMO transmitter–receiver pairs is given by

$$\mathbf{H} \triangleq \begin{bmatrix} \mathbf{H}_{1,1} & \mathbf{H}_{1,2} & \cdots & \mathbf{H}_{1,K} \\ \mathbf{H}_{2,1} & \mathbf{H}_{2,2} & \cdots & \mathbf{H}_{2,K} \\ \vdots & \vdots & \ddots & \vdots \\ \mathbf{H}_{N,1} & \mathbf{H}_{N,2} & \cdots & \mathbf{H}_{N,K} \end{bmatrix}. \tag{9.71}$$

Each entry $\mathbf{H}_{n,k}$ in $\mathbf{H}$ is actually the CIM channel from the kth transmitter unit to the nth detection block, as

$$\mathbf{H}_{n,k} \triangleq \begin{bmatrix} h_{r,r}^{n,k} & h_{r,g}^{n,k} & h_{r,b}^{n,k} \\ h_{g,r}^{n,k} & h_{g,g}^{n,k} & h_{g,b}^{n,k} \\ h_{b,r}^{n,k} & h_{b,g}^{n,k} & h_{b,b}^{n,k} \end{bmatrix}, \tag{9.72}$$

where $h_{p,q}^{n,k}$, $p, q \in \{r, g, b\}$ represents the transceiver channel gain between the electrical current input of the color-q LED of the kth LED unit and the color-p output photocurrent of the nth detection block, given by

$$h_{p,q}^{n,k} \triangleq \eta^{n,k} \int S_q^k(\lambda) R_p^n(\lambda) d\lambda, p, q \in \{r, g, b\}, \tag{9.73}$$

where $\eta^{n,k}$ is the path loss between the kth transmitter and nth receiver, $S_q^k(\lambda)$ and $R_p^n(\lambda)$ denote the spectral response for color-q in the kth transmitter and the optical filter-p at the nth receiver, respectively. It is observed that the elements of $\mathbf{H}_{n,k}$ vary with transmitted symbols due to the transceiver nonlinearity and color crosstalk.

At the receiver, if the nth detection block has W pixels, then the three average output values corresponding to red/green/blue channels are $Y_n^p = \frac{1}{W} \sum_{u=1}^{W} y_{n,u}^p$, $p \in$

$\{r, g, b\}$. For notational convenience, we stack all $3K$ input currents into vector $\mathbf{X}$ and all $3N$ outputs into vector $\mathbf{Y}$. In practice, the elements of $\mathbf{H}$ vary with transmitted symbols due to the nonlinearity effect. To make the problem tractable, $\mathbf{H}$ is assumed to be static over transmitted symbols in the near-linear dynamic range. Note that, since the crosstalk and the received flicker noise are typically strong, it is difficult to reduce these noise signal. Hence, the noise in the CIM-MIMO system can be assumed to be signal-independent. The received vector signal can be expressed as

$$\mathbf{Y} = \mathbf{H}\mathbf{X} + \mathbf{Z}. \tag{9.74}$$

For the equiprobable CIM symbols with Gaussian noise, a maximum likelihood (ML) detector is the optimal detector. If the received noise is assumed to be signal-independent Gaussian noise, ML detection can be achieved through the minimum Euclidean distance rule. Given all observations and the estimated channel matrix $\mathbf{H}$ through training, the symbol can be estimated as

$$\begin{aligned} \hat{\mathbf{X}} &= \arg\max_{\mathbf{X}} \ f_{\mathbf{Y}|\mathbf{X}}(\mathbf{Y}|\mathbf{X}, \mathbf{H}) \\ &= \arg\min_{\mathbf{X}} \|\mathbf{Y} - \mathbf{H}\mathbf{X}\|_2, \end{aligned} \tag{9.75}$$

where $\mathbf{X}$ is the possible transmitted symbol.

9.5.1.2 Principles of system design

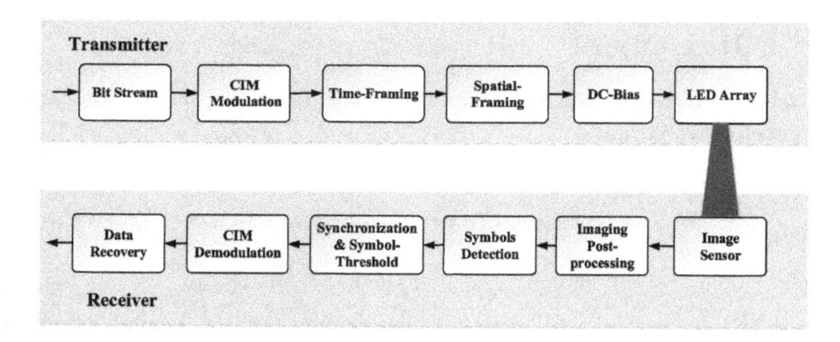

Figure 9.29 The CIM-MIMO OCC transceiver diagram.

The CIM-MIMO OCC system diagram is shown in Fig. 9.29. At the transmitter, after CIM modulation, the synchronization sequence, training sequence, and transmission signal are encapsulated into data frames in the time domain. And then, the time frames are arranged as the spatial frame format and are used to modulate the LED array via a micro-controller unit (MCU). At the receiver, an image sensor captures the images in a successive manner. After adaptive anchor detection, perspective correction, symbol detection, frame synchronization, and signal estimation, each CIM symbol is demodulated in real-time.

In a practical system design, the following challenges must be addressed.

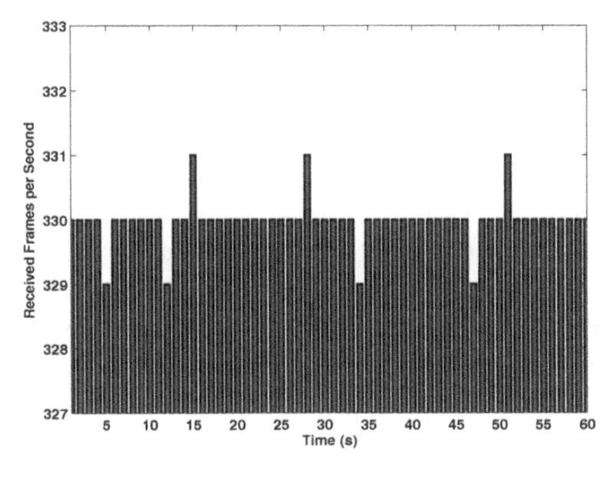

(a) The received effective frames

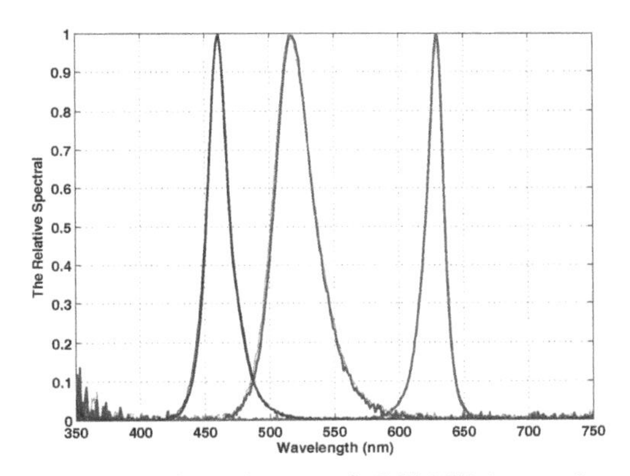

(b) The measured optical spectrum for RGB LEDs in transmitter

Figure 9.30 Observed nonideal issues in CIM-MIMO OCC design.

Unstability: The unstable frame rate for most commercial image sensors will introduce sampling drifts and sampling errors. Figure 9.30(a) shows the statistics of effective received frames per second for the employed image sensor.
Nonlinearity: The nonlinearity exists not only between electrical current and optical power at the transmitter, but also between received optical power and converted electrical current at the receiver.
Flicker noise: The main source of noises in the indoor case is flicker noise caused by background lamp.
Crosstalk: Due to optical spectrum overlaps between red/green/blue colors at the transmitter as shown in Fig. 9.30(b) and mismatches with the Bayer-pattern filters at the receiver, the crosstalk is inevitable.
Sampling: Although the sampling time can be optimally set for the kth pixel-based receiver, it might be the worst sampling time for the lth pixel-based receiver due to

different signal arrival times.

To tackle the challenges mentioned above, some solutions can be employed to realize a successful OCC communication link.

Optimal CIM symbol design:

If the transmitted symbols have equal probability and the ML detector is adopted, at high SNR, the symbol error rate (SER) is dominated by the minimum pairwise Euclidean distance between received symbols. With intensity constraint and color constraint, the symbol set $\mathcal{S} = \{s_1, s_2, \cdots, s_M\}$ that maximizes the minimum distance for a given SNR is given by [39]

$$\begin{aligned} \mathcal{S} = \arg \max_{s_m} \min_{m \neq k} &\|\mathbf{H}(s_m)s_m - \mathbf{H}(s_k)s_k\| \\ \text{subject to} \quad & L_{min} \leq \mathcal{L}(s_m) \leq L_{max} \\ & \mathbf{0} \preceq s_m \preceq [I_r \;\; I_g \;\; I_b], \end{aligned} \tag{9.76}$$

where $\mathbf{H}(s_m)$ is the crosstalk matrix corresponding to symbol s_m, which varies with symbol s_k due to the nonlinearity in each color channel. I_r, I_g, and I_b are the peak currents of the red, green and blue LED, respectively. $L_{\min}$ and $L_{\max}$ are the minimum and maximum allowable total luminous flux, respectively.

The joint nonlinearity and crosstalk effect between red–green–blue channels with different region of interest (ROI) sizes are shown in Fig. 9.31. The average received values in the 30×30 ROI size for red/green/blue channels when only one of red/green/blue LED is illuminated are shown in Fig. 9.31(a). For example, with only red LED activated, not only the red-filter pixels but also the green-filter and blue-filter pixels can perceive signal intensity. Thus, color crosstalks are inevitable in each color channel. The nonlinearity between the transmitted signal intensity and the received gray value, as well as the intensity fluctuations due to rolling shutter, can be observed in Fig. 9.31(a). Furthermore, it is seen that the nonlinearity varies across color channels. To make an effective and low complexity demodulation in real-time, several pixels, which are close to the center pixel of the LED image, are selected to group as a block receiver. Figure 9.31(b) shows the average received gray value in the 3×3 ROI size for red/green/blue channels when only one of red/green/blue LEDs is illuminated.

It is seen that, the crosstalk and nonlinearity still exist even though the ROI size is reduced. Meanwhile, the near-linear dynamic range decreases as the ROI size decreases. Due to the joint nonlinearity and color crosstalk effects, each CIM symbol experiences a unique symbol-dependent channel. More importantly, it is impossible for the receiver to obtain complete channel knowledge through a training procedure. Thus, a sub-optimal CIM constellation is pursued by assuming that all symbols experience a symbol-independent channel in the near-linear dynamic range. Hence, in CIM symbol optimization, the channel matrix $\mathbf{H}_8$ corresponding to the transmitted intensity level in a near-linear dynamic range is assumed as a static channel matrix.

Spatial frame format and packet format design:

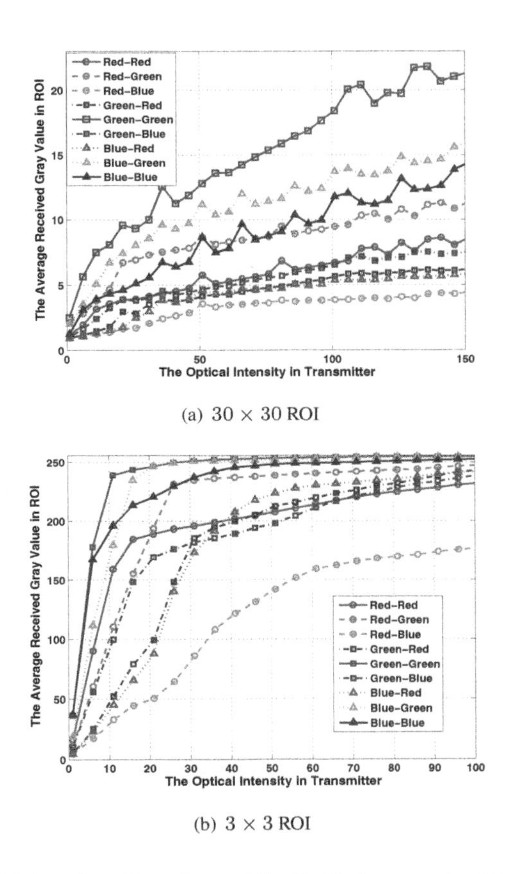

(a) 30×30 ROI

(b) 3×3 ROI

Figure 9.31 The joint nonlinearity and crosstalk effect between red–green–blue channels.

Figure 9.32(a) shows the actual design of the 16×16 LED array in the proposed CIM-MIMO system. The four outmost corner LEDs are used as reference anchors for corner detection. If the coordinates of these anchors are estimated, then all the coordinates of LEDs in the array can be calculated according to the spatial format layout. The remaining 196 LEDs in the inner area of the array are used for data transmission except for four LEDs from the second to the fifth LED in the second row of the array, which are used as a frame cyclic counter, similar to the COBRA system. For robust transmission, OOK modulation is adopted for both frame synchronization and anchor detection. Note that, different signals may arrive at different detection blocks at different times and are possibly sampled asynchronously. Meanwhile, the nonlinearity and crosstalk vary for different LEDs. Thus, the demodulation thresholds and detection time need to be adjusted for different detection areas. To tackle this problem, preamble sequences are inserted in each data packet for each LED unit in CIM-MIMO OCC systems. The packet format is shown in

Fig. 9.32(b), where a Barker code sequence of length 13 is selected as synchronization header due to its robust time synchronization performance. In order to achieve robust synchronization performance, we use the OOK modulation, where the Barker codes "0" and "1" are represented by the transmitted signals s_1 and s_M, respectively. Moreover, to tackle the joint rolling shutter and flicker noise issue, G preamble sequences $s_{1,1}, \cdots, s_{1,M}, \cdots, s_{G,1}, \cdots, s_{G,M}$ are repetitively transmitted and used as training sequences. According to the actual measurement, about $G = 4$ preamble sequences can yield improved and stable BER performance.

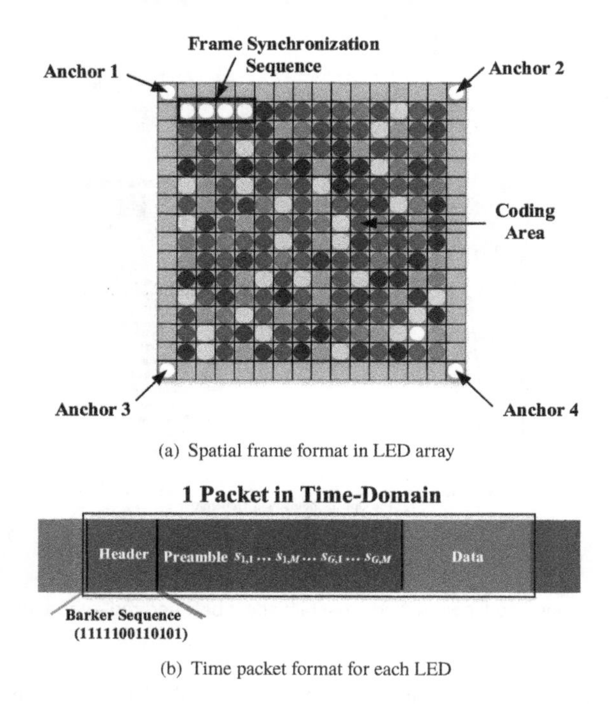

Figure 9.32 The spatial frame layout and packet format for each LED.

As mentioned above, in order to tackle the critical issues in a practical OCC system, the pre-designs, including a redundant transmission, CIM constellation optimization, iterative training sequence design, and spatial frame format design, are employed at the transmitter. And, the corresponding key signal processing techniques toward successful data recovery are needed at the receiver.

Adaptive array anchoring and symbol detection:

To extract and demodulate the signal from the captured images of the LED array, the LED array should be located first under perspective distortion. By using Hough circle detection, the coordinates of the four outmost anchor LEDs can be obtained. To eliminate the background radiation, differential image processing is used to estimate

LED locations reliably. Moreover, to make a smooth communication, a redundant transmission and unsampled reception technique is applied, and the refresh rate of LED array and the frame rate of the camera are set as 82.5 Hz and 330 fps, respectively. Thus, every four captured images in a successive manner should be differentiated to obtain purified intensity copies of each LED signal free of background noise. And then, two pairs of spurred points with the longest distance in between can be found from those images, corresponding to four anchor LED positions. After circle detection, we assume that the centers of these detected anchor circles are the center coordinates of four anchor LEDs if the relative rotation angle of the LED array to the vertical direction is within $-45°$ to $+45°$. Then, we can correct the perspective distortion of the LED array by perspective transformation, as shown in Fig. 9.33. After that, we can calculate the coordinates of all LEDs and color pixels according to the spatial format layout and Bayer-pattern layout, respectively.

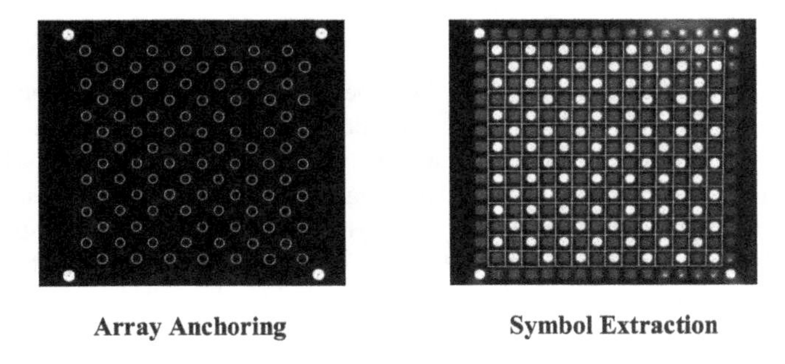

Figure 9.33 LED array and symbol detection.

Synchronization and iterative training:
To make a reliable symbol demodulation, a reasonable training signal threshold is needed. For a practical CMOS image-sensor-based OCC in the indoor case, the flicker noise and rolling shutter effect will inevitably introduce intensity fluctuation and degrade the system performance. The flicker noise can be approximated as a harmonic intensity fluctuation. Moreover, for the rolling-shutter-based receiver, the accumulated charge flows out row-by-row, which will induce bright and dark strips in the image. To tackle the flicker noise and rolling shutter issue, G (G is set as 4) preamble sequences are inserted in time packet before data transmission. Thus, iterative training and recursive estimation can be performed to obtain the average intensity under fluctuation.

For M-CIM, G preamble sequences, each of which contains M kinds of training signals $s_1, \cdots, s_M$, are inserted in the time packet for each LED. Moreover, due to the unstable frame rate for the CMOS image sensor, upsampled signal processing is performed at the receiver to alleviate the frame drift issue. Thus, the image sensor will capture Q (3, 4, or 5) successive images, which contain at least one complete and valid preamble sequence during the training period. Then, the most probable one is selected as the training signal threshold. For example, during the transmission period of the training symbol s_m, the image sensor can capture Q groups of red/green/blue signals corresponding to each LED unit. The 3-D Euclidean distance $E^m_{i,j}$ between the ith group and jth group in the color space can be expressed as

$$E^m_{i,j} = \|\mathbf{P}^m_i - \mathbf{P}^m_j\|_2,\ i \neq j,\ i,j = 1,2,\cdots,Q;\ m = 1,2,\cdots,M, \quad (9.77)$$

where $\mathbf{P}^m_i, \mathbf{P} = [R,G,B]$ is the gray value for red/green/blue colors in ith captured group corresponding to the training symbol s_m. After obtaining the Euclidean distance, we compare it with the Euclidean distance threshold E^m_{th}. If $E^m_{i,j} \leq E^m_{th}$, we conclude that these two groups are similar in the color space. For these similar groups, F^m_i is defined as

$$F^m_i = \sum_{j=1,j\neq i}^{Q} u\left(E^m_{th} - E^m_{i,j}\right), \quad (9.78)$$

where $u(\cdot)$ is the step function. And then, we select the most probable one as the training signal threshold based on F^m_i. Specifically, if $F^m_i \neq 0$, which indicates that it has successfully acquired at least one effective training symbol s_m, we select the one with the largest F^m_i as the obtained training symbol corresponding to transmitted symbol s_m, and then mark the $\text{Flag}^m = 1$. However, if $F^m_i = 0$, which indicates that it fails to detect the training symbol s_m, we mark $\text{Flag}^m = 0$ and choose the middle group signal as the training symbol. The selected m-CIM training symbol threshold T^m is given by

$$T^m = \begin{cases} s^m_{i_0}, i_0 = \max\limits_{i=\{1,\cdots,Q\}} \{F^m_i\} & \text{if} \max\limits_{i=\{1,\cdots,Q\}} \{F^m_i\} \neq 0, \text{Flag}^m = 1, \\ s^m_{i_0}, i_0 = \lceil \frac{Q}{2} \rceil & \text{if} \max\limits_{i=\{1,\cdots,Q\}} \{F^m_i\} = 0, \text{Flag}^m = 0. \end{cases} \quad (9.79)$$

Following the similar processing, we can acquire M kinds of training symbols for each LED unit.

Note that, G preamble sequences are inserted in time packet to average the intensity fluctuation induced by the flicker noise and rolling shutter issue. If we define $U(m,k,g)$ as the renewed received signal corresponding to the m-CIM training symbol for the kth LED in the gth preamble sequence, it is renewed iteratively as

$$U(m,k,g) = \frac{U(m,k,g-1) \times \text{Flag}(m,k,g-1) + U(m,k,g)}{\text{Flag}(m,k,g-1) + 1}, 1 \leq g \leq G. \quad (9.80)$$

Then the final estimated signal corresponding to the m-CIM training symbol for the kth LED is $U^k_m = U(m,k,G)$.

CIM symbol demodulation:

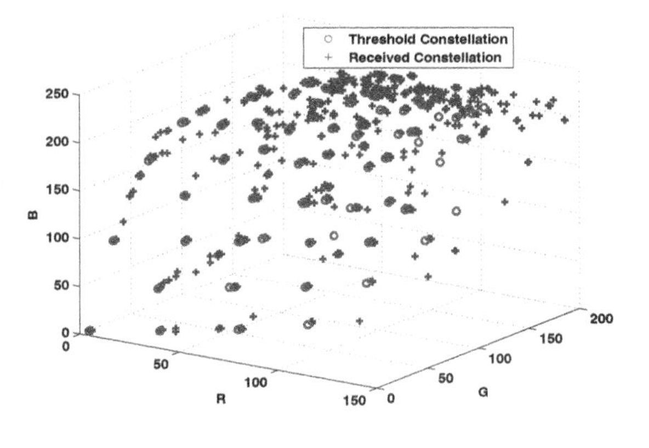

(a) Un-optimal

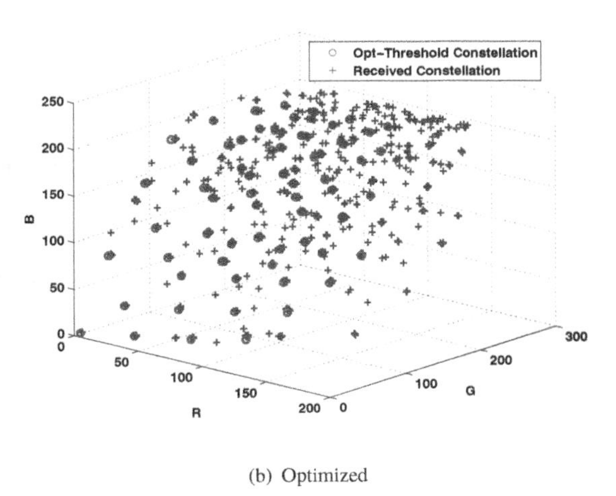

(b) Optimized

Figure 9.34 The 3-D 64-CIM constellation at the receiver.

After obtaining the valid training symbols thresholds $U_m^k, m \in \{1, 2, \cdots, M\}$ of the kth LED, we compare them with the received signals from Q frame images for symbol demodulation, which is given by

$$m_q^k = \underset{m \in \{1,2,\cdots,M\}}{\text{argmin}} \|r_q^k - U_m^k\|_2, \quad q = 1, \cdots, Q. \tag{9.81}$$

The final demodulated symbol for the kth LED is represented by $s_{m,q}^k$ whose index satisfies

$$m_q^k = m_{q-1}^k \text{ or } m_q^k = m_{q+1}^k \text{ or } m_{q-1}^k = m_q^k = m_{q+1}^k. \tag{9.82}$$

Figure 9.34 shows the received constellation points at the receiver for the conventional and optimized CIM symbols. The optimized constellation points in 3-D color space distribute more uniformly than non-optimized constellation points. Aside from the few abnormal constellation points away from threshold constellation points due to rolling shutter, most of the constellation points are close to the thresholds, which illustrates that the M-CIM system can achieve high SNR. The N_c LEDs in the coding domain can transmit $N_c \log_2 M$ bits per transmitted frame. Successive new data frames continue to be demodulated until the frame synchronization sequence value becomes one. Following this decoding mechanism, we can perform an effective and reliable symbol demodulation, and avoid missing demodulation or duplicate modulation.

9.5.1.3 **OCC experimental platform and performance evaluation**

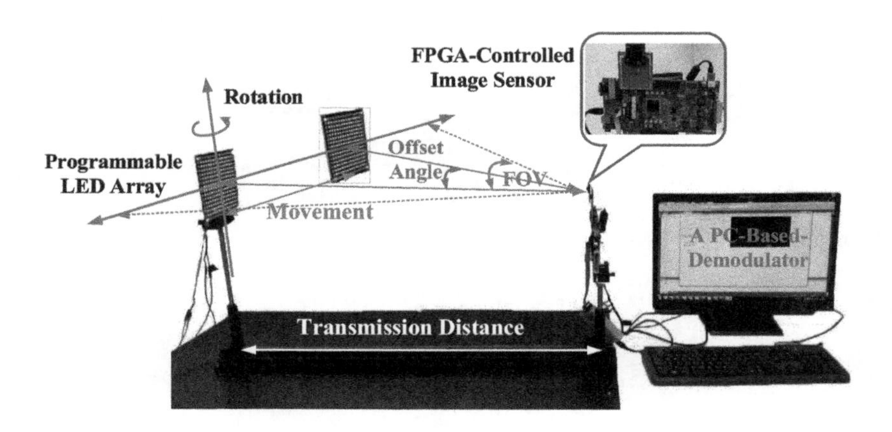

Figure 9.35 The OCC experimental platform.

The OCC experimental platform is shown in Fig. 9.35. A programmable 16×16 LED array is employed as the transmitter and a smartphone image sensor controlled by the FPGA is employed as the receiver. All modules are mounted on the top of an optical breadboard to control the transmission distance accurately. Moreover, to test the robustness to the perspective distortion, we can move or rotate the LED array to vary offset angle and rotation angle. No external lens is used except a small integrated lens on the close top of the sensor. The image sensor can support the refresh rate of 330 fps and LED array can support up to 400 Hz refresh rate. If the array refresh rate is chosen as 330 Hz, the synchronization mismatch between the LED array and the image sensor will dramatically degrade the BER performance. To minimize the synchronization error and tackle the frame rate unstability of the sensor, we set the LED array refresh rate to 82.5 Hz. Table 9.1 lists the specifications of the image sensor receiver and the experimental parameters. The input data stream is continuously fed to the MCU from the computer at the transmitter and data is demodulated in real

Table 9.1 The key specifications and experimental parameters.

Manufacture process	CMOS	LED power	0.5 W
Output formats	10 bits RAW	LED array size	16 × 16
Pixel size	2 μm × 2 μm	LED array refresh rate	82.5 Hz
Frame rate	330 fps	LED pixel pitch	1 cm
Resolution	672 × 380 pixel	LED dimension	5 mm
Lens size	1/3 inch	Communication distance	50–150 cm

time at the receiver.

The actual measured BER performances for CIM-MIMO OCC systems with different modulation orders and transmission distances are shown in Fig. 9.36(a). Note that $M = L_r \times L_g \times L_b$ and we set $(L_r = L_g = L_b = 4)$ for $M = 64$, and $(L_r = L_g = 8, L_b = 4)$ for $M = 256$. For a fixed distance, it is clear that the BER decreases as modulation order M decreases, since the Euclidean distance in the color space becomes larger.

For a given modulation order, the BER performance can be dramatically improved if iterative training and constellation optimization are performed. The effectiveness of iterative training and constellation optimization to trackle the critical issues in OCC systems is experimentally verified. Note that, without external lens assistance, the system is able to achieve a data rate of 126.72 kbps over a range of up to 1.4 m. It is anticipated that much longer distance can be achieved if an extra lens with larger focusing length is used in front of the sensor. We further verify the system robustness to the perspective distortion with different rotation angles and offset angles. Figure 9.36(b) shows the BER performance versus rotation angle when offset angle is fixed to be 0° and the BER performance versus offset angle when rotation angle is fixed to be 0°, with 60 cm separation distance. It is seen that, when the offset angle is within the field of view (FOV) of the image sensor, and the rotation angle is smaller than the threshold value, which can be corrected by perspective transformation, reliable symbol detection and demodulation are unaffected.

9.6 Conclusion

In this chapter, OCC system design principles are presented. First, modulation schemes spanning time, frequency, space, color, and intensity domains are introduced. Subsequently, corresponding nonideal factors and mitigation techniques are discussed. In particular, synchronization problems are tackled by per-line tracking and inter-frame coding, and rateless coding. Then issues in a real-time OCC system design are discussed. Using color and intensity jointly, methods to increase data rate and improve BER performance are proposed. In addition, solutions to deal with

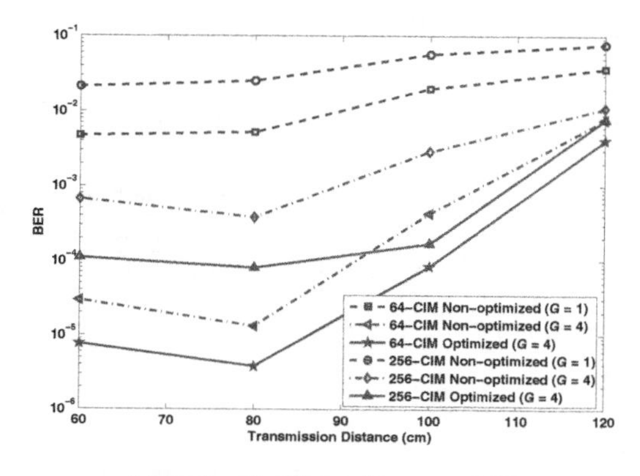

(a) BERs with different distances and M

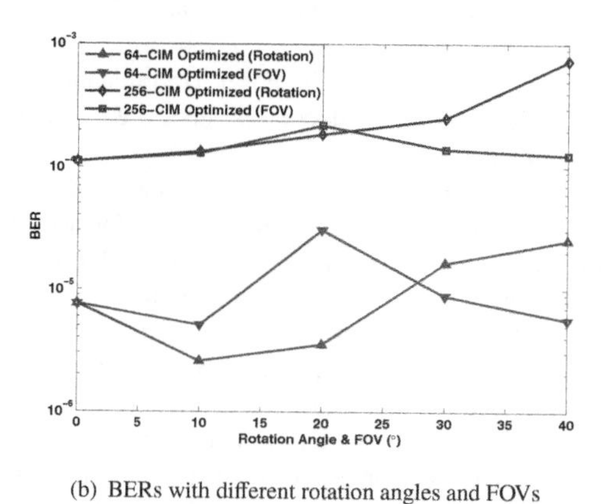

(b) BERs with different rotation angles and FOVs

Figure 9.36 BER performance of CIM-MIMO OCC systems.

frame rate unstability, system nonlinearity and crosstalk, flicker noise, and rolling shutter in the system design are presented.

The preliminary results in this chapter show that it is feasible to use a commercial CMOS camera as a real-time OCC receiver. The corresponding system can be used in the near-field screen–camera communication and indoor visible light positioning. Equipped with an external optical lens, the system is also suitable for other applications, for example, capturing signals from a distant traffic light, or information broadcasting display in a public area such as the shopping mall and the transportation hub.

References

1 R. G. Gallager, *Information Theory and Reliable Communication*, New York, NY, USA: Wiley, 1968.
2 E. A. Ratzer and D. J. MacKay, "Sparse low-density parity-check codes for channels with cross-talk," in *Proc. IEEE Information Theory Workshop 2003* (Paris, France), Mar. 32–Apr. 4, 2003, pp. 127–130.
3 E. Ratzer, "Sparse data blocks and multi-user channels," in *Proc. IEEE International Symposium on Information Theroy (ISIT) 2003* (Yokohama, Japan), Jun. 29–Jul. 4, 2003, pp. 314–314.
4 A. A. Farid and S. Hranilovic, "Channel capacity and non-uniform signalling for free-space optical intensity channels," *IEEE J. Sel. Areas Commun.*, vol. 27, no. 9, pp. 1553–1563, Dec. 2009.
5 J. Cao, S. Hranilovic, and J. Chen, "Capacity and nonuniform signaling for discrete-time poisson channels," *J. Opt. Commun. Netw*, vol. 5, no. 4, pp. 329–337, Apr. 2013.
6 S. Hranilovic and F. R. Kschischang, "A pixelated MIMO wireless optical communication system," *IEEE J. Sel. Topics Quantum Electron.*, vol. 12, no. 4, pp. 859–874, Jul./Aug. 2006.
7 R. D. Roberts, "Undersampled frequency shift on-off keying (ufsook) for camera communications (camcom)," in *Proc. Wireless and Optical Communication Conference (WOCC) 2013* (Chongqing, China), May 16–18, 2013, pp. 645–648.
8 P. Luo, Z. Ghassemlooy, H. Le Minh, X. Tang, and H. M. Tsai, "Undersampled phase shift on-off keying for camera communication," in *Proc. International Conference on Wireless Communications & Signal Processing (WCSP) 2014* (Hefei, China), Oct. 23–25, 2014, pp. 1–6.
9 P. Luo, Z. Ghassemlooy, H. Le Minh, H. M. Tsai, and X. Tang, "Undersampled-pam with subcarrier modulation for camera communications," in *Proc. OptoElectronics and Communications Conference (OECC) 2015* (Shanghai, China), Jun. 28–Jul. 2, 2015, pp. 1–3.
10 P. Luo, M. Zhang, Z. Ghassemlooy, H. Le Minh, H. M. Tsai, X. Tang, and D. Han, "Experimental demonstration of a 1024-QAM optical camera communication system," *IEEE Photon. Technol. Lett.*, vol. 28, no. 2, pp. 139–142, Jan. 2016.
11 N. Rajagopal, P. Lazik, and A. Rowe, "Visual light landmarks for mobile devices," in *Proc. International Symposium on Information Processing in Sensor Networks 2014* (Berlin, Germany), Apr. 15–17, 2014, pp. 249–260.
12 C. Danakis, M. Afgani, G. Povey, I. Underwood, and H. Haas, "Using a cmos camera sensor for visible light communication," in *Proc. IEEE Global Communications Conference (GLOBECOM) Workshops 2012* (Anaheim. CA), Dec. 3–7, 2012, pp. 1244–1248.
13 K. Jo, M. Gupta, and S. Nayar, "Disco: Display-camera communication using rolling shutter sensors,"*ACM Trans. Graphics.*, vol. 35, no. 5, pp. 150:00–150:13, Jul. 2016.
14 P. Hu, P. H. Pathak, X. Feng, H. Fu, and P. Mohapatra, "Colorbars: Increasing data rate of LED-to-camera communication using color shift keying," in *Proc. ACM Conference on Emerging Networking Experiments and Technologies 2015* (Heidelberg, Germany), Dec. 1–4, 2015,

pp. 1–12.

15 H. Y. Lee, H. M. Lin, Y. L. Wei, H. I. Wu, H. M. Tsai, and K. C. J. Lin, "Rollinglight: Enabling line-of-sight light-to-camera communications," in *Proc. International Conference on Mobile Systems, Applications, and Services 2015* (Florence, Italy), May 18–22, 2015, pp. 167–180.

16 J. Gu, Y. Hitomi, T. Mitsunaga, and S. Nayar, "Coded rolling shutter photography: Flexible space-time sampling," in *Proc. IEEE Computational Photography 2010* (Cambridge, MA), Mar. 29–30, 2010, pp. 1–8.

17 A. Dabbo and S. Hranilovic, "Receiver design for wireless optical MIMO channels with magnification," in *Proc. International Conference on Telecommunications (ICT) 2009* (Zagreb, Croatia), Jun. 8–10, 2009, pp. 51–58.

18 S. D. Perli, N. Ahmed, and D. Katabi, "Pixnet: Interference-free wireless links using LCD-camera pairs," in *Proc. International Conference on Mobile Computing and Networking 2010* (Chicago, IL), Sept. 20–24, 2010, pp. 137–148.

19 M. R. H. Mondal, K. R. Panta, and J. Armstrong, "Performance of two dimensional asymmetrically clipped optical OFDM," in *Proc. IEEE Global Communications Conference (GLOBECOM) Workshops 2010* (Miami, FL), Dec. 6–10, 2010, pp. 995–999.

20 E. Katz and Y. Bar-Ness, "Two-dimensional (2-d) spatial domain modulation methods for unipolar pixelated optical wireless communication systems," *J. Lightw. Technol.*, vol. 33, no. 20, pp. 4233–4239, Oct. 2015.

21 M. R. H. Mondal and K. Panta, "Performance analysis of spatial OFDM for pixelated optical wireless systems," *Trans. Emerging Telecommun. Technol.*, vol. 28, no. 2, May 2015.

22 K. M. Wong, J. Wu, T. N. Davidson, and Q. Jin, "Wavelet packet division multiplexing and wavelet packet design under timing error effects," *IEEE Trans. Signal Process.*, vol. 45, no. 12, pp. 2877–2890, Dec. 1997.

23 H. Nikookar, *Wavelet Radio: Adaptive and Reconfigurable Wireless Systems Based on Wavelets*, Cambridge University Press, 2013.

24 W. Huang, C. Gong, and Z. Xu, "System and waveform design for wavelet packet division multiplexing-based visible light communications," *J. Lightw. Technol.*, vol. 33, no. 14, pp. 3041–3051, Jul. 2015.

25 T. Nagura, T. Yamazato, M. Katayama, T. Yendo, T. Fujii, and H. Okada, "Improved decoding methods of visible light communication system for its using LED array and high-speed camera," in *Proc. IEEE Vehicular Technology Conference (VTC Spring) 2010* (Taipei, Taiwan), May 16–19, 2010, pp. 1–5.

26 H. Okada, T. Ishizaki, T. Yamazato, T. Yendo, and T. Fujii, "Erasure coding for road-to-vehicle visible light communication systems," in *Proc. IEEE Consumer Communications and Networking Conference (CCNC) 2011* (Las Vegas, NV), Jan. 9–12, 2011, pp. 75–79.

27 M. R. H. Mondal and J. Armstrong, "Impact of linear misalignment on a spatial ofdm based pixelated system," in *Proc. Asia-Pacific Conference on Communications (APCC) 2012* (Jeju Island, Korea), Oct. 15–17, 2012, pp. 617–622.

28 A. Ashok, S. Jain, M. Gruteser, N. Mandayam, W. Yuan, and K. Dana, "Capacity of pervasive camera based communication under perspective distortions," in *Proc. IEEE International Conference on Pervasive Computing and Communications 2014* (Budapest, Hungary), Mar. 24–28, 2014, pp. 112–120.

29 M. Rubaiyat, H. Mondal, and J. Armstrong, "The effect of defocus blur on a spatial OFDM optical wireless communication system," in *Proc. International Conference on Transparent Optical Networks 2012* (Coventry, England), Jul. 2–5, 2012, pp. 1–4.

30 M. R. H. Mondal and J. Armstrong, "Analysis of the effect of vignetting on MIMO optical wireless systems using spatial OFDM," *J. Lightw. Technol.*, vol. 32, no. 5, pp. 922–929, Mar. 2014.

31 W. Hu, H. Gu, and Q. Pu, "Lightsync: Unsynchronized visual communication over screen-camera links," in *Proc. International Conference on Mobile Computing & Networking 2013* (Miami, FL), Sept. 30–Oct. 4, 2013, pp. 15–26.

32 T. Hao, R. Zhou, and G. Xing, "Cobra: Color barcode streaming for smartphone systems," in *Proc. IEEE International Conference on Distributed Computing Systems 2012* (Low Wood Bay, UK), Jun. 25–29, 2012, pp. 85–98.

33 W. Du, J. C. Liando, and M. Li, "Softlight: Adaptive visible light communication over screen-camera links," in *Proc. IEEE International Conference on Computer Communications (INFOCOM) 2016* (San Francisco, CA), Apr. 10–14, 2016, pp. 1–9.

34 J. W. Byers, M. Luby, M. Mitzenmacher, and A. Rege, "A digital fountain approach to reliable distribution of bulk data," *ACM SIGCOMM Computer Communication Review*, vol. 28, no. 4, pp. 56–67, Sept. 1998.

35 M. Luby, "Digital fountain, inc.luby@digitalfountain.com," 2002.

36 A. Shokrollahi, "Raptor codes," *IEEE Trans. Inf. Theory*, vol. 52, no. 6, pp. 2551–2567, Jun. 2006.

37 W. Huang, P. Tian, and Z. Xu, "Design and implementation of a real-time CIM-MIMO optical camera communication system," *Opt. Exp.*, vol. 24, no. 21, pp. 24567–24579, Oct. 2016.

38 P. Tian, W. Huang, and Z. Xu, "Design and experimental demonstration of a real-time 95kbps optical camera communication system," in *Proc. IEEE International Conference on Communication Systems, Networks and Digital Signal Processing (CSNDSP) 2016* (Prague, Czech Republic), Jul. 20–22, 2016, pp. 1–6.

39 E. Monteiro and S. Hranilovic, "Design and implementation of color-shift keying for visible light communications," *J. Lightw. Technol.*, vol. 32, no. 10, pp. 2053–2060, May 2014.

40 R. Singh, T. O. Farrell, and J. P. David, "An enhanced color shift keying modulation scheme for high-speed wireless visible light communications," *J. Lightw. Technol.*, vol. 32, no. 14, pp. 2582–2592, Jul. 2014.

41 W. Huang, C. Gong, P. Tian, and Z. Xu, "Experimental demonstration of high-order modulation for optical camera communication," in *Proc. IEEE Global Conference on Signal and Information Processing (GlobalSIP) 2015* (Orlando, FL), Dec. 14–16, 2015, pp. 1027–1031.

10
Index

IEEE PRESS SERIES ON DIGITAL AND MOBILE COMMUNICATION

John B. Anderson, *Series Editor*
University of Lund

---

1. *Future Talk: The Changing Wireless Game*
   Ron Schneiderman
2. *Digital Transmission Engineering*
   John B. Anderson
3. *Fundamentals of Convolutional Coding*
   Rolf Johannesson and Kamil Sh. Zigangirov
4. *Mobile and Personal Communication Services and Systems*
   Raj Pandya
5. *Wireless Video Communications: Second to Third Generation and Beyond*
   Lajos Hanzo, Peter J. Cherriman, and Jürgen Streit
6. Wireless Communications in the 21st Century
   Mansoor Shafi, Shigeaki Ogose, and Takeshi Hattori
7. *Introduction to WLLs: Application and Deployment for Fixed and Broadband Services*
   Raj Pandya
8. *Trellis and Turbo Coding*
   Christian B. Schlegel and Lance C. Perez
9. *Theory of Code Division Multiple Access Communication*
   Kamil Sh. Zigangirov
10. *Digital Transmission Engineering,* Second Edition
    John B. Anderson
11. *Wireless LAN Radios: System Definition to Transistor Design*
    Arya Behzad
12. *Wireless Broadband: Conflict and Convergence*
    Vern Fotheringham and Chetan Sharma
13. *Millimeter Wave Communication Systems*
    Kao-Cheng Huang and Zhaocheng Wang
14. *Channel Equalization for Wireless Communications: From Concepts to Detailed Mathematics*
    Gregory E. Bottomley
15. *Handbook of Position Location: Theory, Practice and Advances*
    Seyed A. Reza Zekavat and R. Michael Buehrer
16. *Digital Filters: Principles and Applications with MATLAB*
    Fred J. Taylor
17. *Resource Allocation in Uplink OFDMA Wireless Systems: Optimal Solutions and Practical Implementations*
    Elias Yaacoub and Zaher Dawy

18. *Non-Gaussian Statistical Communication Theory*
David Middleton
19. *Frequency Stability: Introduction & Applications*
Venceslav F. Kroupa
20. *Mobile Ad Hoc Networking: Cutting Edge Directions*, Second Edition
Stefano Basagni, Marco Conti, Silvia Giordano, and Ivan Stojmenovic
21. *Surviving Mobile Data Explosion*
Dinesh C. Verma and Paridhi Verma
22. *Cellular Communications: A Comprehensive and Practical Guide*
Nishith Tripathi and Jeffrey H. Reed
23. *Fundamentals of Convolutional Coding*, Second Edition
Rolf Johannesson and Kamil Sh. Zigangirov
24. *Trellis and Turbo Coding*, Second Edition
Christian B. Schlegel and Lance C. Perez
25. *Problem-Based Learning in Communication Systems Using MATLAB and Simulink*
Kwonhue Choi and Huaping Liu
26. *Bandwidth Efficient Coding*
John B. Anderson
27. *Visible Light Communications: Modulation and Signal Processing*
Zhaocheng Wang, Qi Wang, Wei Huang, and Zhengyuan Xu